Cooking for Geeks

Real Science, Great Hacks, and Good Food

Jeff Potter

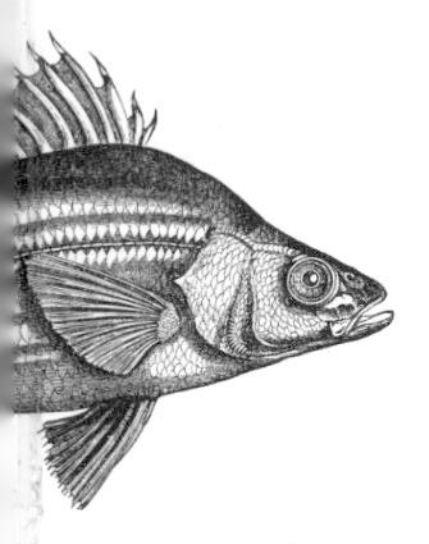
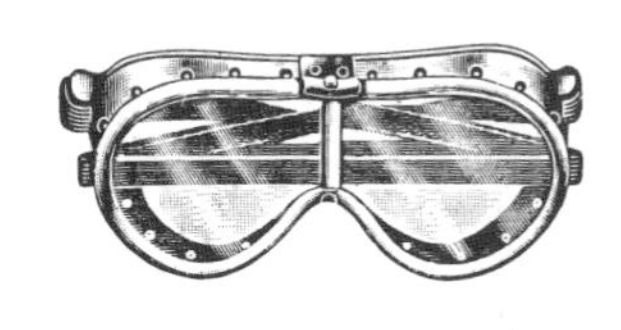

O'REILLY®
Beijing · Cambridge · Farnham · Köln · Sebastopol · Tokyo

Cooking for Geeks
by Jeff Potter

Published by O'Reilly Media, Inc., 1005 Gravenstein Highway North, Sebastopol, CA 95472.

O'Reilly books may be purchased for educational, business, or sales promotional use. Online editions are also available for most titles (*http://my.safaribooksonline.com*). For more information, contact our corporate/institutional sales department: (800) 998-9938 or *corporate@oreilly.com*.

Editors: Brian Sawyer and Laurel R.T. Ruma
Production Editor: Rachel Monaghan
Copyeditor: Rachel Head
Proofreader: Rachel Monaghan
Indexer: Lucie Haskins
Cover Designer: Mark Paglietti
Interior Designer: Edie Freedman
Illustrator: Aaron Double

Printing History:

July 2010: First Edition.

Revision History for the First Edition:

2011-03-25 Sixth release
2011-06-24 Seventh release
2011-11-04 Eighth release
2011-12-22 Ninth release
2012-08-03 Tenth release
2012-11-30 Eleventh release
2013-01-04 Twelfth release

Back cover photograph by Matthew Hrudka.

ISBN: 978-0-596-80588-3
[QG]

Contents

Chapter 3. Choosing Your Inputs: Flavors and Ingredients 79

Chapter 4. Time and Temperature: Cooking's Primary Variables 147

Chapter 5. Air: Baking's Key Variable 217

Recipe Index

Breakfast

Drinks

Breads

Appetizers and Sides

Salads

Soups

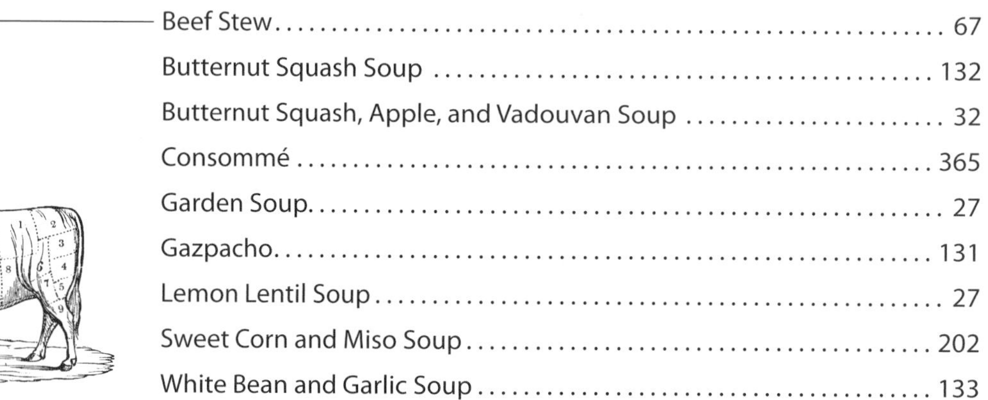

Sauces and Marinades

Mains

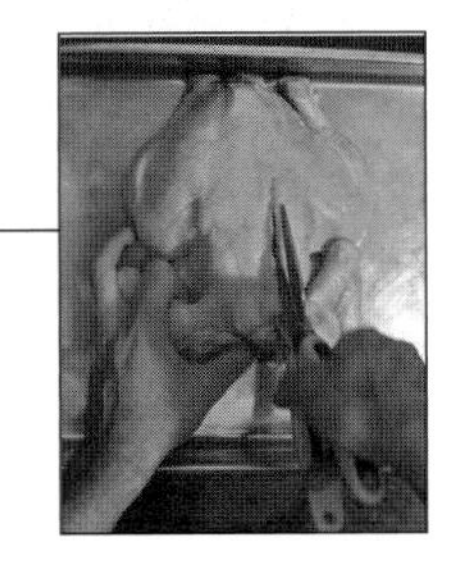

Desserts

Components & Ingredients

List of Interviews

Preface

Hackers, makers, programmers, engineers, nerds, techies—what we'll call "geeks" for the rest of the book (deal with it)—we're a creative lot who don't like to be told what to do. We'd rather be handed a box full of toys or random electronic components, or yarn, or whatever, and let loose to play.

But something happens to some geeks when handed a box full of spatulas, whisks, and sugar. Lockup. Fear. Foreign feelings associated with public speaking, or worse, *coulrophobia*. If you're this type, this book is for you.

Then there's another type of geek: the über-nerd, who's unafraid to try anything...maybe a bit *too* unafraid, but hasn't had that Darwin Award moment (yet). The type of geek who is either "all on or all off," who addresses every aspect of the perfect cup of coffee, down to measuring the pressure with which the grinds are tamped into the espresso machine's portafilter. This kind of geek is always on the search for the next bit of knowledge. If you're this type, this book will inspire you.

And then there's everyone else: the everyday geek, normal, inquisitive, and looking to have more fun in the kitchen. Maybe you're comfortable in the kitchen and would like new ideas, or perhaps you're not quite sure where to start but are ready to give it a go. This book will show you easy ways of trying new things.

Regardless of which type of geek you are, as long as you have "the courage of your convictions" to pick up the spatula and try, you'll do fine. The goal of this book is to point out new ways of thinking about the tools in that box full of kitchen gear.

Of course, I have plenty of tips and secrets to share ("spill the beans," as they say), so I hope you'll buy this book and take it home with you. Scribble notes in the margins about bits that you like (or just star—upvote?—those paragraphs). Write in questions on things that leave you perplexed or wondering. Learning to cook is about curiosity, learning to ask questions, and figuring out how to answer those questions.

When you're done with the book, pass it along to a friend (although my publisher would rather you buy that friend a new copy!). If you've received this book from a friend, I hope it's because they think you'll enjoy it and not because your cooking is lousy. Cooking is about community, and sharing knowledge and food is one of the best ways to build community.

If you're the $(N+1)^{th}$ person to have received this book—if it's dog-eared, worn, and beat up, and by the time you're done with it there aren't any more spots left to write comments in the margins—then I have a favor to ask of you: send me the marked-up book when you're done. In return, I'll send you back something random (possibly only pseudorandom). See the book's companion website for information on how to do this:

> *http://www.cookingforgeeks.com/book/feedback/*

How to Use This Book

This book is designed for use in a couple of different ways.

If you want to "just cook," flip to the recipe index, pick a recipe, and skip straight to that page. The surrounding text will explain some aspects of the science behind the recipe. While the recipes in this book are chosen to complement and provide examples of the science, they're also recipes that are fantastic in and of themselves. Most of the recipes are for single components—say, beef short ribs—without accompanying sides. This allows the various components of a meal to be covered in appropriate science sections, and also keeps each recipe short and easy.

If you're more interested in curling up with a cup of `$favoriteBeverage`, pick a chapter based on your interests and tuck in.

The first portion of this book covers topics you should think about before turning on the oven: how to approach the kitchen and how to think about taste and smell. The middle portion covers key variables in cooking (time and temperature) and baking (air), as well as some secondary variables. The final two chapters address some of the more creative things you can do in the kitchen, either with "software" (chemicals) or "hardware" (blowtorches!). Recipes and experiments are sprinkled throughout the book, along with interviews of scientists, researchers, chefs, and food bloggers. Here's a taste of what you'll find in this book:

Chapter 1, Hello, Kitchen!

What does success in the kitchen mean? How do you pick a recipe, and then how do you interpret it correctly? This chapter considers these questions and also touches briefly on nutrition (really, the all-pizza diet has got to go).

Chapter 2, Initializing the Kitchen

This chapter covers the basic must-haves, but it is ultimately up to you to experiment, adapt, and modify these suggestions to fit your needs and tastes. *Use common sense.* In addition to the essentials, this chapter also touches on storage tips, kitchen organization tricks, and things to keep in mind if you're new to cooking.

Chapter 3, Choosing Your Inputs: Flavors and Ingredients

This chapter explains the physiology of taste and smell and shows how to improve your understanding of flavor combinations, giving ideas on how to stir up new ideas.

Chapter 4, Time and Temperature: Cooking's Primary Variables

This chapter explains the chemical reactions that occur when heating foods, so that you'll know what to look for when cooking. We start with a discussion of heat, looking at the differences between various ways of cooking, how the temperature choice impacts the outcome, and what chemical reactions are taking place. The rest of the chapter then examines a range of temperatures, starting with the coldest and ending with the hottest, discussing the importance of each temperature point and giving example recipes.

Chapter 5, Air: Baking's Key Variable

This chapter takes a brief look at gluten and then examines baking's key variable, air. It covers the three primary methods of generating air—mechanical, chemical, and biological—giving common techniques for creating air and notes on how to work with the associated ingredients.

Chapter 6, Playing with Chemicals

This chapter takes a look at cooking techniques that use food additives, both traditional and modern. Some recent culinary techniques, falling under the genre termed *molecular gastronomy* or *modernist cuisine*, rely on chemicals. Some of these chemical-based techniques are covered in the second portion of this chapter. Even if you're not the type who wants to use food additives, understanding the chemistry and purposes of various food additives makes recovering from kitchen errors quicker and decoding ingredient lists at the grocery store easier.

Chapter 7, Fun with Hardware

Here we cover some of the commercial and industrial tools used in preparing foods, such as sous vide, and throw in a few, uh, "crazy" (and fun!) things that one can do in the kitchen as well. Modern commercial kitchens, most likely including the high-end ones in your area, use a variety of tools that consumers rarely encounter but that can help create some absolutely stellar meals.

As is so often the case with science, what we don't know about cooking seems to be increasing at a faster rate than what we do know. And then there's the difference between theory and practice (in theory, they should be the same; in practice, hahaha). One research paper will find that myosin (a protein in muscle) denatures in fish at 104°F / 40°C, while another reports 107°F / 41.7°C, and yet another at an entirely different temperature. Maybe it's the type of fish that matters (lean versus fatty does make a difference), or maybe it's just *that fish*. Biology does not confine itself to simple models, so when you're trying to combine the various pieces of information into a uniform picture, some discrepancy is unavoidable.

On the Web

So much of cooking is about sharing, community, and discussion. Beyond this book, here are a few places to share your creations, comments, and questions.

- For videos, more recipes, and additional interviews, see this book's companion website, at *http://www.cookingforgeeks.com/*.
- If you use Facebook, see *http://facebook.com/cookingforgeeks*.
- If you tweet, follow @cookingforgeeks; use the hashtag "#c4g" for general discussion.
- Many of the photos included in this book are available under a Creative Commons license on Flickr; see the photosteam at *http://www.flickr.com/cookingforgeeks*. If you post photos of dishes you've made from recipes in this book, tag them with "cookingforgeeks" so that they show up on the Community section on *http://www.cookingforgeeks.com*.

Acknowledgments

I extend my thanks to my good friends Mark Lewis and Aaron Double. Mark suffered through the first versions of both the food and the chapters and provided invaluable feedback on both. Aaron spent too much time turning my chicken-scratch sketches into the charts and diagrams that appear throughout the book. (Aaron is an amazing industrial designer—see *http://www.docodesign.com*—and he can use Illustrator faster than I can use paper and pencil.)

Barbara Vail and Matt Kiggins helped dig up research papers on everything from the aforementioned myosin proteins to average weight gains during the holidays (about 0.5 lbs—not much, but it turns out we don't tend to lose it the following spring), while Quinn Norton fed me congee and helped in more ways than one with the interviews and text.

I'm extremely grateful to the many chefs, bloggers, researchers, and scientists who took time out of their often insanely busy schedules to speak with me. Their insights helped shape the way I think, and I hope the resulting interviews in the book are not just informative, but fun.

Thanks to all those who joined me for the weekly Book Club and Test Club dinners while working on the book, and, finally, thanks to Marlowe, Laurel, Brian, Edie, and the team at O'Reilly and the tech reviewers whose feedback has made this a better book. And of course, thanks to my parents for being so supportive and encouraging. I'll try to not splatter duck fat on the ceiling next time I'm home.

I hope you have as much fun trying out the various ideas in this book as I did putting them together!

My first cookbook, circa 1984.

How to Contact Us

Please address comments and questions concerning this book to the publisher:

O'Reilly Media, Inc.
1005 Gravenstein Highway North
Sebastopol, CA 95472
800-998-9938 (in the United States or Canada)
707-829-0515 (international or local)
707 829-0104 (fax)

We have a web page for this book, where we list errata, examples, and any additional information. You can access this page at:

http://www.oreilly.com/catalog/9780596805883/

To comment or ask technical questions about this book, send email to:

bookquestions@oreilly.com

For more information about our books, conferences, Resource Centers, and the O'Reilly Network, see our website at:

http://www.oreilly.com

Safari® Books Online

Safari Books Online

Safari Books Online is an on-demand digital library that lets you easily search over 7,500 technology and creative reference books and videos to find the answers you need quickly.

With a subscription, you can read any page and watch any video from our library online. Read books on your cell phone and mobile devices. Access new titles before they are available for print, and get exclusive access to manuscripts in development and post feedback for the authors. Copy and paste code samples, organize your favorites, download chapters, bookmark key sections, create notes, print out pages, and benefit from tons of other time-saving features.

O'Reilly Media has uploaded this book to the Safari Books Online service. To have full digital access to this book and others on similar topics from O'Reilly and other publishers, sign up for free at *http://my.safaribooksonline.com*.

1

Hello, Kitchen!

WE GEEKS ARE FASCINATED BY HOW THINGS WORK, AND MOST OF US EAT, TOO.

The modern geek is more than just a refined version of the stereotypical movie geek from the '80s. True, there's a contemporary equivalent, who have swapped *Star Wars* posters, pocket protectors, and large glasses held together by tape for really, *really* smart phones, hipster glasses, and social websites running on virtual machines. The Internet has given the computer geek a new challenge. For most of us techies, the largest obstacle in building something great has changed from a technical to a social one. The question is no longer *can you build it*, but *will people want it*? We're becoming a different kind of community, one that has to relate to a half a billion Facebook users, Twitterers, and lolcats. (I can has cheezburger? See page 169.)

But what it means to be a geek today can also be broader. Overly intellectual. Obsessed with details. Going beyond the point where a mainstream user would stop, often to the bemusement of those who don't "get it." Physics geeks. Coffee geeks. Almost-anything geeks. A geek is anyone who dwells with some amount of obsession on why something works and how to make that something better. And it's become a badge of honor to be a geek.

At our core, though, all of us geeks still share that same inner curiosity about the *hows* and *whys* with the pocket-protector crowd of yesteryear. This is where so many cookbooks fail us. Traditional cookbooks are all about the *what*, giving steps and quantities but offering little in the way of engineering-style guidance or ways of helping us think.

Unfortunately, there's no way (yet) to download a program for kitchen techniques and experience straight into your brain. Don't expect to walk away after finishing this book (or any other) knowing how to make a perfect four-course meal. That'd be like saying, "Hey, I want to learn how to program, so maybe I should start by writing my own operating system!"

But don't despair. Learning to cook is not so much about rote memorization or experience as it is about curiosity, and that's something us geeks have way more of than your average "random." With the right mindset and a few "Hello, World!" examples, you can crack the culinary code and be well on your way to having a good time in the kitchen.

In this first chapter, we'll cover how to approach the kitchen. What does it mean for a geek to hack in the kitchen? What things should a beginner keep in mind? What does success in the kitchen mean? How do you pick a recipe, and then how do you interpret it correctly?

We'll also briefly touch on nutrition. If you're already comfortable in the kitchen, you might want to skim this and the next chapter and dig right in with Chapter 3.

Always read through the entire recipe, top to bottom, before starting.

Think Like a Hacker

hack: 1. n. Originally, a quick job that produces what is needed, but not well. 2. n. An incredibly good, and perhaps very time-consuming, piece of work that produces exactly what is needed.

—Eric S. Raymond's *Jargon File*

hacker: a person who delights in having an intimate understanding of the internal workings of a system, computers and computer networks in particular. The term is often misused in a pejorative context, where "cracker" would be the correct term.

—RFC1392, Internet Engineering Task Force

My microwave has no 3 key, but I can enter 2:60.

—As tweeted by Tom Igoe, @tigoe

Cooking has the same types of hard constraints that code, hardware, and most science disciplines do. Processes (chemical or virtual), reactions, allocation of resources (more veggies!), and timing all matter. And while each discipline has standard techniques for solving these constraints, invariably there are other clever alternatives. Hacks don't have to be quick and dirty (that'd be a hack job), or overly involved works of perfection.

Some of the best hacks start out as safe and stable ways of solving unexpected problems, and being able to see those solutions is what it means to think like a hacker. It's rare to see a hack called for in a spec. Imagine a programmer coding a script that needs to count the number of lines in a text file. Standard method? Open, read a line, ++, close. Five minutes until demo? `` `wc -l "$file"` ``. While the hack is easier and faster to write, you should probably understand open/read/close first and know how and when to use them.

If you're new to the kitchen, buckle down and be prepared to learn the system from the inside out before breaking out the blowtorch, methylcellulose, or centrifuge. Every one of the well-respected chefs and instructors interviewed for this book has a thorough grasp of the fundamentals of cooking. Those who use tools like centrifuges and ingredients like methylcellulose use them as ways of extending those cooking fundamentals, not merely for the sake of novelty. To the pros, these newer techniques and ingredients simply expand their repertoires, taking their place alongside olive oil, flour, and other pantry staples.

Spraying a muffin tin on a dishwasher door.

Mug as plastic bag holder.

Strainer as splatter guard.

Roasting peppers in a toaster.

What does it mean to take the hacking mindset into the kitchen? Sometimes it's technique. Rolling pizza or pie dough to a uniform thickness by eye can be tricky, but slap a few rubber bands on each end of the rolling pin, and you've got an instant guide. Need to pour spices or coffee grinds into a plastic bag? Drop the plastic bag in a mug or cup and fold the edge of the bag over the edge. "Hacking" can apply to the ingredients as well, as you'll see in Chapter 3.

Metal bowl as double boiler.

Ways of doing things become obvious once you see them. The challenge in the kitchen is to see where you want to go and then find a path that gets you there. Thinking like a hacker means thinking of an end state and then figuring out how to get there in a time- and space-optimal (and fewest-dishes-possible) way.

How does one go about discovering hacks and tricks in the kitchen? Here's a thought experiment: imagine you're given a candle, a book of matches, and a box of nails, and asked to mount the candle on a wall. Without burning down the house, how would you do it?

Functional Fixedness

The problem just described is called Duncker's Candle Problem, after Karl Duncker, who studied the cognitive biases that we bring to problems. In this example, things like the paper of the matchbook have a "fixed function" of protecting the matches. We don't normally think of the matchbook cover as a piece of thick cardboard that's been folded over; we just see that as part of the matchbook. Recognizing the object as capable of serving other functions requires mental restructuring, something that the scriptwriters for *MacGyver* excelled at.

This mental restructuring is something that most geeks are naturally good at. All those interview puzzles common in the tech industry? You know: how would you start a fire with a can of soda and bar of chocolate?* Or, you're given 12 gold coins and a balance scale, but wait! One of the coins is fake, either lighter or heaver than the others, and the balance scale will magically break after exactly three uses. Problems like these almost invariably come down to breaking functional fixedness and overcoming confirmation bias (here, in the sense of being blinded to new uses by knowing previous uses). The obvious solutions to the candle problem—pushing the nails through the candle or melting the candle so that it sticks to them—will either split the candle or leave it too close to a wall to be safe. The solution, or at least the one Duncker was looking for, involves repurposing the box that had been holding the nails into a shelf. (I'm dreading all the emails I'm going to get with photos of this being done in other ways.)

Approaches for overcoming functional fixedness in puzzles, code, or the kitchen are the same. Understand what you actually have and what you're asked to do, break it down into individual steps, and explore different possibilities for each discrete step. Take the quest for the perfect cup of coffee: can you isolate the variables for bean grind, temperature, pressure, etc. and then explore the combinations in a controlled way, varying just one variable at a time? Think about the ingredients you're starting with and the end state you want, as opposed to the straight execution of a recipe. This way, when the execution inevitably veers off course, you can understand the step you are at and how to catch and correct the exception. Of course, be open to other possible outcomes—the way a meal turns out will sometimes be different than what you originally conceived.

Thinking about the end state will also help broaden how you think about cooking more generally. Cooking is not just about food in a pan; it's about health and well-being, community and giving. Why do you want to cook? Watching your waist or your wallet? Health and finances are common considerations. Building community? Potlucks, shared meals, and barbeques can be fun social activities and even spur friendly competition. Expressing love? Cooking can be an act of giving, both in the literal sense of sustenance and in the spiritual sense of sharing time and breaking bread together.

* "Right" answer: use the chocolate to polish the bottom of the can to a mirror-like finish, and then use the can as a parabolic reflector to focus sunlight onto a dry twig. My answer: trade the can of soda for a light from the nearest smoker; eat the bar of chocolate.

Cooking also allows you to try new things—there are plenty of foods that you can't order in a restaurant. Perhaps you want to get closer to your source of food, in which case learning how simple it can be to put together many common dishes will bring you at least one step closer. And then there's getting yet another step closer: I happen to eat meat, but what I buy at the store is so far removed from a living, breathing animal that I find it hard to identify with the life of the critter. (The English language doesn't help. We eat *beef*, but it's a cow. We eat *pork*, but it's a pig. Fish don't seem to be smart enough to merit a clear separation.) To properly respect an animal's life, to understand where my food comes from and to be mindful of not wasting it, I feel that at some point I should have to butcher an animal myself. (You could try lobster, but I've yet to get teary-eyed over one.) For me, cooking is also as much about escaping from work as it is about satisfying hunger, not to mention having fun trying new things with friends and knowing that what I'm putting in my body is healthy.

Regardless of your reason for wanting to cook, realize that there's more to cooking than just following a recipe. When looking at the end goal, think beyond the cooking stage. If your reason for cooking is to express affection, you should consider the sensations that your food brings your guests and the perceptions and reactions they have to it as much as the cooking itself. On the other side, if you're cooking primarily for health or financial reasons, the quality and price of ingredients will be much more important.

If your goals are social, the end state isn't the food on the plate; it's the perceptions that are brought about by the experience of eating. If you're making a meal for a romantic interest, give thought not just to the work done in cooking, but also the experience at the table. While you can't control your guest's perceptions, you do have control over the inputs, cooking, and sensations, all of which inform and shape those perceptions. Even something as simple as preheating your plates so that hot food remains hot can make an impact. (Cold sautéed fish and vegetables? Yuck.) For some, the extra effort of setting the table with nice flatware or festive plates can be a powerful signal of attention and affection.

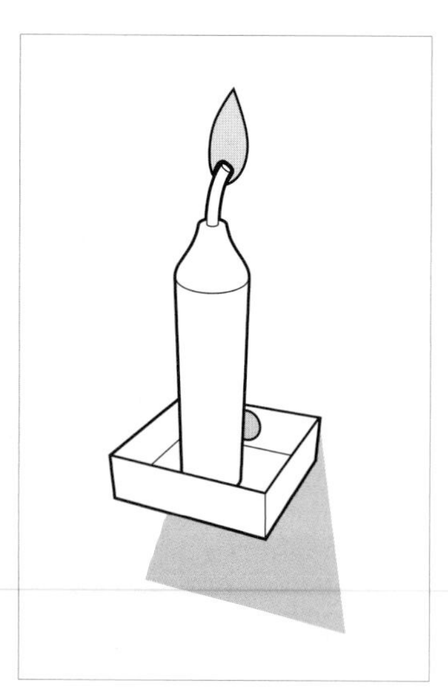

The solution to Duncker's Candle Problem, at least according to Duncker, is to use the box holding the nails as a makeshift shelf to hold the candle.

Here's a way of thinking about this visually:

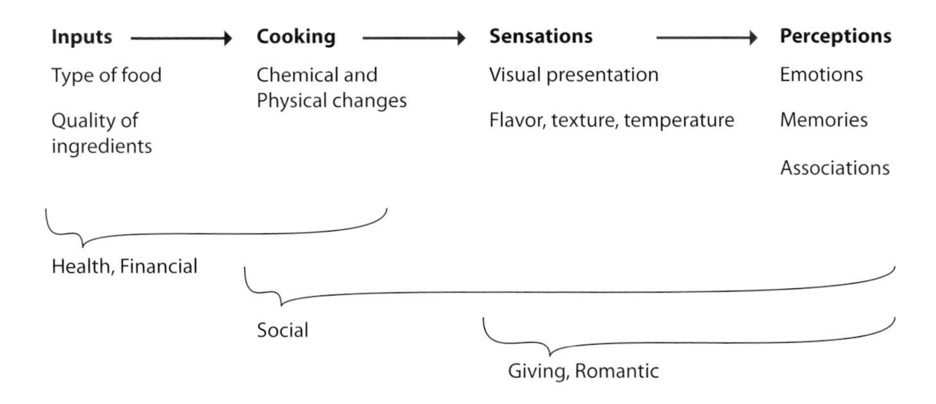

Stages and reasons for cooking.

We'll cover the first column of the stages and reasons for cooking diagram, *Inputs*, in Chapter 3 and give *Cooking* its due respect in Chapters 4 and 5. Some elements from the final two columns, *Sensations* and *Perceptions*, are covered indirectly in Chapters 6 and 7, since playing with textures and presentations is a great way to evoke memories. But the essence of sensations and perceptions is much more in the personal domain. If your reasons for cooking include being social, giving, and romantic, consider how to draw upon these aspects as you try things from this book.

Finally, for those who say presentation doesn't matter, think about dining-hall food, and then check out Fancy Fast Food (*http://www.fancyfastfood.com*). How we approach food, from a food psychology and consumer behavior perspective, impacts our experience much more than we are typically willing to admit, even when confronted with hard data. See the interview with Brian Wansink on the next page for a story about his graduate students and Chex Mix to get a sense of just how far this denial goes!

Brian Wansink on Cooking Styles

PHOTO USED BY PERMISSION OF BRIAN WANSINK

Brian Wansink is a professor at Cornell University, where he studies the way we interact with food. His book, Mindless Eating *(Bantam), examines how we make our choices about how much and what kinds of foods we eat.*

Tell me a bit about the styles of cooking that you have found.

We find that the nutritional gatekeeper, what we call the person who purchases and prepares the food in their home, controls about 72% of all the food their family eats. They do it either positively or negatively: positively if they serve fruit bowls, negatively if they serve candy dishes.

We did a study of 1,004 North Americans. These were *good cooks*, people considered by themselves and by at least one of their family members to be a far above average cook. We asked probably about 120 questions of all different aspects. We found that about 80% to 85% could be categorized in one of five different ways.

The *giving cooks* are the people who see the food they make as giving love. They tend to be great bakers, very traditional in the recipes they make. There's not a lot of changing or tweaking of the recipes. They're the ones who all the families go to on Thanksgiving or Christmas.

The second one of these good cooks is the *healthy cook*. This shouldn't be that surprising, but these are the people who will sacrifice taste to make something healthy. They eat lots of fish and tend to be most likely an exerciser of all these groups. They're more likely than the others to have a garden as well.

The third group is the *methodical cook*. The methodical cook can pretty much make anything, but she or he has to have that cookbook right in front of them the entire time. After they finish making something, it's going to look exactly like it looks in the cookbook. Their kitchen is going to look pretty much like Iwo Jima. They have some of the skills, but they don't have the familiarity, the "second nature" of cooking that would make them facile in the kitchen.

The fourth group is an *innovative cook*. They cook by second nature. They seldom use cookbooks and if they do they just look at the picture and say, "Yeah, I can do that!" These people are pretty creative in a lot of other areas of their life, too. Cooking for them is almost like painting might be for an artist or messing around with music might be for a musician. It's not just a hobby; it's sort of an expressive release. Innovative cooks are interesting, because very little of their ego is involved in the food they make. If something goes wrong, they're not going to be shattered and cry in the corner for the rest of the day. They're just going to be like, "Eh, tried it and it didn't work, no big deal."

The fifth group of good cooks are also very fun. These are *competitive cooks*. They cook to impress other people. You can kind of consider them to be the Iron Chefs of the neighborhood.

They'll try new things. They'll try weird things, but not because they like new weird things; they want you to leave that night going, "That guy is incredible! Man, that was great!"

If you have two people who are in a relationship, I have to imagine there are combinations of style that lead to some amount of conflict, like a giving cook baking for someone trying to get fit. Do you have any advice for couples or families where the nutritional gatekeeper is doing something that's antagonistic without knowing it?

In most people's lives, about five meals per week can cause conflict. First, breakfasts are eaten at staggered times and since they're kind of low-prep things, the person can do whatever they want to or skip it. Lunches are often eaten offsite and you can pack your own. The action is usually at dinner, but even with dinners, there could be a night or two where people might eat somewhere else, or a night or two when one of them is away. It leaves a very small handful of meals where there is potential conflict. That can be further reduced by having one night when the nondominant cook ends up doing the cooking for everyone else. So, for instance, on Mondays I usually cook for my family to give my wife a break.

What are the easiest habits that a geek can change to help them eat healthier?

This depends on what their issues are. One would be simply to use a smaller plate. A smaller plate leads you to eat 22% less than a larger plate. That only works with fresh food, because if all you're doing is heating up frozen food, you're going to eat all you can heat. Another one would be to serve off the counter. What we find is a person ends up being about two-sevenths as likely to go back for seconds or thirds or fourths if a dish is simply six or more feet away.

How much of those biases and those cues can we counteract by knowing about them? Do they tend to go away once we're aware of them?

I took 60 graduate students and for 90 minutes told them that if I give them a big serving bowl of Chex Mix they will take and eat a lot more than if I give them a slightly smaller bowl. I demonstrated and showed videos. They broke into groups to figure out how they could let this not happen to them. Then I invited them to a big Super Bowl party. We had huge bowls of Chex Mix in one room and slightly smaller bowls of Chex Mix in the other. Those in the room with the big bowls ate about 200 more calories over the course of the night. When they were leaving, I asked "You ate about 50% more than the group in the other room. Do you think the size of the bowl had anything to do with it?" They said no. They would make random stuff up like, "I didn't have breakfast last Tuesday!" Mindful eating might work for some people. For those of us who have 10 things going on, I don't know how we could be much more mindful 21 times a week just for meals, let alone snacks.

We've got wives, we've got kids screaming, we've got lists that we're making in our heads in the middle of dinner, we have four phone calls we need to make when we finish. We're way too busy to do mindful eating, unless you live in a castle by yourself. So for most of us, the solution isn't information; the solution is simply to change our environment so that it works for us. We find that if you give people short wide glasses rather than tall skinny glasses with the same volume, people end up pouring about 32% more into them. Even bartenders pouring a shot will pour more in a short wide glass than a tall skinny glass. They never look at the width of the glass; they just look at the height. You could say, "I must not over pour every time I have a short wide glass." That's ridiculous. We're not going to do that. A better solution is just to get rid of all of your short wide glasses. Now that I'm aware that that could happen it's not going to happen; of course it's going to happen. Just change that cue.

Change the environment. That sounds like it's the secret.

The first line of *Mindless Eating* says the best diet is the diet you don't know you're on. I actually started the last chapter by saying it's easier to change your environment than to change your mind.

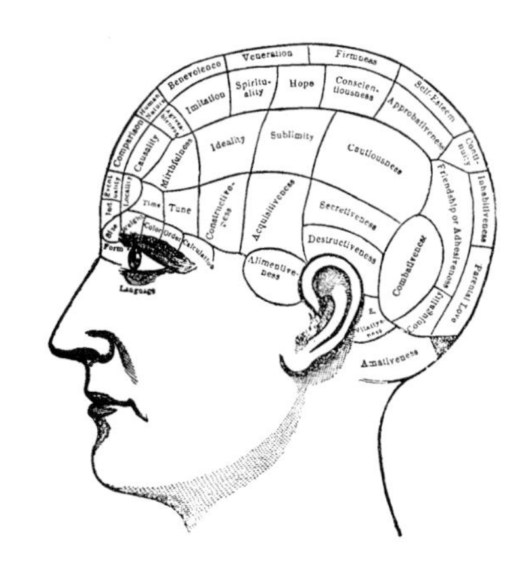

A Few Words on Nutrition

Most of this book deals with cooking for pleasure and enjoyment, but at the end of the day, it *does* come down to making sure your body has the nutrients it needs to keep you going and to keep you healthy. Okay, okay; I know... You probably don't want yet another rant on how you should be eating more veggies. I promise to keep it short.

While our bodies are amazingly adaptive systems, able to tolerate just about anything in the short term, the general consensus is that you'll be happier, healthier, and live longer if you eat the right things. What's "right" isn't the same for everyone, since genetics, metabolisms, and activity levels differ so much. Nor is it easy to prescribe an exact diet, a "perfect" diet, because the human body seems to adapt so well. Just talk to anyone who's tried to lose weight. The human body can adapt to a wide range of eating patterns. After all, we have evolved to survive under less-than-ideal situations. There's even a *New York Times* article about one guy who lives primarily on candy, and it seems to work for him!

Still, there are two general rules for nutrition that you should keep in mind: eat the right amount and eat healthy foods. Portion control is a big issue for many Americans, especially as restaurant meals tend to be way larger than they need to be. And it's easy to plop down on the couch in front of the TV and eat away. You should eat until you're just full, not until you're stuffed or your plate is empty. And while there's no perfect list of foods, you should eat "whole" foods—grains, vegetables, fish, and moderate amounts of meats—and restrict the amount of processed foods, especially those high in sugar, fats, and salt. Personally, I believe eating foods that would have been recognized a century ago is a good rule of thumb.

We eat for two physiological reasons: to provide our bodies with food to break down into energy (via *catabolism*), and to provide our cells with the necessary building blocks to synthesize the chemicals that cells need to function (called *anabolism*). At the simplest level, there are *macronutrients* (proteins, fats, and carbs) and *micronutrients* (trace elements, vitamins). Both provide the chemical compounds that your body needs for anabolism, but it's the macronutrients that provide the energy necessary to read, grocery shop, and cook. As long as you're ingesting enough (but not too much) of each type of nutrient, your body will be good to go. If you're regularly cooking balanced meals for yourself, you probably don't need to worry too much about micronutrients.

When it comes to measuring the amount of energy in food, the standard unit of measurement used in the United States is the *food calorie*, equal to 1,000 *gram calories* (the amount of energy needed to heat 1 gram of water by 1°C). In nutrition, "food calorie" is sometimes capitalized to "Calorie" to distinguish it from a gram calorie and is abbreviated as kcal or C.

(Other parts of the world use joules and kilojoules.) How many calories your body needs depends upon both your body's base caloric requirements and your activity level. If you're a desk jockey, you're probably not burning as many calories as a student running back and forth between classes and the lab. If you routinely eat more calories than your body burns, your body will convert the excess calories into fat, even if the source of those calories wasn't fat. (Sugary fast-food items labeled "low fat" are *not* "low fattening.") Eat too few calories, and your body will either lose weight or slow down your metabolism; that is, slow down the rate of chemical reactions related to anabolism and catabolism, leaving you with less energy. Eat far too few calories for an extended period of time and your body will suffer from malnutrition.

While a calorie is a calorie in the sense of energy, the food you ingest isn't just about energy. Your body needs various types of nutrients for specific purposes. Protein, for example, provides essential amino acids necessary for building and repairing muscle. If you ate only carbs, you wouldn't last long! Unless you have particular dietary needs—say, for athletic training or pregnancy—you're probably getting sufficient quantities of proteins and fats.

Not all fats are created equal, nor are all carbs the same. As a general rule, you want your fats to be liquid at room temperature (good: olive oil, canola oil; bad: lard, shortening), and you want your carbs to not be white (that is, cut down on white rice, white flour, and sugar). As with so many things in food, it's the dosage that matters. A little bit of salt won't hurt you; too much will kill you. Avoid processed foods as much as possible. Most processed foods are engineered for consistency and shelf stability, which usually results in trade-offs of nutritional benefit. Even white flour has its drawbacks: the wheat germ and bran are removed (the oils located in the germ and bran go rancid, so removing them extends shelf life), but the germ and bran are beneficial to our health. Still, if your body needs calories, processed food is better than no food, and the occasional brownie isn't going to hurt.

If all of this still leaves you wondering what to cook for dinner, consider what Michael Pollan wrote in the *New York Times* ("Out of the Kitchen, Onto the Couch," August 2, 2009):

> *I asked [food-marketing researcher Harry Balzer] how, in an ideal world, Americans might begin to undo the damage that the modern diet of industrial prepared food has done to our health.*
>
> *"Easy. You want Americans to eat less? I have the diet for you. It's short and it's simple. Here's my diet plan: Cook it yourself. That's it. Eat anything you want—just as long as you're willing to cook it yourself."*

Tips for Newbies

Knowing how to overcome functional fixedness problems such as Duncker's Candle Problem requires understanding how to read a recipe and break it down into the individual steps, so that you can control and vary the discrete stages. As with any protocol, understanding the structure is critical; you have to understand a system before you can hack it. Here are a few tips for getting yourself in the right state of mind to learn the kitchen equivalents of programming's "open, read, close":

- Have fun! Learning is about curiosity, not work.
- Know your type. Like to grill? Then grill. Rather bake? Then bake.
- Read the whole recipe before starting, and make sure you understand each step.
- Take time to taste things, both to adjust seasoning and to learn how the taste changes during cooking.
- Don't be afraid to burn dinner!

Have fun!

I was talking with a friend of mine, a fellow geek who was just starting to learn to cook, when he said:

> *I was never that curious about cooking, so I thought that buying* The Joy of Cooking *and going through it would be the right approach. That's probably like sitting down with Donald Knuth's* The Art of Computer Programming *in order to learn to program, when really all you should be doing at first is trying to make something you like.*

He's right: make something you like, give yourself enough time to enjoy the process, and have fun doing it. Slaving through the *Joy* or Knuth will work, but it's not the way most people learn to cook or write code. It'd be like picking up a dictionary to learn how to write. The culinary equivalent of *The Oxford English Dictionary* or *The Art of Computer Programming* is Harold McGee's *On Food and Cooking* (Scribner). It's a fantastic reference and a substantial contribution to our understanding of the everyday processes in food, and you should make space for a copy on your bookshelf. But it is not a book for learning how to cook.

If there's one secret about learning how to cook, it's this: have fun in the kitchen. Go experiment. Play. Take that hacker curiosity that you use in front of the keyboard and bring it with you into the kitchen, to the grocery store, and on your next meal out. Cook to please yourself. Doing someone else's work is nowhere near as much fun as working on your own projects, and it's no different in the kitchen: pick something *you* want to learn how to cook and try making it.

Caught between two different ways of cooking something? Do an A/B test: make it one way, then a second way, and see which works better.

Don't cook a new dish for an important guest. If you're nervous about how it'll turn out, cook for just yourself, so you don't have to worry about trying to impress someone (especially a potential romantic interest!). It's entirely okay to screw up and toss it in the trash; it's no different than a programmer refactoring code. Most people's first drafts of software, food, or books need refinement before they're ready to ship. Sure, it'll hurt a little on the wallet, but it's not wasted: you *did* learn something, yes? Success!

Finally, don't expect your cooking to taste exactly like restaurant or packaged foods. For one, a lot of commercial cooking is designed to appeal to the palete via a salty, fatty, or sugary assault on the senses. Tasty? Yes. Healthy? Not exactly. Learning to cook is a great way to control what you eat and, by extension, your health.

Know your type

There are two kinds of people in the world: those who divide people into two kinds, and those who don't. Or is it 10 types: those who know binary, and those who don't? All joking aside, "binning" yourself into the right category will make the learning process a whole heck of a lot easier. And in case you're the irreverent type who insists you don't fit into any standard category, work with me here. Consider the following: *vi* or *emacs*? Windows or Mac? PHP or Python? Sure, you might not have strong preferences, but it's still clear that divisions exist.

The culinary world has its divisions, too. The biggest one in the professional world is that of *cooks* versus *bakers*. Cooks have a reputation for an intuitive, "toss it into the pot" approach, adding a pinch of this or a dash of that to "course-correct" along the way. Bakers are stereotypically described as precise, exact in their measurements, and methodically organized. Even culinary schools such as *Le Cordon Bleu* split their programs into cooking ("cuisine") and baking ("patisserie"). But this is probably due to the differences in technique and execution. Cooking is split into two stages: prep work and then an on-demand, line-cook portion. Pastry and baking is almost always done production-style, completed in advance of when the order comes in.

This isn't to say that professional cooks hate baking, or that bakers don't enjoy cooking. But if you find yourself dipping your finger into the cake batter and being tempted to add more of this or that, pay attention to what it means. If you're the type who really likes to have an exact set of instructions to follow, taking the guesswork out of the process, learn to relax and develop your kitchen instincts when whipping up a dinner. Give yourself permission to dislike some parts of cooking. For most of us, it's a hobby, not a profession, so it's okay to skip the culinary equivalent of documenting your code.

What Type of Cook Are You?

When I prepare a meal, I typically:

a. Rely on classic dishes my family has always enjoyed

b. Substitute more healthful ingredients

c. Follow a recipe step-by-step

d. Rarely use recipes and like to experiment

e. Go all out and try to impress my guests

Some of my favorite ingredients are:

a. Lots of bread, starches, and red meat

b. Fish and vegetables

c. Beef and chicken

d. Vegetables, spices, and unusual ingredients

e. A trendy ingredient I saw on the Food Network

In my free time I like to:

a. Visit with friends and family

b. Exercise or take a fitness class

c. Organize the house

d. Take part in creative or artistic pursuits

e. Be spontaneous and seek adventure

My favorite things to cook are:

a. Home-baked goodies

b. Foods with fresh ingredients and herbs

c. Casseroles

d. Ethnic foods and wok dishes

e. Anything that lets me fire up the grill

Other people describe me as:

a. Really friendly

b. Health-conscious

c. Diligent and methodical

d. Curious

e. Intense

There may be overlap in the answers you give, but is there one letter that you picked most often? Here's what your answers say about your cooking style:

a. **Giving:** Friendly, well-liked, and enthusiastic, giving cooks seldom experiment, they love baking, and like to serve tried-and-true family favorites, although that sometimes means serving less healthful foods.

b. **Healthy:** Optimistic, book-loving, nature enthusiasts, healthy cooks experiment with fish, fresh produce, and herbs. Health comes first, even if it means sometimes sacrificing taste.

c. **Methodical:** Talented cooks who rely heavily on recipes. The methodical cook has refined tastes and manners. Their creations always look exactly like the picture in the cookbook.

d. **Innovative:** Creative and trend-setting, innovative cooks seldom use recipes and like to experiment with ingredients, cuisine styles, and cooking methods.

e. **Competitive:** The "Iron Chef" of the neighborhood, competitive cooks have dominant personalities and are intense perfectionists who love to impress their guests.

Used by permission of Brian Wansink, author of Mindless Eating

Avoid PEBKAC-type errors: RTFR!

Avoid Problem Exists Between Knife And Chair–type errors by Reading The F'ing Recipe! Recipes *are* code, although they require some interpretation, so read the recipe, top to bottom, before starting. One interviewee, Lydia Walshin, explains:

> *The biggest, biggest piece of advice that I can give any cook starting out, and even a lot of experienced cooks, is to take a minute, breathe deeply, read the recipe first, and know from the beginning where you think you want to end up. Don't start out thinking you're making a soup and halfway through you find out you're making a stew, because it's a recipe for disaster.*

Every. Word. Matters. I've watched geeks with PhDs in chemistry skip right over steps that say "turn off heat" in the middle of a recipe that involves melting chocolate in simmering port. Turn off heat? But melting things requires heat! In fact, the residual heat from the port will melt the chocolate, and this way you don't accidentally burn it.

It's okay to go "off recipe." In fact, it's a great way to learn; just do it intentionally. Maybe you don't have all the ingredients and want to substitute something else. Perhaps the recipe is poorly written or has errors. Or, as in programming, you can see there's more than one way to do it and you want to do it differently. A recipe isn't a strict protocol, but do understand the suggested protocol before deviating.

There's a lot of room for personal preference in cooking. Just because a recipe for hot chocolate might say "½ cup heavy cream, 1 cup milk," that doesn't mean you *must* use those quantities. As another interviewee put it, "Please, let's get off the recipes!" I couldn't agree more. If you're following a recipe and think it needs more or less of something, or could benefit from an extra spice, go for it. I usually stick to the recipe the first time I make something, but after that, all bets are off. I'll pull out a pencil, make notes, change quantities, drop and add ingredients. I encourage you to do that to this book! After making something, take a pencil and make notes as to what you'd do differently next time. That way, when you next pick up the book, you'll remember how to tweak the recipe to your taste. (And if there are any errors in the text, you won't repeat them.)

If a brownie recipe calls for walnuts, but you really like almonds, yes, it'll still work! Out of vanilla extract? Those chocolate chip cookies will be fine. Your timer says the chicken has been in for the prescribed time, but it's still got that gross, raw chicken look? Pop it back in the oven. (Better yet, use a probe thermometer, as explained on page 63 in Chapter 2.)

In most modern cookbooks, recipes are laid out in two sections: ingredients and methods. The ingredients section lists the quantities and prep steps for each of the ingredients, and the methods section describes how to combine them. Recipes in this book are laid out in a more conversational format that walks you through the recipes with ingredients listed as they come up. Pay attention to the notes, as they show where you can do things differently.

To get started, consider the recipe for hot chocolate on the following page.

The recipes in this book give both weight in grams and standard U.S. volume-based quantities. Sometimes, the weights are rounded up or down a bit. 1 cup milk, for example, actually weighs 256 grams (1 cup = 237 ml). We'll cover when to use weight and when to use volume in Chapter 2, but be aware that the conversions given between the two are sometimes rounded for convenience's sake.

What kind of milk? Whole milk? Skim? If a recipe doesn't specify, it shouldn't matter too much, although as a general rule I tend to split the difference and grab lowfat/semi-skimmed milk. Sometimes the choice is governed by taste preference, so if you're used to that watery stuff or are the stick-of-butter type, go for it. Some cookbooks will specify defaults in their introductions, perhaps defining milk as whole milk. The most common generic term is flour. When it's called for, you can assume that what you need is AP (all-purpose) flour. AP flour really isn't all-purpose; it just has a moderate amount of gluten (10–12%) as compared to cake flour (6–8%) or bread flour (12–14%).

When a recipe calls for something "to taste," add a pinch, taste it, and continue adding until you think it is balanced. What constitutes balanced is a matter of cultural background and personal preference

Hot Chocolate

In a saucepan, gently heat over low heat until hot, but do not boil:

1 cup (250g) milk

½ cup (125g) heavy cream

Once the milk and cream are hot, turn off heat, add, and whisk until completely melted:

3 tablespoons (40g) chopped bittersweet chocolate

Salt to taste. (Why add salt? See page page 97 in Chapter 3.)

Notes

- *Try adding a few pinches of cinnamon or cayenne pepper. For a smoky version, use powdered chipotle peppers. For a lighter version, use just milk.*
- *Be careful not to burn the chocolate! Adding chocolate to hot liquid with the heat off will prevent this.*

Oaxacan Drinking Chocolate

Heat until hot (stovetop or microwave) in a bowl that you can whisk in:

½ cup (125g) whole milk

½ cup (125g) water

Once hot, remove from heat and add 2 to 3 small squares, about 20 to 30 grams, of Oaxacan chocolate ("Mexican chocolate"). Thoroughly whisk to melt the chocolate and combine.

If your mug is wide enough, you can use a whisk directly in the mug, rolling it back and forth between your palms.

Notes

- *The Oaxacan (roughly "o-a-hawk-an")—who live in a region of Mexico known for chocolate production—use a chocolate that has skipped the conching process, which Europeans use to create a smoother chocolate. Oaxacan chocolate has sugar and sometimes cinnamon added into it as well.*
- *Pre-Columbian Oaxacan would have used only water rather than including milk. Try making this variation and compare. For a modern dairy-free version, replace both the milk and water with hazelnut milk, which makes for a fantastically light taste.*

for some ingredients, especially seasoning ingredients such as salt, lemon juice, vinegar, and hot sauce. There's some evidence that some of these preferences are actually a matter of biological differences between the way different people taste, as discussed in Chapter 3.

f(g(x)) != g(f(x)) Translation? Order of operations is important! "3 tablespoons bittersweet chocolate, chopped" is *not* the same thing as "3 tablespoons chopped bittersweet chocolate." The former calls for 3 tablespoons of chocolate that are then chopped up (taking up more than 3 tablespoons), whereas the latter refers to a measure of chocolate that has already been chopped. When you see recipes calling for "1 cup nuts, chopped," measure the nuts, *then* chop; likewise, if the recipe calls for "1 cup chopped nuts," chop the nuts and *then* measure out 1 cup.

Taste == Feedback

Learn to *really* taste things. The mechanical aspects of cooking—combining ingredients, applying heat—come down to smell and taste. Pay attention to your sense of smell and see if you can notice a change in the odors just as the food finishes cooking. Take time to taste a dish and ask yourself what would make it better. And taste things throughout the process of cooking to see how the flavors evolve over time.

One of the first things I was taught in painting class was to be comfortable scraping the still-wet oil paint off the canvas. We were told to paint a still life; a few hours later, our instructor said, "Great, now take the palette knife and scrape the paint off. All of it." Talk about frustration! But it's a good lesson: becoming attached to the current state of something prevents you from being able to see better ways of doing it. In writing, it's called "killing your babies": deleting favorite bits of the text that no longer serve their original purpose. (These are usually pieces of text that are older and have survived rewrites due to emotional attachment.) "Killing your babies" is about getting beyond the current version, about getting from point *A* to a better point: *B*.

How does all this relate to cooking? Given a sauce, stew, cookie dough, whatever food you're working with, its "current state" is *A*. If you taste it and think it's not quite right, how do you get to *B*? Start with *A*, taste it, take a guess at what might make it better, and try version *B*. Turning out great food isn't about following a recipe exactly and getting it right on the first pass; it's about making many small guesses and picking the better choice with each guess.

Try making a guess with a small side portion if you're unsure. Making stew? Put a few spoonfuls in a bowl and season that. Making cookies? Bake just one cookie, see how it comes out, and tweak the dough before making the next one.

How Many Milliliters in a Cup?

It depends. In a standard U.S. cup, 237 ml. But if you're talking about a U.S. "legal" cup, as used on nutrition labels, it's 240 ml. Live in Canada, eh? That'll be 250 ml, please. Or are we British? An imperial cup is 284 ml. This leaves me wondering: is a pint of Guinness actually larger in Ireland?

Randall Munroe of xkcd (*http://www.xkcd.com*) has kindly provided the following guide to converting to metric.

THE KEY TO CONVERTING TO METRIC IS ESTABLISHING NEW REFERENCE POINTS. WHEN YOU HEAR "26°C," INSTEAD OF THINKING "THAT'S 79°F" YOU SHOULD THINK, "THAT'S WARMER THAN A HOUSE BUT COOL FOR SWIMMING." HERE ARE SOME HELPFUL TABLES OF REFERENCE POINTS:

TEMPERATURE

60°C	EARTH'S HOTTEST
45°C	DUBAI HEAT WAVE
40°C	SOUTHERN US HEAT WAVE
35°C	NORTHERN US HEAT WAVE
30°C	BEACH WEATHER
25°C	WARM ROOM
20°C	ROOM TEMPERATURE
10°C	JACKET WEATHER
0°C	SNOW!
-5°C	COLD DAY (BOSTON)
-10°C	COLD DAY (MOSCOW)
-20°C	F**KF**KF**KCOLD
-30°C	F************CK!
-40°C	SPIT GOES "CLINK"

PTOO

CLINK!

LENGTH

1 cm	WIDTH OF MICROSD CARD
3 cm	LENGTH OF SD CARD
12 cm	CD DIAMETER
12.5 cm	WIFI WAVELENGTH
15 cm	BIC PEN
80 cm	DOORWAY WIDTH
1 m	LIGHTSABER BLADE
170 cm	SUMMER GLAU
200 cm	DARTH VADER
2.5m	CEILING
5m	CAR-LENGTH
16 m 4 cm	HUMAN TOWER OF *SERENITY* CREW

SPEED

kph	m/s	
5	1.5	WALKING
13	3.5	JOGGING
25	7	SPRINTING
35	10	FASTEST HUMAN
45	13	HOUSECAT
55	15	RABBIT
75	20	RAPTOR
100	25	SLOW HIGHWAY
110	30	INTERSTATE (65 MPH)
120	35	SPEED YOU ACTUALLY GO WHEN IT SAYS "65"
140	40	RAPTOR ON HOVERBOARD

VOLUME

3 mL	BLOOD IN A FIELDMOUSE
5 mL	TEASPOON
30 mL	NASAL PASSAGES
40 mL	SHOT GLASS
350 mL	SODA CAN
500 mL	WATER BOTTLE
3 L	TWO-LITER BOTTLE
5 L	BLOOD IN HUMAN MALE
30 L	MILK CRATE
55 L	SUMMER GLAU
65 L	DENNIS KUCINICH
75 L	RON PAUL
200 L	FRIDGE

SO, WHEN IT'S BLOCKED, THE MUCUS IN YOUR NOSE COULD ABOUT FILL A SHOT GLASS.

RELATED: I'VE INVENTED THE WORST MIXED DRINK EVER.

55+65+75 < 200

MASS

3g	PEANUT M&M
100g	CELL PHONE
500g	BOTTLED WATER
1 kg	ULTRAPORTABLE LAPTOP
2 kg	LIGHT-MEDIUM LAPTOP
3 kg	HEAVY LAPTOP
5 kg	LCD MONITOR
15 kg	CRT MONITOR
4 kg	CAT
4.1 kg	CAT (WITH CAPTION)
60 kg	LADY
70 kg	DUDE
150 kg	SHAQ
200 kg	YOUR MOM
220 kg	YOUR MOM (INCL. CHEAP JEWELRY)
223 kg	YOUR MOM (ALSO INCL. MAKEUP)

MROWL?

P.S. Which weighs more: an ounce of gold or an ounce of feathers? (Hint: 31 grams in a troy ounce; 28 grams in a normal ounce.)

Sure, to be proficient at something you do need the technical skill to be able to see where you want to go and to understand how to get there. And happy accidents do happen. However, the methodical approach is to look at *A*, wonder if maybe *B* would be better, and rework it until you have *B*. ("Hmm, seems a bit dull, needs a bit more zing, how about some lemon juice?") The real skill isn't in getting to *B*, though: it's in holding the memory of *A* in your head and judging whether *B* is actually an improvement. It's an iterative process—taste, adjust, taste, adjust—with each loop either improving the dish or educating you about what guesses didn't work out. Even the bad guesses are useful because they'll help you build up a body of knowledge.

Taste the dish. It's your feedback mechanism both for checking if *A* is "good enough" and for determining if *B* is better than *A*.

Don't be afraid to burn dinner!

Talking with other geeks, I realized how lucky I was as a kid to have parents who both liked to cook and made time to sit down with us every day over a good home-cooked meal. Because of this background, approaching the kitchen has never been a big deal for me. But for some, the simple idea of stepping into the kitchen sets off panic attacks as the primitive parts of the brain take over (you can blame your brain's *locus coeruleus*; it's not your fault).

Here's the thing. Failure in the kitchen—burning something, "wasting" money, and having to order pizza—is actually *success*. Think of it this way: there's not much to learn when things work. When they fail, you have a chance to understand where the boundary conditions are and an opportunity to learn how to save something in the future when things go awry. Made mac 'n cheese from scratch but the sauce turned out gritty? Spend some time searching online and you'll discover that gritty cheese sauce = "broken" sauce, which is caused by too much heat and stirring, or using nonfat cheese. The key to learning how to cook is to define success as a chance to learn rather than as a perfect meal. Even if dinner does end up in the trash, if you learned something about what went wrong, that's success. Failure in the kitchen is a better instructor than success.

Fear of failure is another thoroughly modern American phenomenon. We're bombarded with images of the perfect Thanksgiving turkey (they probably used a plastic one during the photo shoot), photos of models sporting impossible physiques (thanks, Photoshop), and stories of triumph and success (where they don't disclose the sad parts and trade-offs). Then when we go to try something, we often find it doesn't work for us the way it seems to for others. Setbacks. Negative feedback. No wonder there's so much fear of failure: we've set ourselves a bar so high that it simply doesn't exist.

There's a generation of Americans hung up on being perfect. The perfect white teeth, the perfect clothing, the perfect "carefree" tossed-together wardrobe. Helicopter parents. Overly critical Yelp.com reviews that rag on everything, down to who cuts our hair and the food we eat. Insane expectations in reviews on Amazon.com about the books we read. (A good book is one that gives you more value than the cost of the book and your time. Be kind. ;-)) No wonder why some parts of American society seem to match the DSM-V criteria for schizophrenia: we're literally going insane trying to be perfect when it just isn't possible. It's much easier to love yourself for who you are than to try to be perfect (the latter will never bring true happiness), and it's much easier in the kitchen to aspire to "fun and tasty" than the *perfect* 16-course gourmet meal (although attempting it can be fun on occasion).

Be okay with being "just" good enough. Part of the appeal of Julia Child was her almost-average abilities and her "nothing special" aura. The reason some people fear Martha Stewart is because her cooking looks perfect and always comes out perfectly on the first try. (I have the world of respect for Ms. Stewart.) Given her background—starting a catering business out of her basement—she had to be perfect to succeed. (Wedding days have to be perfect, no?) This quest for perfection comes at a real cost, though; even if it's achievable for a day, it isn't practical day-in, day-out.

Set reasonable goals, and expect to get frustrated on occasion. Cooking well takes practice. Play around with various ingredients and techniques, and come up with projects you want to try. (*Mmm, bacon and egg breakfast pizza.*) It's like learning to play the guitar: at first you strive just to hit the notes and play the chords, and it takes time to gain command of the basic techniques and to move on to the level where subtle improvisation and nuanced expression can occur. If your dream is to play in a band, don't expect to get up on stage after a day or even a month; start by picking up a basic book on learning to play the guitar and practicing somewhere you're comfortable.

A beta tester for this book commented:

> *While there are chefs with natural-born abilities, people have to be aware that learning to cook is an iterative process. They have to learn to expect not to get it right the first time, and proceed from there, doing it again and again.*

What about when you fubar (foobar?) a meal and can't figure out why? Think of it like not solving a puzzle on the first try. When starting to cook, make sure you don't pick puzzles that are too difficult. Start with simpler puzzles (recipes) that will allow you to gain the insights needed to solve the harder ones. And give it time. You might have days when you feel like you've learned nothing, but the cumulative result will lead to insights.

If a recipe doesn't work as well as you'd have liked, try to figure out why and then try it again. It might also be the fault of the recipe, or that the recipe is simply too advanced. I know some newbies who have gotten stuck trying to perfect one dish. They usually burn out in frustration. If you're not happy with the results of your early attempts, try a different source of recipes. Some books, especially those from top-tier restaurateurs such as Chefs Thomas Keller or Grant Achatz, are highly technical and complicated. Don't begin with these recipes; instead, pick recipes that limit the number of variables to just a handful that you can manage.

"A Kinder, Gentler Philosophy of Success"

Celebrity Chef Gordon Ramsay has carved out a niche as a raging culinary maniac. (Secretly, I bet he's "tough but gentle on the inside," and that the TV series *Hell's Kitchen* has edited the footage to exaggerate his hot temper.) Getting results doesn't have to be about fear and intimidation, though. There's a great TED talk (TED is an annual conference loosely related to "Technology Education Design") by Alain de Botton available online, called "A kinder, gentler philosophy of success"; see *http://www.ted.com/talks/lang/eng/alain_de_botton_a_kinder_gentler_philosophy_of_success.html.*

Picking a Recipe

I hope by now I've convinced you that it's okay to burn the meal, to read the entire recipe before starting, and that xkcd is awesome. (Maybe you already knew all these things...) You're ready to venture into the kitchen and want to make your favorite dish. Where to start?

If you're new to cooking, do what experienced programmers do when encountering a new language: look at a few different examples. Don't just print out and follow the first recipe you find; that'd be like downloading a random executable and running it. Recipes should be treated as reference implementations, especially if you're cooking a dish that you've never made before. Pull up a couple of examples, consider what the authors were doing, and ask yourself what makes their "code" (recipe) work and what parts of their "code" apply to what you want to do.

Let's use pancakes as an example. They're quick to make, and the ingredients are cheap, so if you want to try variations, you're not risking financial meltdown. (One friend told me about learning to de-bone animals in culinary school. It basically amounted to "do it 100 times, and by the time you're done, you'll know how to do it." That's gotta add up.) Start by going online and searching for "pancake recipe."

Here's the thought process that runs through my head. I look at three or four different recipes, reading through at least the ingredient list. Of the recipes, usually one of them will be an outlier with odd instructions or require something I don't have on hand—say, yogurt—or call for ingredients that I'm not in the mood for (nothing against peanut butter).

Of the remaining few, I look at the ratio of ingredients and pick one that looks reasonable. Don't worry if it turns out not to *be* reasonable; you'll learn afterward that it wasn't—that's the point!

Then, it's off to the kitchen. The first time I make something from a recipe, I try to follow it as precisely as I can, even if I think it should be altered. For pancakes, I might think the batter is too runny (add more flour) or thick (add more milk). Or maybe the batter looks fine, but it comes out too thin (add more baking powder for cakier pancakes). Regardless, on the first pass, I remain true to the recipe because sometimes it'll surprise me. I love it when this happens; it means there's something I don't understand and it gives me a chance to correct my mental model of how things work.

If you really want to geek out, print out a handful of recipes and figure out the ratios between the ingredients in each recipe, and then look at the differences in the ratios between the recipes. Why would some ingredients remain at a relatively constant percentage of the recipe while others differ? Even if you can't answer the *why*, you'll have a huge clue as to what's critical in the recipe. If buttermilk pancake recipes always call for baking soda, there's probably a chemical reaction going on with the baking soda. Compare those recipes to nonbuttermilk ones. Besides leaving out the buttermilk, there's no baking soda. From this, you can infer that the baking soda is reacting with the buttermilk. Sure enough, buttermilk has a pH of 4.4–4.8, while regular milk has a pH of ~6.7, so it follows that baking soda will buffer and neutralize the more acidic buttermilk. (See page 239 in Chapter 5 for more about the chemistry of baking soda.)

The following chart is the breakdown for eight pancake recipes that came up for me on an Internet search for "pancake recipes."

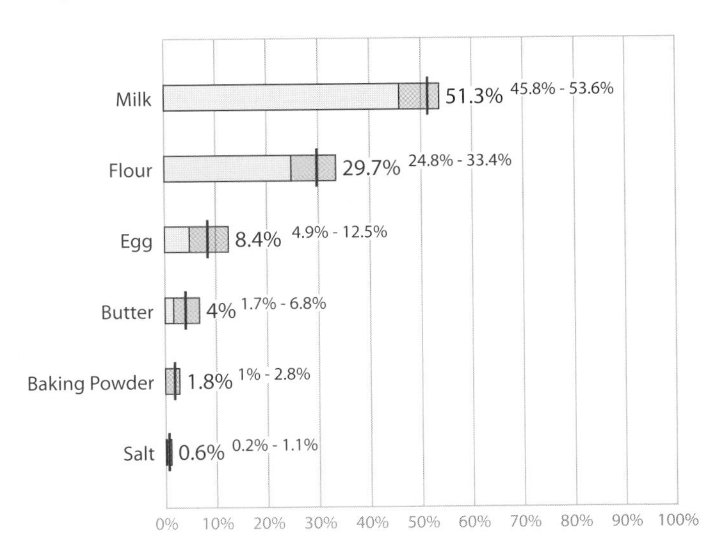

Eigen Pancakes: The Hello, World! of Recipes

No one's ever wrong on the Internet, so the average of a whole bunch of right things must be righter, right? The quantities here are based on the average of the eight different pancake recipes from an online search. For each ingredient, I converted the measurement to grams and then calculated that ingredient's percentage of the total weight of the recipe. (This is somewhat like a "baker's percentage," in which ingredients are given as a percentage of the weight of flour in a recipe.)

In a mixing bowl, measure out and whisk together:

> **1½ cups (190g) flour**
> **2 tablespoons (25g) sugar**
> **2 teaspoons (10g) baking powder**
> **½ teaspoon (3g) salt**

In a separate, microwave-safe bowl, melt:

> **2 tablespoons (25g) butter**

By default, assume that the order of ingredients in a recipe indicates the order in which you should add the ingredients into your bowl. It doesn't always matter, of course, but in this case you should add the milk before the eggs to prevent the eggs from cooking in the hot butter.

Add to the butter and whisk to combine thoroughly:

> **1¼ cups (330g) milk**
> **2 small or 1 extra-large (80g) eggs**

Pour the dry ingredients into the liquid ingredients and mix them together with a whisk or spoon until just incorporated. Little pockets of flour are okay; you want to avoid overstirring the batter to minimize the amount of gluten formed from two proteins, glutenin and gliadin, present in flour (they crosslink and bond together to create a stretchy, net-like matrix—think French bread).

Place a nonstick frying pan on a burner set to medium-high. Wait until the pan is hot. The standard test is to toss a few drops of water into the pan and see if they sizzle; the geek test is to take an IR thermometer and check that the pan is around 400°F / 200°C. Use a ladle, measuring cup, or ice cream scoop to pour about half a cup of batter into the pan. As the first side cooks, you'll see bubbles forming on the top surface of the pancake. Flip the pancake after those bubbles have started to form, but before they pop (about two minutes).

Wolfram|Alpha (*http://www.wolframalpha.com*) is a great resource for converting standard measurements to metric. Enter 1T sugar and it'll tell you 13g; enter the entire Eigen Pancake recipe—use "+" between individual ingredients—and it'll tell you 38 grams of fat, 189 grams of carbs, and 46 grams of protein.

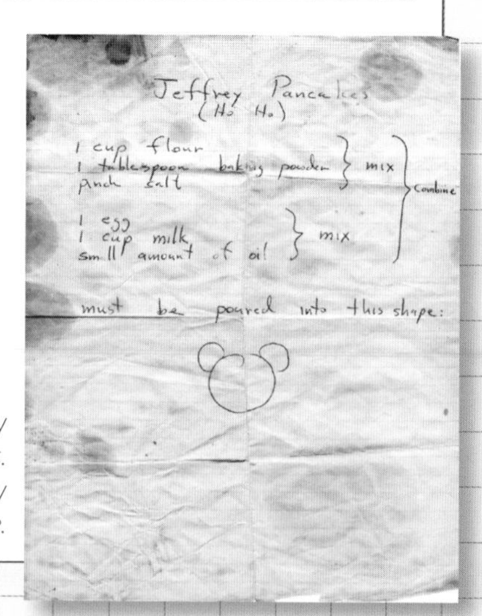

Jeffrey Pancakes
(Ho Ho)

1 cup flour
1 tablespoon baking powder } mix
pinch salt

1 egg
1 cup milk } mix
small amount of oil

combine

must be poured into this shape:

Pancakes are a great way to teach cooking to kids. This is the first recipe my parents taught me.

Notes

- *If you use a nonstick frying pan, you don't need to butter the pan first. If you're using a regular sauté pan, butter it and then wipe out as much of the butter as you can with a paper towel. Too much butter on the surface of the pan will prevent parts of the pancake from reaching temperatures hot enough for browning reactions, as you can see here in the pancake on the left.*

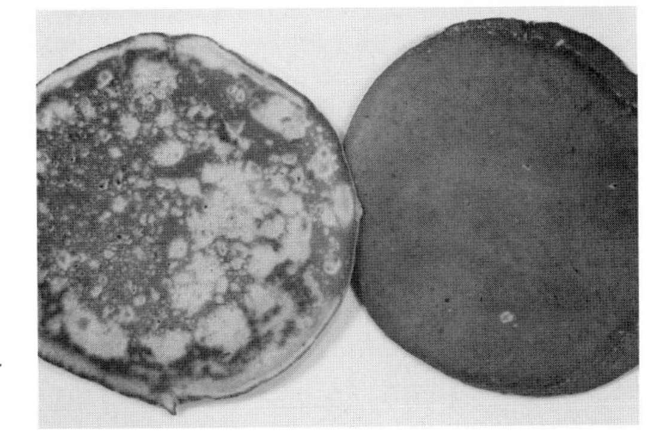

Using a scale with gram weights? Hit the "tare" button after adding each ingredient.

- *When a recipe calls for an egg, what size should you use? By default, use large eggs, unless you're in the EU, in which case use medium eggs. (Fun egg trivia: quail eggs weigh about 9 grams on average, while duck eggs weigh around 70 grams. I was happy to learn that ducks themselves also weigh about 8 times more than quails.)*

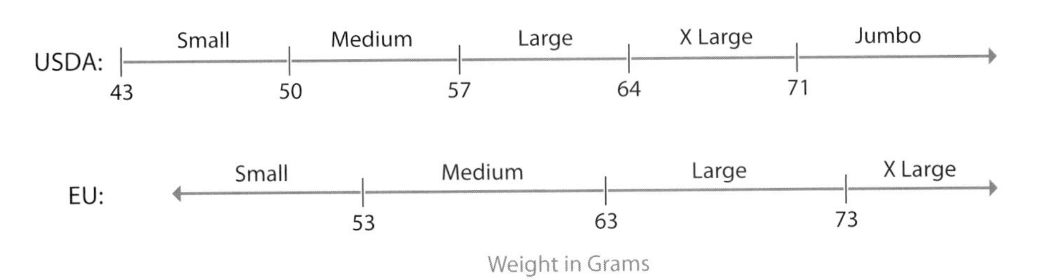

- *Cracking an egg? Tap it on the counter, not the edge of a bowl. The shell of an egg cracked on a flat surface will have larger pieces that aren't pushed into the egg. Eggs cracked on a sharp lip are much more likely to have little shards of shell poked into them that then end up in the bowl and have to be fetched out.*

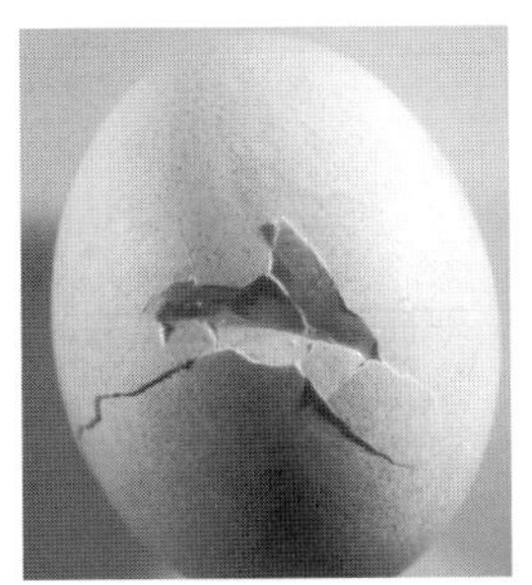

Egg cracked on edge of bowl.

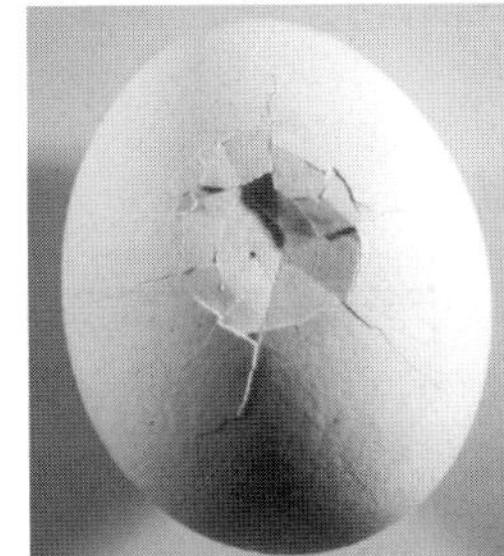

Egg cracked on flat surface.

Reading Between the Lines

If you're still with me and haven't already skipped to "the fun part," here are a few more thoughts on recipes, plus my favorite dish, duck confit sugo.

Recipes are, by definition, documentation of what works for their authors. When reading a recipe, realize that it's only a suggestion, and it's also abbreviated. Unlike software, where given the same code a machine will execute it the same way regardless of the hardware (at least in theory), the same recipe given to a dozen different experienced chefs will yield a dozen variations.

The main reason for differences in execution isn't inexperience or error; it's that recipes themselves are only notes, like a score or script. The oldest known recipe dates back to around 1,700 BC and reads more like a tweet than a scientific protocol.

For a good history on food, see *http://www.foodtimeline.org*.

It wasn't until sometime in the 1800s that cookbooks began to give more precise measurements (for an example, search Google Books for "Boston Cooking-School Cook Book"). Even with today's precise measurements, variability in ingredients is beyond our control. A teaspoon of your dried oregano won't necessarily be the same strength as a teaspoon of the recipe author's dried oregano, due to age, breakdown of the chemicals (carvacrol in this case), and variations in production and processing. To a good chef, recipes aren't about exact reproductions of the original work; they're reminders of combinations, ratios, and steps.

In most recipes, the measurement accuracy exceeds the error tolerances; that is, if you're off by a few percentage points on quantity of flour, it won't make a marked difference in the results. In cooking, does it matter that this egg has slightly more of the compound lecithin than that egg, or that one onion has slightly more water than a second one? Probably not. But in baking, the error tolerances are tighter than they are in cooking. It's a small difference between the right amount of water to hold a dough together and too much water, making the dough too wet to dry when baking. In these sorts of situations, the recipe should give you hints such as "drizzle water into food processor until the dough forms a ball." When you see these kinds of things, you'll get better results if you think to yourself, "why would the recipe say to hold back some of the water and drizzle it in?"

Maureen Evans' Twitter Recipes

PHOTO BY BLAINE COOK; USED BY PERMISSION OF MAUREEN EVANS

Maureen Evans posts recipes on Twitter (@cookbook) limited to the length of a single tweet (140 characters). See her book Eat Tweet *(Artisan) for more recipes.*

How did you come up with the idea to tweet recipes?

I noticed there wasn't a place for home cooking in people's lives anymore, but there was still a lot of passion about the idea. I just on a whim started writing these recipes on Twitter because a lot of my friends are geeks and they were the early adopters.

Fitting something down into 140 characters has got to be a real challenge.

I approach cooking as a metaprocess. I've never been one with a lot of patience for long-winded recipes. When I'm in the kitchen, I look at the overall processes and steps. I don't find it too tricky to apply the way I think about recipes into these tiny clauses. I've sort of invented a grammar around cooking based on punctuation in order to cram as much information in as possible. Even a semicolon can convey a break in the cooking steps.

It sounds like the recipes are written for somebody who already knows how to perform the individual steps?

The first step when I write the recipe is trusting the intelligence of the reader, even if they're just a beginning cook. People have the capacity to figure things out and to know when something looks right or wrong. I think the most geeky aspect of the recipe is the spirit that anyone can try things and tinker with them and learn by doing.

Any final thoughts?

Really, really old recipes tended to be about one line. I have one here from *Famous Old Receipts* by Jacqueline Harrison Smith from 1906 for popovers and all it says is "Popovers: 1 egg well beaten, 1 cup sweet milk, 1 cup sifted flour, salt to taste. Drop in hot tins and bake quickly." That's it. People today are kind of freaked out by making that, but obviously back in the day, 112 characters was enough to explain stuff.

RT @cookbook...

Hummus: soak c dry chickpea 8h. Replace h2o; simmer3h@low. Drain. Puree/season to taste+⅓c tahini&lem&olvoil/½t garlic&salt/cayenne. Chill.

Garden Soup: brwn ½lb/225g chopd chickn/onion/3T oil; +t s+p/2T wine&herbs/bay/2c pep&carrot&tom/c orzo/5c h2o. Simmr15m. Top w parm&olvoil.

Yam Leek Soup: saute leek&onion/T buttr/t piespice. Simmr20m+4c stock/3c yam/tater/bay. Rmv bay; blend+½c yogurt/s+p. Top w tst pumpkinseed.

Lemon Lentil Soup: mince onion&celery&carrot&garlic; cvr@low7m+3T oil. Simmer40m+4c broth/c puylentil/thyme&bay&lemzest. Puree+lemjuice/s+p.

Spice Cookies: cream 8T butter&sug; +egg/t vanilla. Mix+2c flour/t cinnamon/½t bkgpwdr&salt/dash cayenne. Chill/roll/cut~25. 15m@350°F/175°C.

TWEET RECIPES USED BY PERMISSION OF MAUREEN EVANS

Duck Confit Sugo

Prepare four duck legs, confit-style, as described in Chapter 4 (see page 192). This can be done days in advance and stored in the fridge. If you're not in the mood to wait a day or don't have a slow cooker, check if your grocery store sells prepared duck confit.

In a large pot, bring salted water to a boil for making pasta.

Prepare the duck meat by pulling the meat off four legs of duck confit, discarding the bones and skin or saving them for stock. In a pan, lightly sauté the duck leg meat over medium heat to brown it.

Add to the pan:

28 oz (1 can, 800g) diced tomatoes

8 oz (1 can, 225g) tomato sauce

¼–½ teaspoon (0.25–0.5g) cayenne pepper

Simmer the tomatoes and tomato sauce for five minutes or so. While the sauce is simmering, cook the pasta per the directions on the package:

⅓ pound (150g) long pasta—ideally, *pappardelle* (an egg-based noodle with a wide, flat shape) or spaghetti

Once the pasta is cooked, strain (but do not rinse) the pasta and add it to the sauté pan. Add and stir to thoroughly combine:

2 tablespoons (2g, about 12 sprigs) fresh oregano or thyme leaves (dried is nowhere as good)

½ cup (100g) grated Parmesan cheese

¼ cup (50g) grated mozzarella cheese

Want to see a video of me making this?
Visit http://www.cookingforgeeks.com/book/duck-confit-sugo/

Notes

- *You might find it easier to transfer the duck mixture to the pasta pot and stir in there, because your frying pan might not be big enough. When serving, you can grate Parmesan cheese on top and sprinkle on more of the oregano or thyme leaves.*
- *The secret to duck confit sugo is in its combination of ingredients: the heat of the capsaicin in the cayenne pepper is balanced by the fats and sugars in the cheese, the fats in the duck are cut by the acids in the tomatoes, and the aromatic volatile compounds in the fresh thyme bring a freshness to this that's just plain delicious. If the world were going to end tomorrow, I'd want this tonight.*

So, what can go wrong in making this dish?

- *Hot or cold pan? Any time you see a recipe call for something to be sautéed, that means you should be browning the food. Maillard reactions begin to occur at a noticeable rate at around 310°F / 154°C, and sucrose (sugar) caramelization and browning start to occur at around 356°F / 180°C. (We'll cover these two reactions in Chapter 4.) You'll have a hard time getting those reactions to occur when putting cold duck into a cold pan. On the other hand, you don't want an empty pan to overheat, especially if you're using a nonstick frying pan, which can offgas chemicals when too hot. When sautéing, heat the pan empty, but keep an eye on it to make sure it doesn't get too hot. (You can hover your hand above the surface to check for radiant heat.)*
- *When separating the duck meat from the duck fat, skin, bones, and gelatin (the clear gloppy stuff that's culinary gold), how do you determine what's good and what's not? Duck fat will be whitish and slippery; the meat will be darker and more strand-like. When in doubt, if it looks yummy, it probably is. And yes, the duck confit is already cooked, so feel free to sample the goods. Since the meat is to be browned, you want to avoid the gelatin, as it will melt and then burn as the water boils off.*
- *When pulling fresh thyme off the stem, be careful not to get the actual stem in the food. It's woody, chewy, and not enjoyable. Pinch the top of the stem with one hand and run the fingers of your other hand down the stem, against the direction the leaves grow in, to strip them off.*

Start by gripping near the bud end of the plant.

To strip the leaves, run your fingers down toward the base of the stem.

Lydia Walshin on Learning to Cook

PHOTO USED BY PERMISSION OF LYDIA WALSHIN

Lydia Walshin is a professional food writer who also teaches adults how to cook. She founded Drop In & Decorate, which supports and hosts cookies-for-donation events to benefit local community agencies and shelters in more than 30 U.S. states and across Canada. Her blog, The Perfect Pantry, is at http://www.theperfectpantry.com.

Tell me a bit about your blog.

The Perfect Pantry is a look at the 250 ingredients that are in my pantry with recipes for how to use each ingredient. Every time I would open my refrigerator or go into the cupboard, there were things I used every day in the front, and then things in the back that I'd bought for one particular recipe and never used again. The blog is about all of the ingredients in the pantry and giving people different ways to use the things they've already bought.

How do you go about learning what to do with unfamiliar ingredients?

The best way to learn how to use something new is to substitute it in something familiar. So, for instance, I have a great butternut squash soup that I make in the fall and winter. When I get a new spice that I think might have similar characteristics to something in that soup, I start by making a substitution. First, I'll substitute part of the ingredient for part of another ingredient, and I'll see how that tastes. And then maybe I'll substitute that ingredient entirely.

Using the butternut squash soup as an example, my recipe uses curry powder, which in itself is a blend of many ingredients. Recently I discovered an ingredient called vadouvan, a French curry powder. How do I learn the way vadouvan behaves? I put it into something I already know, and I say, if I take half the curry powder and substitute it for vadouvan, how does that change the taste? And then the next time I make it, I substitute vadouvan for all of the curry powder, and how does that affect the taste?

Once I understand the difference between something that's familiar and something that's unfamiliar, then I can take that into other kinds of recipes. But if I start with a recipe that I don't necessarily know, and it uses an ingredient I don't know, then I don't know what the ingredient has done in the recipe, because I can't isolate that ingredient from the recipe as a whole.

You speak of isolating ingredients almost like how a programmer would when writing code: isolating one variable at a time and going through and changing just one thing to see how the system changes. One thing I think a lot of geeks and techies forget to do is to take that set of skills that they have in front of the keyboard and go into the kitchen with that same set of skills, to apply that same methodological approach to food.

Except that the outcome might not be as quantifiable or predictable when you're cooking, but that's the nonscientific part of me saying that cooking, to me, is both an art and a science. You need to know some fundamentals. It only takes one time making tomato sauce in a cast iron pan to realize that it's not a good idea from a science point of view and from a taste point of view—it's pretty darn awful to watch your sauce turn green and bubbly. So you need to understand the basic fundamentals of science in order to cook, but you don't need to be a scientist in order to cook, and you need to accept the fact that your outcome might be a bit more random than if you were sitting in a computer lab.

What are the basic science fundamentals that you see people failing to understand?

I can tell you the place that I fail constantly is in baking. I'm not a baker, I'm not a measurer. I still don't really understand the difference between baking powder and baking soda. I think understanding how acids react with all food is important. And understanding how salt reacts with foods, understanding things that are opposites. If something is too salty, you don't necessarily put in something sweet in order to balance the flavor. Having a fundamental knowledge of those things enables you to cook without recipes and also to look at a recipe that might be faulty and figure out where the faults are.

One of the assumptions that many cooks make, especially when they're starting out, is that if it's published in a cookbook, it must be true. Well, not true. Cookbooks are subject to bad editing, bad writing, and bad recipe testing. You can faithfully follow a recipe in a cookbook and at the end have something that goes right into the garbage disposal, and it's not your fault. It might be your fault, but there's an equally good chance that it's a flaw in the recipe. If you know enough basics, you can look at a recipe and say, "Wait a minute, there must be a typo here, this just isn't going to work."

The first thing that I tell students when they come to cooking classes with me is to read the recipe all the way through. This is the biggest reason that people's cooking goes awry; they do not read the recipe. You can't start cooking without reading to the end and knowing where you're going. The biggest, biggest piece of advice that I can give any cook starting out, and even a lot of experienced cooks, is to take a minute, breathe deeply, and read the recipe first. Know from the beginning where you think you want to end up. Don't start out thinking you're making a soup and halfway through find out you're making a stew, because it's a recipe for disaster. That's not science; that's common sense.

I'm surprised at how often the people I interview say that it seems like people just aren't using common sense.

I think there are several reasons for that. One is a lack of confidence that comes from not having grown up around cooks so that you're afraid to trust your instincts, whereas if you've grown up making cookies with your grandmother for your whole life, then a cookie recipe wouldn't scare you. I think there's some fear of failure there.

I think there's another thing that probably relates to science and common sense more than people realize: don't poke at your food while you're cooking it. If you have something cooking in a hot pan, and the recipe says either sauté it or cook until the onions are wilted or whatever it is, if you get in there with a spoon and you keep moving it around so that the food doesn't have any chance to come into contact with the heat, whether you're actually stirring or, as more often happens, poking, that food doesn't ever cook. People see a recipe that says stir constantly; that doesn't actually mean that you have to stir it so much that the food never gets hot. I have had to confiscate spatulas and spoons from my cooking students, set a timer and say, "Until the timer rings, you cannot stir the food again!"

I teach adults, so these are people who have been feeding themselves for their whole lives, and yet when I take a look at where their downfall is on a recipe, it's always two things: they haven't read the recipe, and they're not giving the food a chance.

It does seem like there's an American fetish with overstirring and poking.

Judging by what I see in my classes, absolutely.

Why do you think there's a fear of cooking?

Honestly, I see this more in younger people, people in their 20s and 30s. I think our entire way of raising kids, educating kids, all of the pressures that we read about to succeed, and whatever punishment there seems to be for failure, seems to have translated to the kitchen. Not only are you supposed to be great at your job and great at being a parent, but you're supposed to be a gourmet cook; and if you're not a gourmet cook, then the fault is yours. I think that's really kind of sad. Julia Child had the right idea: you drop a chicken on the floor, you pick it up, you wipe it off, and you just carry on. We have come to take cooking too seriously. We've come to take ourselves too seriously.

For me, once it stops being fun, I'm going to give it up, because I really do think that you should have a good time in the kitchen. I think you should make a mess in the kitchen. I think you should put some things down the disposal if nobody really should eat them, and then you should go out for pizza, and it's all okay. We don't let it be okay anymore. That's me. That's my rant.

I'm surprised at how often the line of "you can always order pizza" comes up. That seems to be the universal go-to for when dinner ends up down the garbage disposal: just order pizza. So the secret to learning to cook in the kitchen is that it should be fun?

I think it's critical that it should be fun, and I understand that there is survival cooking and there's the weekday cooking that I used to do when we had young kids at home, but even that should be fun, and the better you are at it—not in terms of creating gourmet meals but in terms of understanding how to cook, how the fundamentals work, and making it not a painful experience for the cook—the better it is for everybody. If you've created your food with fun, and created it with or for people you care about, or just because you want to sit down and watch *Top Chef* on Wednesday night or whatever it is, if you've had a good time doing it, your food reflects that.

Butternut Squash, Apple, and Vadouvan Soup

- 1 medium (750g) butternut squash, peeled and cut into 2-inch / 5 cm cubes
- 1 small (70g) red onion, peeled and chopped
- 1–2 tablespoons (12–25g) olive oil
- 1 large (150g) tart apple such as Macoun, cored but not peeled, roughly chopped
- 1 tablespoon (6g) vadouvan
- 1 tablespoon (6g) hot curry powder (if your curry powder isn't hot, add ½ teaspoon hot sauce in addition)
- 2 cups (475g) chicken broth or vegetable broth
- 2–3 teaspoons (14–20g) honey, to taste
- 2–3 teaspoons (10–15g) lemon juice, to taste
- ½ teaspoon (1g) fresh ground black pepper, or more to taste

In a large stockpot, place the squash, onion, and olive oil. Over medium heat, cook, stirring frequently, until the onion is translucent and the edges of the squash are starting to soften. Add the apples, and cook two minutes more. Add vadouvan and curry powder and cook, stirring constantly, for two minutes, until the spices are lightly toasted and fragrant. Add the chicken broth. Bring to a boil, then reduce heat to low and cook for 20 minutes or until the squash is quite soft.

Remove the pot from heat and, using an immersion blender, purée the soup until smooth. If you don't have an immersion blender, purée the soup in batches in a blender or food processor and return the soup to the pot. Stir in honey, lemon juice, and black pepper. Return the pot to the stove, and cook over low heat for five minutes. Taste, adjust seasoning, and serve hot.

RECIPE USED BY PERMISSION OF LYDIA WALSHIN

Cooking for One

What about us geeks who eat dinner solo? Cooking for one presents a number of challenges, especially if you don't want to spend too much time or money. Without someone to help share in the cooking and cleanup work, more complicated recipes become less attractive. And the cost of ingredients doesn't scale down linearly, meaning that recipes with longer ingredient lists become less affordable. On the plus side, cooking for yourself has the great advantage of allowing you to truly experiment and improvise without worrying about what others think. Pasta and fish? Chicken in a red wine sauce? Chocolate and beets? The sky's the limit.

Preparing a batch of a particular common ingredient can also save you a lot of time over the course of a week. If you're trying to save money or watching what you eat, try cooking a large batch of chicken breasts or stir-fried tofu on the weekend. Having a batch of precooked ingredients can help challenge you, too. This can be a great way to play with flavors and learn about new combinations as well, since chicken or tofu day-in, day-out by itself can get pretty darn boring. You'll end up being driven to experiment with seasonings!

One way you can reduce the price of ingredients is to amortize it: plan a number of meals in a row that use common perishable ingredients. Unused tomatoes and parsley purchased for a chicken dish can be used with eggs the next morning or in a lunchtime salad. Sticking to specific types of cuisine such as Italian also increases the amount of overlap in ingredients between recipes, since the regional variation in ingredients is much smaller. Another trick: if your grocery store has a salad bar, you can sometimes find the ingredient you're looking for. If I'm making a pizza for myself, I'll sometimes skip buying a whole red bell pepper and yellow bell pepper and snag just the amount I need from the salad bar at my grocery store. The best part? Presliced and already roasted. And because of the "buffet" pricing at the salad bar, I've sometimes found things to be cheaper!

Look for oven-safe plates and bowls. You can cook items like chicken tenderloins directly in the bowl, meaning fewer dishes to wash. Be careful with the hot bowl, though!

Leftover sauces, and sometimes entire dishes, can be recycled as components in entirely new dishes. (School cafeteria food!) Chicken and vegetables from one dinner can be recycled into chicken noodle soup. Tomato sauce made for a pasta dish can be reused in lasagna the next night, and the lasagna can be reused as an unexpectedly delicious filling in omelets. Leftover cake scraps or bread can be turned into bread pudding. Sandwiches are a great vehicle for odds and ends. Using dinner leftovers for breakfast can be a huge, untapped resource for creativity as well. A slice of pizza can be turned into a breakfast pizza by cracking an egg on top and putting it under the broiler for a few minutes. Next time you're in the kitchen, open the fridge door and scrounge around for leftovers, doing your best to see past the functional fixedness that we talked about earlier.

If you do find that cooking for one ends up being too expensive or time consuming, consider finding a cooking buddy with whom you can split the cost of groceries and cooking duties. Getting together with someone on a regular basis to spend a few hours cooking a few days' worth of meals can also help ensure that you eat and socialize regularly, especially for the busy geek.

Cooking for Others

Anytime you have someone else in your home, you're the host and are responsible for taking care of their comfort. This doesn't mean stuffy formality. Even a quick "Hey, good to see you, help yourself to a drink in the kitchen!" goes a long way toward telling the guest what's acceptable. Your responsibility starts the moment you extend the invitation and even includes those times when you're *not* cooking. Inviting people over for a party with "food and drinks"? Make it clear whether you're serving an actual meal or just appetizers.

The most important responsibility you have when cooking for others is keeping them safe from allergic reactions and foodborne illnesses. Ask ahead of time if your guests have any food allergies. Being aware of dietary restrictions and of food aversions or intolerances in advance will spare you last-minute surprises. If you are cooking for someone with a true food allergy, you should take extra precautions to avoid triggering an allergic reaction.

> Take a look at the appendix for information on food allergies and common substitutions.

You might find some guests are vague regarding whether they have an aversion or an allergy, as some people think of food sensitivity as an allergy and don't realize the burden they put chefs through by overstating their needs. I've known individuals who get gas from consuming too much bread. This doesn't make them gluten intolerant, however! Likewise, lactose intolerance is different from lactose allergy: a small amount of lactose will not hurt lactose-intolerant individuals (depending upon their tolerance, they might even be able to taste a milk-based dish without discomfort), whereas those with a true allergy might go into anaphylactic shock and die. When a guest tells you that she is allergic to something, check if it is an intolerance or an outright allergy.

Closely related to food allergies is food preference. Sometimes, you'll be cooking for guests who are following restricted diets, either limiting certain types of foods—e.g., vegetarians (no fish or meat), vegans (no animal products), lacto-ovo-pescetarians (milk, eggs, and seafood okay, but no other meats)—or limiting certain classes of foods—e.g., avoiding saturated fats, simple carbs, or salty foods. Either way, in these cases, pick a menu that keeps various side dishes in different serving containers—putting grilled veggies in one bowl, bread in another, roasted chicken in another—as opposed to, say, making a stew or casserole. You don't need to plan the entire meal around the one individual on a restricted diet, but you should have at least one dish that is suitable. This allows you the freedom to choose anything from your standard repertoire for most of your guests, while still showing consideration and addressing the needs of that individual.

Beyond the actual menu planning, think about doing something special to show that you care. Even a minor touch—a tablecloth, special plates, anything beyond your daily ritual—will communicate thoughtfulness. One of the easiest ways of showing consideration is to have a few appetizers on hand for your guests to snack on before starting the meal. Simple things like bread and olives, pita and hummus, or fresh fruit are quick, easy, and useful for guests who are hungry before the meal is ready or while you wait for remaining guests to arrive—or, in the event of a kitchen meltdown, the pizza delivery guy.

Speaking of kitchen meltdowns, don't try out new recipes with first-time guests. When it comes to picking a recipe, choose something that you're comfortable making. You'll be more relaxed cooking a dish that you're familiar with, which will translate to a more relaxed atmosphere for everyone. This isn't to say that experimenting with dishes needs to be limited to those times when you're cooking alone. I certainly enjoy trying out new things with friends, because their feedback helps me understand how others react to new dishes.

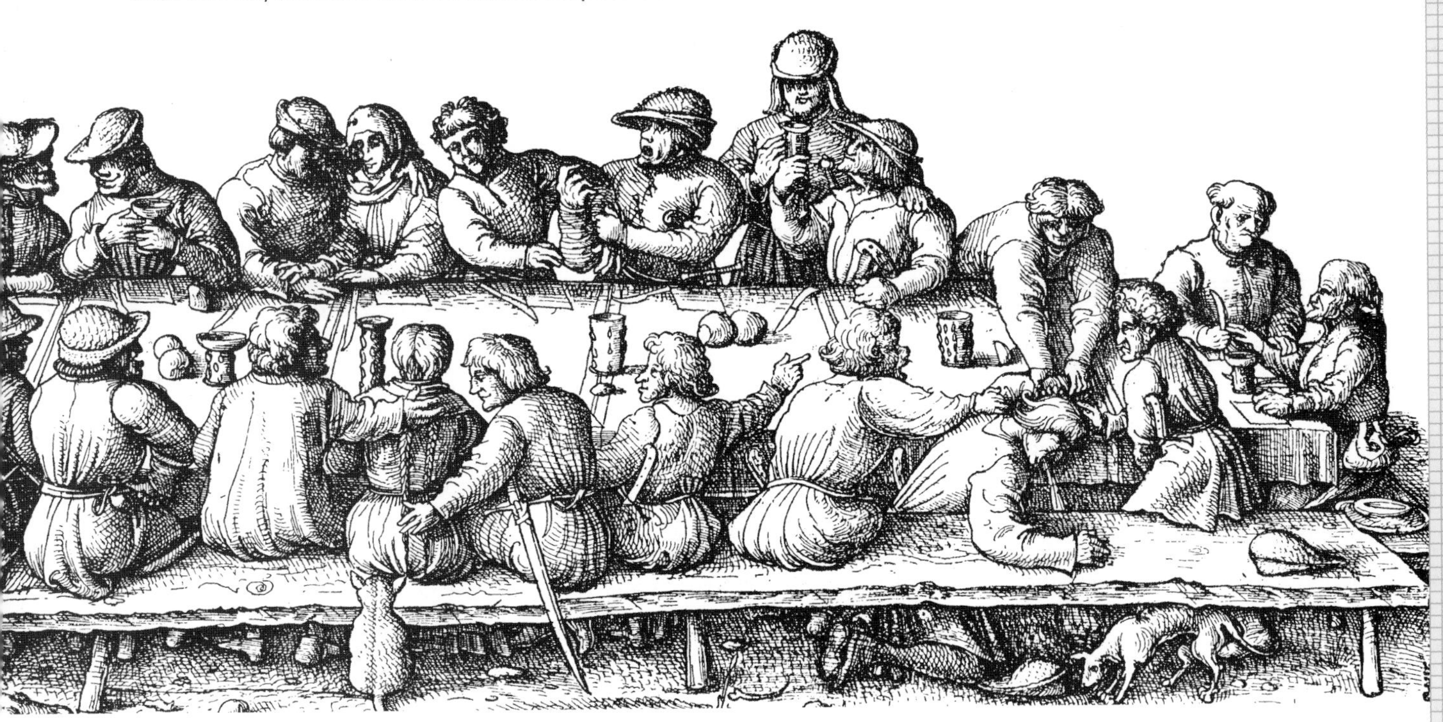

Restaurants do the same thing, using their staff as beta testers: one cook will use leftover bits from the normal meal service to make a "family meal," and the better experiments can end up on the menu. Just keep in mind who your guests are and their general openness to experiments. If you're not sure, stick with the familiar.

Pick a recipe that's in line with what's expected by your guests. Making sushi for someone who likes his meat well done is probably too much of a stretch. Some dishes lend themselves well to a more casual, family-style meal (e.g., lasagna), while others are better suited to being plated in the kitchen, where you can spend time on the presentation.

Finally, choose recipes that leave you time to spend with the guests. After all, they're there to see you! Depending upon the complexity of the meal and the number of people you are cooking for, try to pick recipes that have a distinct prep phase that you can do in advance of your guests arriving. I survived hosting a four-course dinner party for 40 by prepping individual servings of duck confit sugo (see page 28, earlier in this chapter) and individual chocolate cakes in small ramekins. As people arrived, I snatched the appropriate number of ramekins from the fridge and tossed them into the oven. This left me time to work on the other courses and still hang out with my guests. (I used the ramekins a few weeks later for making Christmas fruitcakes that I sent off to my friends.) This attention to planning will limit the amount of attention you need to give to the preparation of food and will free you up to interact and socialize with your guests. After all, it should be fun!

Adam Savage on Scientific Testing

Adam Savage is co-host of Discovery Channel's MythBusters, *a popular science program that examines rumors, myths, and conventional wisdom, "putting them to the test" with a scientific approach.*

How do you go about testing a myth?

One of the earliest things we realized on the show is that you always have to have something to compare to. We would try to come up with an answer like: is this guy dead, is this car destroyed, is this an injury? And we would be trying to compare it to an absolute value, like X number of feet fallen equals dead. The problem is the world is very spongy and nonuniform, and trying to nail down a value like that can be really difficult. So we always end up doing relative tests. We end up doing a control under regular circumstances and then we test the myth under identical circumstances, and we compare the two things. In that comparison, we get to see our results.

We did one where we were testing whether or not you could tenderize steaks with explosives. We had to figure out what tenderness is. The problem is you can give two different people each a piece of steak from the same cut compared to a piece of steak from a different cut, and they might come up with two different assessments of which one is more tender. We actually did a whole day of testing that didn't end up on film because we realized we were using the wrong parameters for assessing steak tenderness. The USDA actually has a machine for testing the tenderness of steak that measures the pounds of force it takes to punch a hole through a steak. We replicated that machine and to our great surprise, it worked exactly as it was supposed to. Coming up with something for $50 that equals the USDA testing equipment: that was thrilling!

How can testing a myth translate into learning more about cooking?

Changing one variable is probably the single hardest thing for people to understand. Change only one variable. It's not like changing only a small number of variables; it's really changing one variable at a time, because only then do you know what caused the change between your first test and your second test. You get so much clarity from the process that way.

I'm an avid cook. My wife and I both cook a lot of elaborate things, and we really do love playing around with single variables, changing things and learning how things work. We were reading Thomas Keller, and he talked about how salt is a flavor enhancer, and he mentioned that vinegar does a similar thing. It doesn't add a new taste, but it often alters the taste that's there. My wife was making a cauliflower soup, and it was kind of bland. I didn't want to put any more salt in it, because I could tell it was about to go in the wrong direction. We tossed in a little bit of vinegar and the whole thing just woke up. It was thrilling! I love that.

Have you done other myths related to food?

We have—certainly a whole bunch of drinking myths. We did poppy seed bagels to see if eating a poppy seed bagel causes you to test positive for heroin, which is absolutely true. In fact, parolees are completely forbidden from eating poppy seed bagels. They're told if you test positive for drugs, we are not going to wonder why. You are just going to go back to jail, so make it easy, don't eat poppy seeds.

I had a whole episode written called the surreal gourmet, which ended with tenderizing steak with dynamite, but it had all those other things like poaching fish on your catalytic converter or cooking eggs in your dishwasher. Jamie loves the idea of tenderizing meat in the dryer.

I thought of roasting almonds in a dryer, but not tenderizing meat.

Also, the idea of *is it safe to eat fresh road kill*. We think that would be just hilarious and gross.

The problem-solving aspect of the show is really fascinating. Do you have any advice on how to get to where you want to be when problems arise?

The first thing to realize is that you're not going to end up where you think. The world is smarter than you are. A craftsman isn't somebody who never screws up. A craftsman screws up just as much as you do. They can just see it coming, and can adjust; it's an ongoing process. Everybody's oven heats at a different rate. You open it up to check, the temperature drops. There are all sorts of variables. Maybe it's humid, maybe it's not. Humidity was affecting all sorts of my wife's cookie recipes. People tend to overfocus on the final product, when you've got to be awake to the process. So problem solving doesn't mean doing whatever it takes to get to the end result; it means following the path that you're on. You're going to probably end up changing your definition of what the result is before you're done.

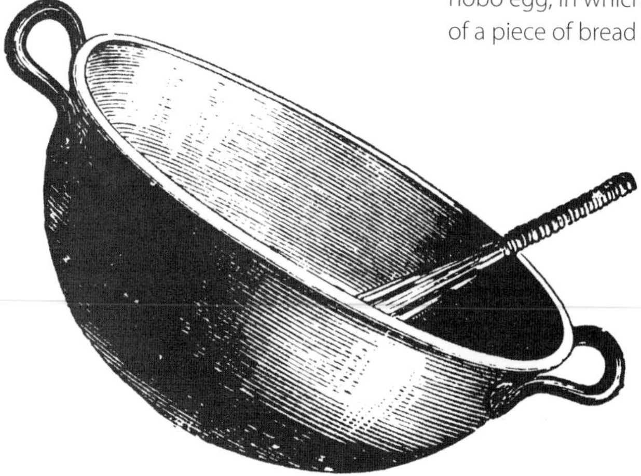

The better you get, the more that things start to turn out like you planned. When my wife started doing really serious baking, I couldn't believe how much of a difference just having all your ingredients at room temperature made in terms of the emulsifying and chemical reactions—getting the doughs flaky, for instance. Just the simple thing of pulling all of your ingredients out of the refrigerator an hour before you start cooking has a massive effect on the final product. Or things like certain kinds of berries in certain kinds of pastries; the acidity of the berries means having to add more baking soda. I love that. You just have to learn as you go.

What do you enjoy cooking?

My favorite thing to cook of all is eggs. After years of practice, I've almost mastered the pan flip for an omelet without the spatula. I've actually held brunches for 15 people where the theme was "come and I'll cook you eggs any way you want." My kids are both really getting into it now. They wake up (they're 10-year-old twins) and they both have their specific ways that they like cooking eggs. My son Addison prefers the hobo egg, in which you cut a hole out of a piece of bread and fry an egg in that hole, and my son Riley likes scrambled eggs. He likes them a little bit on the hard side, but I'm trying to teach him not to cook them too much.

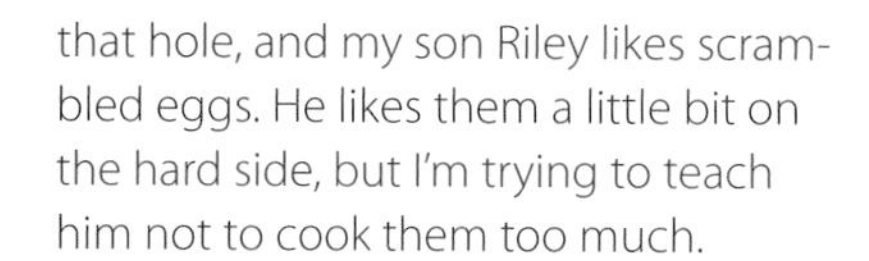

That does seem to be a common affliction, overcooking eggs and getting dry scrambled eggs.

With enough sauce, they'd work, but when you start to cook them right, it turns out that there is this tiny band in which they're unbelievably good. That's why I like the eggs. They're kind of unforgiving in some ways and that's really exciting.

One of the great things about cooking is that, unless you're doing something really specifically unforgiving, most recipes are really quite impressively forgiving. That's a part I really love. You can change all sorts of variables and it still comes out pretty darn good. It's a great test platform.

How do you learn from the things that don't succeed?

I hand-whipped my first whipped cream about six or seven years ago. I whipped it, and the very first thing I did once it was whipped was I whipped it too far on purpose. I thought, "I know this is perfect, but I want to know where the line is," and I just kept on going until I had butter. It was surprisingly fast and taught me a really clear thing about exactly where you can go with whipping cream.

Whipped cream tastes great. Flavoring it and sweetening it is just trivial. If you're good, you can do it almost as fast as it takes to get the mixers and the bowls out and do it all mechanically. It's a lovely thing to sit there and talk to your guests while you're hand-whipping cream.

Making Whipped Cream

You can whisk up whipped cream by hand in less time than it takes to get an electric mixer out. Start with a cold bowl (chill it in the freezer for a few minutes, ideally), add either heavy cream or whipping cream, and whisk until the cream holds its shape. Add a spoonful of sugar and a little vanilla extract for a sweeter-tasting version. For more on what happens to cream during whisking, see page page 254 in Chapter 5.

30 seconds: still liquid; light bubbles.

60 seconds: still liquid; light bubbles.

90 seconds: thin cream, would be great on berries.

120 seconds: whipped, soft peaks. Add some vanilla and sugar, and you're good to go.

150 seconds: overbeaten, a little buttery flavored.

180 seconds: whipped butter.

2

Initializing the Kitchen

FIGURING OUT WHICH TOOLS TO HAVE IN THE KITCHEN CAN BE A DAUNTING TASK, ESPECIALLY IF YOU'RE JUST STARTING OUT. With so many products on the market, the number of decisions to be made can overwhelm anyone, especially overly analytical perfectionists (you know who you are). What type of knife should I buy? Which pan is right for me? Where should I store my cherry pitter?

Take a deep breath and relax. To a newbie, kitchen equipment probably seems like the key to success in the kitchen, but in all honesty, kitchen equipment isn't *that* important. Two sharp knives, two pots, two pans, a spoon to stir and a spatula to flip, and you're covered for 90% of the recipes out there and have a better kitchen setup than 90% of the world. Heck, in some parts of the world it's just one pot and a spatula that's been sharpened on one side to double as a knife. I know one culinary pro who backpacked through New Zealand for a year; she narrowed it down to a paring knife, vegetable peeler, heatproof spatula, and cutting board. Still, while having great kitchen equipment won't make or break you, having the right tool for the job, and one that you're comfortable with, does make the experience more enjoyable.

Back to the list of questions. The right answer to any question on which piece of kitchen equipment to use is: *whatever works for you, is comfortable, and is safe*. This chapter will cover the basic must-haves, but ultimately it is up to you to experiment and to adapt and modify these suggestions to fit your needs and tastes. The one consistent piece of advice I can give is *use common sense*.

In addition to the basic essentials, this chapter also provides some common sense tips on storage, kitchen organization, and other things to keep in mind if you're new to cooking, and maybe a few new ideas for the already initiated.

Approaching the Kitchen

So you've picked out a recipe to start with and you're raring to go. Now what? Beyond the grocery shopping list, there are a few things you can do before putting the knife to the cutting board to avoid mishaps while cooking.

Calibrating Your Instruments

A scientist can only run experiments and make observations up to the level of accuracy that his equipment allows. This isn't to say that you need to approach the kitchen with the same rigor that a scientist shows at the lab bench, but if you're trying to bake cookies or roast a chicken and your oven is off by 50°F / 28°C, your results will be less than desirable. The largest variance in most kitchen equipment is usually the oven, and it can be hard to tell if your oven is running cold or hot just by feel. (Dull knives are also a common misdemeanor; more on that later.) Check and calibrate your oven using an oven thermometer. On the road visiting someone and don't trust their oven? See "The Two Things You Should Do to Your Oven RIGHT NOW," below, for instructions on calibrating an oven using sugar.

The Two Things You Should Do to Your Oven RIGHT NOW

One piece of equipment that you're probably stuck with is your oven. What makes an oven "good" is its ability to accurately measure and regulate heat. Since so many reactions in cooking are about controlling the rate of chemical reactions, an oven that keeps a steady temperature and isn't too cold or too hot can make a huge difference in your cooking and baking.

Improve your oven's recovery time and even out the heat: keep a pizza stone in the oven. Say you're baking cookies: oven set to 350°F / 180°C, cookies on pan, ready to go. In an empty oven, the only thing hot is the air and the oven walls, and opening the door to pop the cookies in leaves you with just hot oven walls. I find I get much better results by keeping a pizza stone on the very bottom rack in my oven, with a rack directly above it. (Don't place the cookie sheet directly on the pizza stone!)

The pizza stone does two things. First, it acts as a thermal mass, meaning faster recovery times for the hot air lost when you open the door to put your cookies in. Second, if you have an electric oven, the pizza stone serves as a diffuser between the heating element and the bottom of your baking tray. The heating element emits a hefty kick of thermal radiation, which normally hits the bottom side of whatever pan or bakeware you put in the oven. By interposing between the heating element and the tray, the pizza stone blocks the direct thermal radiation and evens out the temperature, leading to a more uniform heat. For this reason, you should go for a thick, heavy pizza stone; they're less likely to crack, too. I've turned crappy ovens that burned everything into perfectly serviceable ones capable of turning out even "picky" dishes like soufflés just by adding a pizza stone. Just remember that like any thermal mass, a pizza stone will add lag to heating up the oven, so make sure to allow extra time to preheat your oven.

Calibrate your oven using sugar. I know this sounds crazy, and yes, you should get an oven

Prepping Ingredients

When making a meal, start by prepping your ingredients before you begin the cooking process. Read through the entire recipe, and get out everything you need so you don't have to go hunting in the cupboards or the fridge halfway through. Making stir-fry? Slice the vegetables into a bowl and set it aside before you start cooking. In some cases, you can do the prep work well in advance of when you start cooking the meal. Restaurants wash, cut, and store ingredients hours or even days ahead of when they're needed. The stages of prepping and cooking are like the stages of compiling and executing in software programming. If compiling is looking through all the steps and assembling the instructions into a single stream of optimized commands that are ready to be executed, the prep stage of cooking is similarly "precomputing" as much of the work as possible so that, when it's time to fire off the recipe, you can execute it as quickly and easily as possible.

The *mise en place* technique (French for "put in place") involves laying out all the ingredients and utensils needed to cook a dish before starting. Think of it like cache priming in computer programming: mise en place is equivalent to prefetching the various bits you'll need

thermometer. But how do you know that the oven thermometer is right? My three thermometers—an IR thermometer, a probe thermometer, and the oven's digital thermometer—have registered temperatures of 325°F / 163°C, 350°F / 177°C, and 380°F / 193°C, all at the same time. (They're all designed for accurate readings in different temperature ranges.)

It's common practice to calibrate thermometers with ice water and boiling water because the temperatures are based on physical properties. Sugar has a similar property and can be used for checking the accuracy of your oven thermometer. Sucrose (table sugar) melts at 367°F / 186°C. It turns from a powdered, granulated substance to something resembling glass. (Caramelization is different from melting; caramelization is due to the sugar molecules decomposing—literally losing their composition—and happens over a range of temperatures coincidentally near the melting point.)

Pour a spoonful of sugar into an oven-safe glass bowl or onto some foil on a cookie sheet and place in your oven, set to 350°F / 177°C. Even after an hour, it should still be powdered. It might turn slightly brown due to decomposition, but it shouldn't melt. If it does, your oven is too hot. Next, turn your oven up to 375°F / 190°C. The sugar should completely melt within 15 minutes or so. If it doesn't, your oven is calibrated too cold. Check to see if your oven has either an adjustment knob or a calibration offset setting; otherwise, just keep in mind the offset when setting the temperature. Note that your oven will cycle a bit above and below the target temperature: the oven will overshoot its target temperature, then turn off, cool down, turn back on, etc. It's possible that your oven could be "correctly" calibrated but still melt the sugar when set to 350°F / 177°C due to this overshooting, but it would have to overshoot by about 15°F / 8°C.

Sugar at 350°F / 177°C.

Sugar at 375°F / 190°C.

while executing to avoid cache misses. If you are going to prepare the same dish multiple times (say, omelets for a large brunch), having a bunch of containers ready with the various fillings in them will allow you to work quickly. *Mise en place* isn't an absolute necessity, although it does generally make the cooking process smoother. Measure out the ingredients at this stage whenever possible; this way you'll have a chance to discover if you're short of a critical ingredient (or if it's gone bad!) before committing to the cooking process. It also helps avoid those panicked moments of trying to locate a strainer that's wandered off while a sauce that needs immediate straining cools down. (Happens to me all the time...) Sure, a "just-in-time" approach is fine for simple meals. However, if you're cooking for a large number of people or attempting a particularly complicated menu, keep the mise en place approach in mind.

Kitchen Equipment

Regardless of your needs, a well-equipped home kitchen shouldn't cost much money. I once heard the products sold in consumer kitchen stores described as "kitchen jewelry." Stores like Williams-Sonoma offer beautiful products that make for beautiful gifts, but just because they call their products "professional-quality cookware" doesn't mean that professionals routinely use them. Sure, their kitchenware is beautiful and functional, but if you're willing to settle for just functionality and skip the bling factor, you can save a bundle.

If you live in a large city, look for a restaurant supply store. These stores stock aisle after aisle of every conceivable cooking, serving, and dining room product, down to the "Please wait to be seated" signs. If you can't find such a store, next time you eat out, ask your waiter to ask the kitchen staff. If that fails, the Internet, as they say, "is your friend."

If you do get stuck or want recommendations of which features to look for in a product, look at recent reviews from Adam Ried of *America's Test Kitchen* and *Cook's Illustrated* or Alton Brown of *Good Eats*. Products continually change as manufacturers revise, update, and improve their offerings, so don't be surprised if specific models you read about are no longer available. Common sense and thinking about your requirements are really all you need, though.

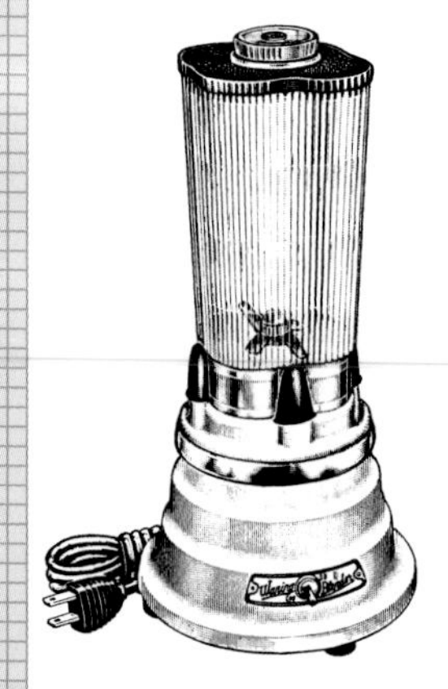

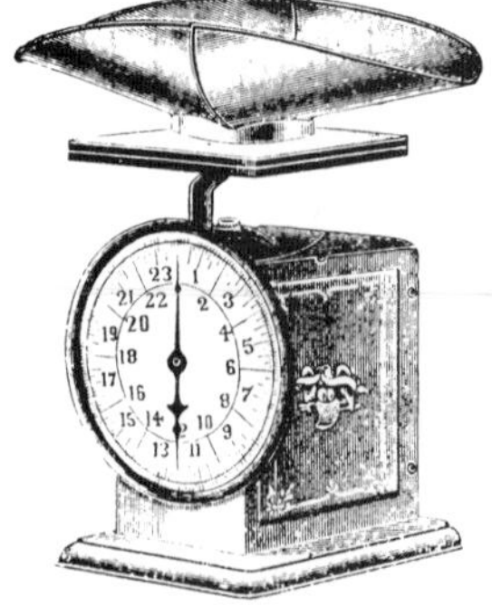

Storage Tips for Perishable Foods

Should you wash your produce when you unpack your weekly groceries or at time of use? And how should you store various other foods? Here's a look at basic storage rules, from most to least perishable.

Seafood. Seafood is the most perishable item you're likely to handle. Ideally, seafood should be used on the day of purchase. A day or two longer is okay, but past that point enzymes and spoilage bacteria begin to break down amine compounds, resulting in that undesirable fishy odor.

Fun science fact: Fish live in an environment that is roughly the same temperature as your fridge. The specific activity of some enzymes is much higher in fish than mammals at these temperatures. Putting seafood on ice buys a bit more time by increasing the activation energy needed for these reactions. Meat is already far enough away from the ideal reaction temperatures that the few extra degrees gained by storing it on ice don't change much.

Meats. Follow the sell-by or use-by date. The *sell-by date* is the point in time until which the store still considers the product safe for sale. (Not that you should push it, but it's not as if the meat will suddenly turn green and smelly at 12:01 a.m. the next day.) The *use-by date*, as you'd imagine, is the recommended deadline to cook the food. If you have a package of chicken whose use-by date is today, *cook* it today, even if you're not ready to eat it. You can store the cooked product for a few more days. If you can't cook the fish or meat you've bought on or before its use-by date, toss it in the freezer. This will affect the texture, but at least the food won't go to waste.

Freezing meat does *not* kill bacteria. It takes being zapped with radiation AND over a month at 0°F / –18°C to render nonviable the bacteria in salmonella-contaminated meats. Nice to know, but not very helpful unless you happen to have a radiation chamber lying around.

Fruits and vegetables. How you process and store fruits and vegetables impacts their ripeness and flavor, and can also delay the growth of mold. When it comes to ripening, there are two types of fruit: those that generate ethylene gas, which causes them to ripen, and those that don't generate it. For those that do ripen when exposed to ethylene, you can speed up ripening by storing them in a paper bag, which traps the gas.

Store raw meats below fruits and vegetables in your refrigerator, because this reduces the likelihood of cross-contamination. Any liquid runoff from the meats won't be able to drip onto other foods that won't be effectively pasteurized by cooking. (Storing meats below other foods is required by health code in commercial establishments.)

Ripen in the presence of ethylene gas

To speed up ripening, store these in a loosely folded paper bag out of direct sunlight, at room temperature.

Apricots, peaches, plums. Ripe fruits will be aromatic and will yield slightly to a gentle squeeze, at which point you can store them in the fridge. Don't store unripe stone fruits in the refrigerator, in plastic bags, or in direct sunlight. If you're lucky enough to be gifted pounds and pounds of these fruits, either freeze them or make jam before they have a chance to go bad.

Avocados. Ripe fruit will be slightly firm but will yield to gentle pressure. Color alone will not tell you if the avocado is ripe. Storing cut avocados with the pit doesn't prevent browning, which is due to both oxidation and an enzymatic reaction, but does stop browning where the pit prevents air from coming in contact with the flesh. Plastic wrap pressed down against the flesh works just as well, or if you have a vacuum sealer, go for overkill and seal them.

Bananas. Leave at room temperature until ripe. To prevent further ripening, store in the refrigerator—the peel will turn brown, but the fruit will not change.

Blueberries. While blueberries do ripen in the presence of ethylene, their flavor is not improved from this. See advice for blackberries et al.

Tomatoes. Store at temperatures above 55°F / 13°C. Storing in the fridge is okay for longer periods of time but will affect flavor and texture. If the ultimate destination for the tomatoes is a sauce, you can also cook them and then refrigerate or freeze the sauce.

Potatoes. Keep potatoes in a cool, dry place (but not the fridge). Sunlight can make the skin turn green. If this occurs, you must peel off the skin before eating. The green color is due to the presence of chlorophyll, which develops at the same time that the neurotoxins solanine and chaconine are produced.* Since most of the nutrients in a potato are contained directly below the skin, avoid peeling them whenever possible.

* While you're unlikely to die from consuming the solanine content present in an average potato that's gone green (~0.4 mg), it appears to be possible to give yourself a rather unpleasant digestive tract experience for the better part of a day. For a more thorough explanation, see *http://en.wikipedia.org/wiki/Solanine*.

Unaffected or negatively impacted by ethylene gas

Store these separately from ethylene-producing produce.

Asparagus. Store stalks, with bottoms wrapped in a damp paper towel, in the crisper section or the coldest part of the fridge. You can also put them in a glass or mug, like cut flowers. Eat as soon as possible because the flavor diminishes with time.

Blackberries, raspberries, and strawberries. Toss out any moldy or deformed berries. Immediately eat any overripe berries. Return the other berries to the original container, or arrange them (unwashed) in a shallow pan lined with paper towels and store in the fridge. To absorb additional moisture, place a paper towel on top of the berries. Wash them just prior to use; washing and storing them adds moisture that aids the growth of mold.

Broccoli, cabbage, collard greens, kale, leeks, Swiss chard. Store in the crisper drawer of the refrigerator or in a plastic bag poked with holes to allow for any excess moisture and ethylene to escape. Ethylene causes florets and leaves to turn yellow.

Carrots. Break off green tops. Rinse carrots, place in a plastic bag, and store in the crisper section of the fridge. Storing carrots in the fridge will preserve their flavor, texture, and beta-carotene content.

Garlic. Store in a cool, dark place (but not the fridge). You can still use cloves that have sprouted, but they will not be as strong in flavor. The sprouts themselves can be cut up like scallions or chives and used in dishes.

Lettuce and salad greens. Check greens bought in bunches for insects. Wash leaves, wrap in a towel or paper towel, and then store in the fridge in a plastic bag.

Onions. Keep in a cool, dry space away from bright light. Onions do best in an area that allows for air circulation. Do not place onions near potatoes, because potatoes give off both moisture and ethylene, causing onions to spoil more quickly.

Questions about other produce?
See *http://www.fruitsandveggiesmatter.gov* and *http://postharvest.ucdavis.edu/Produce/ProduceFacts/*.

Here's what I consider the essential kitchen items. We'll cover each in turn.

Bare Minimum Equipment

- Knives
- Cutting board
- Pots and pans
- Measuring cups and scales
- Spoons & co.
- Thermometer and timers
- Bar towels

Standard Kitchen Equipment

← All that, plus...
- Storage containers
- Strainers
- Mixers & co.

Bare Minimum Equipment

Here's the equipment that you'll need at a bare minimum.

Knives

Knife blades made of steel are manufactured in one of two ways: forging or stamping. *Forged* blades tend to be heavier and "drag" through cuts better due to the additional material present in the blade. *Stamped* blades are lighter and, because of the harder alloys used, hold an edge longer. Which type of knife is better is highly subjective and prone to starting flame wars (or is that flambé wars?), and with some specialty sashimi knives listing for upward of $1,000, there are plenty of options and rationales to go around.

Some people like a lighter knife, while others prefer something with more heft. Personally, I'm perfectly happy with a stamped knife (currently, Dexter-Russell's V-Lo series) for most day-to-day work, although I do have a nice forged knife that I use for slicing cooked meats.

Chef's knife. If I could take only one tool to a desert island, it would be my chef's knife. What size and style of chef's knife is best for you is a matter of preference. A typical chef's knife is around 8" / 20 cm to 9" / 23 cm long and has a slightly curved blade, which allows for rocking the blade for chopping and pulling the blade through foods. If you have a large work surface, try a 10" / 25 cm or larger knife. Or, if you have smaller hands, you might want to look at a Santoku-style knife, a Japanese-inspired design that has an almost flat blade and a thinner cross-section. Keep in mind, though, that Santoku knives are best suited for straight up-and-down cutting motions, not rocking chopping motions or pulling through foods.

Paring knife. A paring knife has a small (~4" / 10 cm) blade and is probably the most versatile knife in the kitchen. I've had some chefs confide to me that their favorite knives are the scalloped paring knives, since they are useful for cutting so many different types of items. They're designed to be held up off the cutting board, knife in one hand, food item in the other, for tasks such as removing the core from an apple quarter or cutting out bad spots on a potato. I find that the almost pencil-like grip design of some commercial paring knives allows me to twirl and spin the knife in my fingers, so I can cut around something by rotating the knife instead of rotating the food item or twisting my arms. Personally, I prefer a scalloped blade—one that is serrated—because I find it cuts more easily.

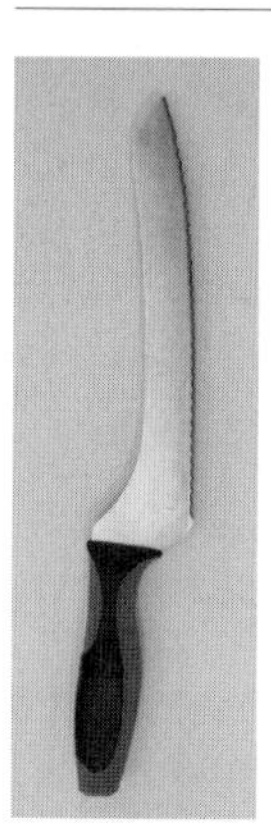

Bread knife. Look for an offset bread knife, which has the handle raised up higher than the blade, avoiding the awkward moment of knuckles-touching-breadboard at the end of a slice. While not an everyday knife, in addition to cutting bread and slicing bagels, bread knives are also handy for cutting items like oranges, grapefruits, melons, and tomatoes because of the serrated blade.

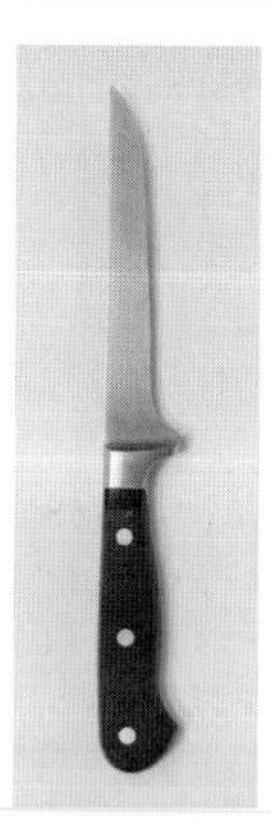

Boning knife. If you plan to cook fish and meat, consider acquiring a boning knife, which is designed to sweep around bones. A boning knife has a thinner, more flexible blade than a chef's knife, allowing you to avoid hitting bones, which would otherwise nick and damage the knife blade. Some chefs find them indispensable, while others rarely use them.

Knife Skills 101

The sound of a failed disk drive grinding itself down is pretty bad, but watching someone use a knife improperly is far worse. I swear, if I were going to develop PTKD (post-traumatic kitchen disorder) over something, it'd be from watching people use knives improperly.

I treat knives as the second most dangerous implements in the average kitchen. (Microplanes and mandolins hold the top spot.) When using a knife, I'm always thinking about the "failure mode." If it slips, or something goes wrong, how is it going to go wrong? Where is the knife going to go if it does slip? How can I use the knife and position myself such that if an exception does occur, it isn't fatal? Of course, getting a good, clean cut and keeping the knife in good working order are also important. Here are my top tips for knife usage:

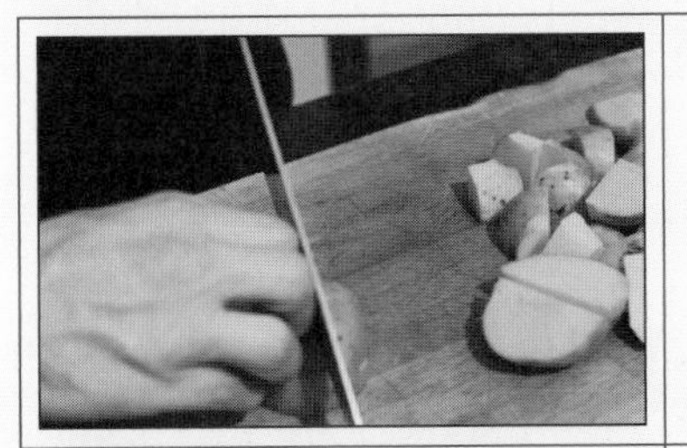

Feed the food into the cutting plane with your fingers positioned so that they can't get cut. Keep the fingers of the hand holding the item curled back, so that if you misjudge where the knife is, or it slips, your fingers are out of harm's way. You can also rub the upper side of the knife against your knuckles to get better control over the location of the knife. Use a smooth, long motion when cutting. Don't saw back and forth, and don't just press straight down (except for soft things like a block of cheese or a banana). Let the knife do the work!

When scraping food off a cutting board, flip the knife over and use the dull side of the blade. This will keep the sharpened side sharper.

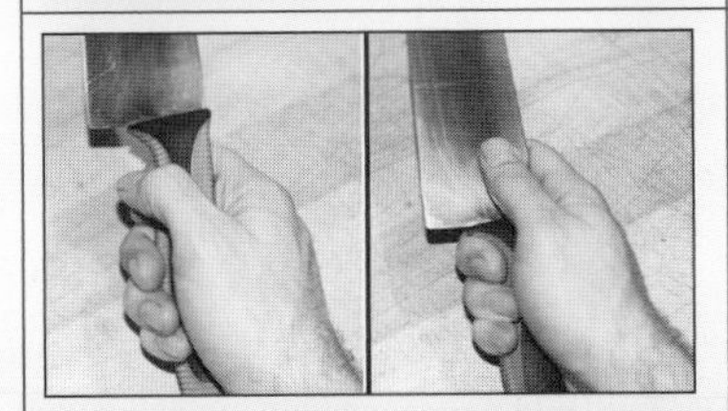

There's more than one way to hold a knife: try using a "pinch grip" instead of a "club grip." A pinch grip allows for more flexibility, as it gives you more dexterity in moving the knife.

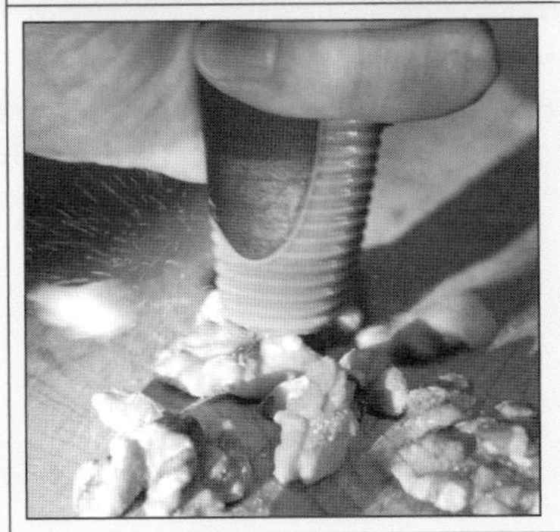

Don't use the edge of the blade to whack or crack hard objects, such as a walnut shell or a coconut; you'll nick it! Repeat after me: *knives are not hammers* (you know who you are). Unless, of course, you have a commercial knife that has a butt that actually is a hammer, in which case, go right ahead… You can, however, use the side of the blade as a quick way to crush garlic or pit cherries or olives. Place food on board, place side of blade on top of food, press down on blade with fist.

Buck Raper on Knives

PHOTOS USED BY PERMISSION OF BUCK RAPER

Buck Raper is the manager of manufacturing and engineering for Dexter-Russell, the largest and oldest cutlery manufacturer in the United States. Above, Buck holds a knife next to an edge sharpness and edge life test apparatus in their metallurgy lab.

How did you come to work at Dexter-Russell?

In a former life, I was working on a doctorate in synthetic organic chemistry.

Wow. What happened?

I got drafted to Vietnam.

And then you came back…

I came back and there weren't many job opportunities for PhD chemists, and I was still looking at two more years in school, and I had a family to support. So I went and got an MBA and got twice the starting salary I would have gotten as a PhD. My family had always been in the cutlery business, my grandfather, and my father, and all I ever heard was knife talk. When I was a Baby Buck, my father would take me into the pocket-knife factory on Saturday mornings and hand me off to a foreman so he could get some work done, and I'd make knives with a foreman.

Did the background in chemistry, combined with your family's history in knife making, complement each other?

To some degree…but it was more of the scientific method and analytical techniques that you learn in a hard science, applying them to manufacturing. I looked at it from a different standpoint than a history major MBA would, or an English major MBA would. Coming from a real science, you take a different approach, an engineering approach.

Can you give me an example?

Much of the heat treating, the grinding, and the choice of steels was done almost by folklore. It's always been done that way and nobody remembers why. Now when we're trying to choose a steel for a particular application, we do some testing, make some blades, and try them out to see what the results are. We have a control sample and record data. That's the type of change that I made. Dexter-Russell is 192 years old, and we still have machinery and tooling that we were using at the turn of the century, from 1900. Those techniques still work and they're still very good, but nobody really knew why we were doing things the way we were doing them.

What surprised you when you were testing the folklore?

We're number one in professional oyster knives, and there's the chronic problem with the tips of oyster knives breaking off. We had a heat-treatment process that we thought was making the steel hard enough to not break. The theory was if the blade is breaking, make it harder, and then the tip won't break off. The reality was what we needed to do was to make tougher steel. So we changed our heat treatment process to create a tougher, softer steel.

What does it means for a steel to be tough versus hard?

It's a trade-off to hold an edge. The harder the steel is, the better it will hold an edge. But you also want to have some flexibility. If you need a flexible bone or fillet knife, a harder steel is more brittle; it would fracture. So you have to trade off the hardness for the toughness that allows you some flexibility. The toughness also gives you wearability, resistance to abrasion. One way an edge fails is that you literally wear away the grains of steel, and to resist that, you're looking for a tough steel.

When you heat-treat steel, you martenize it to the temperature that's going to give you the maximum hardness. But if you underheat it, if you undercook it a little bit, it comes out tougher. If you overcook it, it's also tough, but then it corrodes. In our case, when we're talking heat-treatable stainless 400 series steel, the optimum temperature is 1934°F / 1057°C. If you heat it to 1950°F / 1066°C, you get the same hardness that you would if you heat it to 1920°F / 1049°C, but one is tougher, and the other will corrode.

Steel is formed of grains. If you were to snap a knife blade in half, and look at it with the naked eye, the texture would look like a fine cement inside the knife. What you're seeing is groups of grains. Steel exists in 9 or 10 different phases. Depending on how it's been processed, temperature-wise, it has a mixture of these various phases, and that determines the toughness of the steel. I use the analogy of baking a cake when I'm explaining heat treatment. You have raw dough and expose it to heat. There's a chemical change and a phase change, and you go from slurry to a porous solid once it's baked.

With steel, once it is heated to a critical temperature, cooling—called *quenching*—is also critical. You've probably seen old movies where the blacksmith is pounding away, when he gets the iron hot, he plunges it into the water and there's a hiss of steam. The reason for that is the rapid cooling. In the case of stainless steel, you have to get it below 1350°F / 732°C in less than three minutes in order to maintain the phase that you want. If you cool it slower, you get a different mixture of phases in the steel. So it's not just in bringing it up to temperature, the cooling curve is key.

Steel is also determined by the alloy. There're two or three dozen different types of stainless cutlery steels, and stainless cutlery steels are just a very small subset of alloyed steels. Alloyed steels are a subset of carbon steels. And all the heat-treatment processes are determined by which alloy you're working with.

Are there other types of steels that you would want to use for particular purposes for knife making?

We want to use a stainless steel, although carbon steel makes wonderful knives. Everybody likes their old carbon steel knives, but nowadays, with the National Sanitation Foundation and other regulatory bodies, you can't use carbon steel knives in most restaurants. We choose stainless, which has chromium in it; the chromium makes it stainless. You also have to have carbon in the steel so that you can harden it. You add more carbon if you want to create a harder blade, and more chromium if you need to get more corrosion resistance. When you heat-treat it, you want to come out with a very fine texture, and things like molybdenum, vanadium, tungsten, and cobalt help you get a fine grain. Tungsten and cobalt help make the steel tougher.

What's the rationale prohibiting carbon knives in restaurants?

They rust, and rust is iron oxide. It's dirty, and where the blade has rusted, there are pits that will retain grease. The grease will breed bacteria. It's usually controlled by city or state or county ordinance.

Carbon steel versus stainless steel. Which is better?

That was a classic question that I wondered about for 30 years. I finally had a seminar with a metallurgist from a French steel mill, and he developed a machine to test the sharpness of edges and the life of edges. The answer is that you can get a carbon steel edge about 5% sharper while a stainless steel edge will last about 5% longer. With stainless steel being tougher, it is harder to create the edge, so stainless steel often gets a bad reputation because people can't sharpen it correctly. It is possible to get carbon steel 5% sharper, but you would never perceive that using a knife. You need the scientific apparatus to bring out that difference. The practical difference is it's very easy to bring up an edge on carbon steel, so most people's carbon steel knives are sharper because they're easier to resharpen. A carbon steel knife responds very easily to a butcher's steel; you have to work a little bit more with a stainless steel knife.

I'm going to ask the question that'll probably lead to the gates of Hell: how do I sharpen a knife correctly?

There are lots of ways to do it. Probably the best general-purpose way and what I recommend to people is to use a diamond sharpening steel. The traditional serrated butcher's steel is a ½" or ⅝" rod with ridges running longitudinally. Those are now being replaced by rods that are plated with diamond. The diamond rod brings up an edge very quickly, because it's hard enough to remove metal, creating a new edge.

An edge is actually a whole bunch of little burrs, sort of like hacksaw teeth that are standing up, perpendicular to the back of the blade. When you cut, those little burrs (here we call them *feathers*) roll over. The first thing that happens when you swipe with a butcher's steel is you stand those feathers up, and you have a real good edge. After a time, they bend back and forth. They work-harden and break off, like breaking a wire by twisting it until it work-hardens and snaps. Then you have to create a new edge, new burrs, and the grit on a diamond steel is perfect for that. That's what the long serrations do on your regular butcher's steel, but it's a lot easier to do with a diamond steel.

When you run a knife edge along a steel, you stand up the burrs, and you start thinning down the edge. I can do it with the back of a porcelain plate, or I can rub a knife on a brick wall and bring up the edge, but a diamond steel is best.

I've made a lot of trips to China, and they have very primitive kitchens as far as equipment, tools, and utensils go. They make do with the one basic knife. People call it a cleaver, but it's not really a cleaver. It's a slicer and a spatula and a scraper and everything else, but with that one knife, they stop and squat on the floor and bring the edge back on a brick that's in the floor. They keep those knives very, very sharp. I learned in Chinese cooking how nicely things are sliced up counts as much as the taste, the presentation, and the freshness of ingredients. All of that can be ruined if you have cut raggedy chunks.

I would recommend either a diamond butcher steel or a whet stone. But a whet stone takes more skill, more training to use. I would stay away from electric sharpeners.

At some point the burrs snap off, and I presume that's the point at which one needs to actually grind down the edge of the knife to form a new edge?

With a diamond steel, you're doing grinding at the same time you're straightening up the edge. A traditional butcher's steel isn't hard enough to remove metal. The deal with using a butcher's steel is your steel has to be harder than the metal of the blade you're sharpening. Otherwise, you get nowhere, like using a common file to smooth or shape metal. Your file won't cut the metal if the file isn't harder than the metal it's cutting. If you let your knife get very dull, bringing the edge back is a real bear. If you give it a few strokes on a butcher's steel every other day, or once a week, or every time you go to put the knife in the drawer, then the knife is always ready.

At what point is a knife effectively used up? [Buck shares with me the photo shown below.] I cannot believe how much the bottom knife has been sharpened away compared to the new knife on top. What's the story with this actual knife?

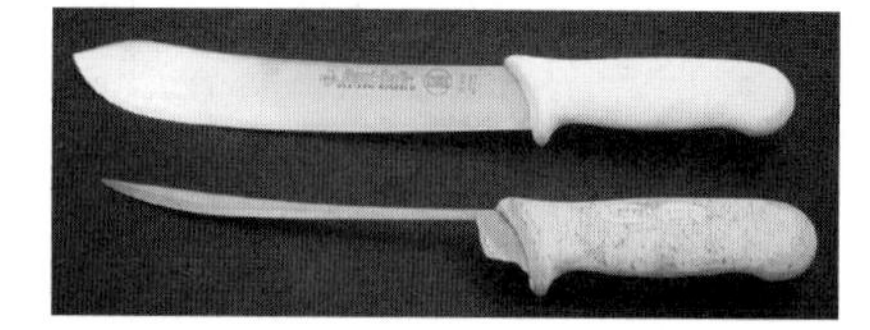

Whoever was resharpening that knife was very, very good. It came back to our customer service people for replacement from a mom-and-pop butcher shop. I train our sales force, and one of the questions they ask is how long is a knife useful. I show them this. That's pushing the ridiculous. I would think that that knife had seen about five or six years of service.

We usually find in a restaurant that a knife is good for six to nine months. With professional cutlery, and in particular with packing houses, they'll need a wide blade for breaking down a side of beef. They need a large curved knife, which we call a *cimeter steak knife*. When it starts out life, it's about 2 1/2" wide, and when it gets down to about 1" or 1 1/4" wide, it's no longer suitable for breaking down the big sides of beef. So then they use it for the smaller cuts, and call it a *breaking knife*. When they wear it down to about under an inch, they use it as a *boning knife*.

So these knives actually go to a series of different lives? As they get smaller from sharpening, they get repurposed and reused?

They get narrower, and they get shorter. People find different applications for them. The poultry industry still does that. What I'm talking about is mostly pre-WWII. After WWII, people started coming to us and saying, "Hey, can't you make this shape from scratch?" So we started to create the same shape as the worn-out knife. You wouldn't have to wear out a giant cimeter; you could just buy a breaking knife off the shelf. A lot of our traditional knife shapes have evolved from large blades that were worn down and used for different applications, and then we started making a blade with that shape.

What advice would you give somebody new to the kitchen?

If I were being a smartass, I would tell you don't run with a knife. Keep your knives out of the dishwasher. Wipe them clean with a damp rag. When you put them in the dishwasher, they bang together and you nick up your edges. If you do put them in the dishwasher, make sure you pull them right out of the basket and dry them off. Keep up with your sharpening; don't let your knife get dull. Maintain the edge every time you use it or every other time you use it. Give it one or two strokes on a steel and sharpening will never be a chore, and you will always have a sharp knife.

Knife Sharpening 101

Keeping your knives sharp is the kitchen equivalent of backing up your files: it's something you should do more often than you think. A sharper knife is safer and easier to use:

- Sharp knives require less pressure for making cuts so there's less force involved—meaning you're less likely to slip and cut yourself.
- Sharp knives cut cleaner; there is less "tear" through whatever you're cutting.
- Sharp knives keep your arm from getting tired because you don't have to muscle through things. Of course, you'd probably need to be slicing and dicing for many hours to notice.

Keeping your knives in good working order involves both keeping the blade "true" (in alignment) and grinding down the blade to reshape the edge if the trued shape is lost. To keep your knives true, use a sharpening steel (those steel rods ubiquitous in celebrity chef photos) as part of your cleanup and wash routine at the end of a cooking session. By running the knife against the sharpening steel, you push any portion of the edge that is out of alignment ("burrs") back into alignment. (*Never* try to true a serrated knife, such as a bread knife—the sharpening steel won't fit against the serrated edge.) Look for a diamond-coated sharpening steel; the diamond coating is harder than the steel, so it can not only realign the burrs but also create a new edge, keeping the knife truly sharp and actually removing the need to reshape the edge.

More serious sharpening involves grinding down the blade to form a new edge and can be done against any hard surface: a sharpening stone, a grinding wheel, even a brick! (See the interview with Buck Raper on the preceding pages for details.) If it comes to that, I find it easier to have my knives professionally sharpened. Grinding down the edge isn't a great thing, though, because creating the new edge removes material. Knives used in restaurants can be "sharpened through" in under a year—that is, sharpened down to a point where the new edge on the knife becomes too thick to hold a sharp edge for long.

Cutting boards

Most cutting boards are made of either hardwoods, such as maple or walnut, or plastics like nylon or polyethylene. Regardless of which type you get, look for ones that are at least 12″ × 18″ / 30 cm × 45 cm. Bigger is better, as long as the board fits in your sink or dishwasher. If you choose a plastic board, consider snagging both a rigid one, which can serve double duty as a serving board, and a thin, flexible one, which can be used as a makeshift funnel (e.g., chop veggies, pick up board, and curl it while sliding the food into your pan).

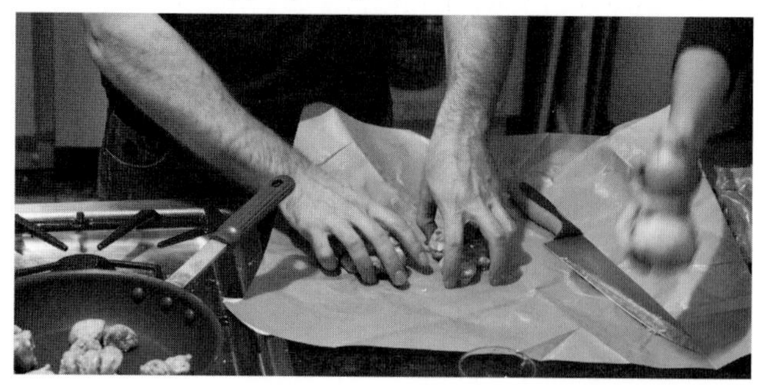

You can use the wrapping paper that some meats come in as an impromptu disposable cutting board if you are just cutting something like a sausage to sauté. One less dish to wash!

Always use two different cutting boards when working with meats: one for raw meats and a second for cooked items. I use a plastic cutting board for raw meats and a wooden one for after cooking because I find the difference in material to be an easy visual reminder. I then toss the plastic cutting board into the dishwasher for cleanup. Since I have more than two boards, I use the plastic one exclusively for raw meats.

Plastic cutting boards have the advantage of being sterilized when washed in a dishwasher because the heated water kills common bacteria. (Don't put your wooden cutting board in the dishwasher, though: the hot water will damage the board.) Note that washing a cutting board in the sink with hot water and soap is *not* sufficient to remove absolutely all traces of bacteria like *E. coli*. Whether wood or plastic is "safer" depends on your habits. Some studies have shown that wood is better than plastic at preventing cross-contamination, possibly due to chemical properties of wood, which suggests that wooden cutting boards are more forgiving to lapses in sanitization. If you don't have a dishwasher, current research suggests that a wooden cutting board is the way to go.

Researchers at UC Davis found that disease-related bacteria such as *E. coli* survived for a longer period of time on plastic cutting boards than wooden ones, and that treating wooden cutting boards with mineral oil did not materially affect the die-off rate. Additional research found that home chefs using plastic cutting boards are twice as likely to contract salmonellosis than those using wooden cutting boards, even when cleaning the board after contact with raw meat.

Here are a few additional tips:

- Place a bar towel or slip mat under your cutting board to prevent it from moving while you're working.
- Some cutting boards have a groove around the edge to prevent liquids from running over the edge. This is handy when you're working with wet items, but it makes transferring dry items, such as diced potato, more difficult. Keep this in mind when choosing which board—or which side of a board—to use.
- You can clean wooden cutting boards by wiping them down with white vinegar (the acidity kills most common bacteria). If your board smells (e.g., of garlic or fish), you can use lemon juice and salt to neutralize the odors.
- Prep vegetables and fruits before starting to work on raw meats. This further reduces the chances of bacterial cross-contamination.

Pots and pans

Which pot or pan is ideal to use for cooking an item, and how the materials in that pot will affect the cooking process, is a topic that could easily be expanded to fill an entire chapter and yet still leave questions unanswered. When it comes to the metals used in making pans, there are two key variables: how quickly the metal dissipates heat and how much heat the metal can retain (see Chapter Metals, Pans, and Hot Spots on page 59). For new cooks, the biggest issues are avoiding hot spots and being careful not to overheat the pan. Avoid hot spots by using pans with materials that conduct heat well (and avoid those really cheap thin pans). Also, don't just automatically crank the heat up to high. Hotter doesn't mean faster! And if you do find yourself with a pan full of ingredients that are starting to burn, dump the food into a bowl to halt the burning. Even off the burner, the pan will still be hot enough to continue cooking and burning its contents.

All that being said, don't obsess over the "perfect" pan for a job. Looking at cladded pans (two types of metals sandwiched together) and can't decide between copper and aluminum? If they're properly made (in terms of the thickness of the metal and the construction), there won't be a huge difference. Same thing when it comes to size and shape.

Sure, to a professional it matters: cooking 10 pounds of onions in a narrower pot will yield more consistent results than cooking them in a wide, shallow pan (the narrower pot will retain water better, which assists in the cooking). But as a home chef, you'll typically achieve similar results in both cases, as long as you use common sense about the amount of heat you're using and keep a watchful eye on the pan.

As with knives, let your preferences and cooking style guide your selection of pots and pans, and be willing to experiment and replace items to suit your needs. Avoid purchasing a set of pots and pans, because sets often come with extra items that aren't quite ideal and end up wasting space and money. Instead, select each pot or pan individually and purchase only the ones that best suit your needs. Browse your local restaurant supply store or search for commercial products online. Commercial frying pans are cheap multitaskers. If you're going to splurge on a pot or pan, spring for an enameled cast iron pan (Le Creuset is the leading maker), a good skillet, or a sauté pan.

A skillet is technically the same thing as a frying pan, but I think of frying pans as being the cheap-but-good commercial aluminum ones and skillets as being stainless steel. A sauté pan is like a skillet, but the inside corners are square instead of rounded up.

When using pots and pans, follow these tips. Unless you're heating a pan to sauté something, don't absentmindedly leave it empty while it's heating on the burner. Overheating a pan, especially the nonstick type, will ruin the pan's finish and possibly warp it. Cast iron is the exception, but you still risk destroying the seasoned finish. Also, if you're anything like me, when you throw a dinner party the dishes often wait until the next morning. Don't leave pots and pans soaking in water overnight. In some cases, the water can get "under" nonstick finishes and blister it. In the case of cast iron, the pans will rust.

Frying pans. A frying pan is a shallow, wide pan with slightly sloped edges. Look for frying pans that have a smooth cooking surface and are as large as your stovetop will comfortably accommodate. If you get one that's too large, the burners on your stove will heat the center but not the outer region, which will lead to uneven cooking.

Nonstick frying pans are useful for sautéing fish and for breakfast items such as eggs, pancakes, or crepes. Using a nonstick pan for eggs or fish also allows you to reduce the amount of butter or oil needed during cooking.

Since nonstick coatings prevent the formation of fond (the bits of food that brown in the bottom of the pan and provide much of the flavor in sauces), you might also want to purchase a stainless steel frying or sauté pan.

How do they get Teflon (polytetrafluoroethylene, PTFE) to stick to the pan if it doesn't stick to anything? By using a chemical that can actually stick to both PTFE and the pan, called an *adhesion promoter* in chem-speak. Perfluorooctanoic acid (PFOA) is the adhesion promoter of choice. Unfortunately, it's rather toxic, but according to the manufacturers it's not present in the finished products. PTFE itself melts at 620°F / 327°C. Most stoves can get pans up above that temperature, which is why nonstick pans shouldn't be used for searing or under the broiler. DuPont says nonstick pans coated with PTFE are fine up to 500°F / 260°C and that the material won't begin to "significantly decompose" until 660°F / 349°C. Still, don't try it: polymer fume fever isn't fun.

I personally use nonstick frying pans as a default for day-to-day cooking because they're easier to clean and well suited to the type of food I eat. My stainless steel frying pan gets used for those times when I am cooking "for real" (not to knock my morning scrambled eggs) and want to deglaze the pan to capture the fond. But you might cook different foods than I do, in which case your default pan might end up being stainless steel or cast iron.

I recommend that you have at least three frying pans on hand: one for searing items such as fish, a second for sautéing vegetables, and a third for those times when you want to reduce a sauce or sweat onions at a lower temperature. I prefer Vollrath's Lincoln Wear-Ever Ceramiguard 10" frying pans (EZ4010): they're cheap, they get the job done, and the silicone handles are oven-safe. If you're lucky enough to have a larger stovetop with burners rated for higher BTUs, snag a 12" / 30 cm frying pan in lieu of a third 10" / 24 cm pan. And, if you're often cooking for one, a smaller 8" / 20 cm frying pan is a useful size for quick dishes like scrambled eggs.

You don't need to completely wash nonstick frying pans every time you use them, unless there's particulate food left behind. Wipe the pan down with a paper towel, leaving a thin layer of oil behind.

I find it useful to have multiple frying pans so that I can cook different components of a dish separately. Onions (left pan), for example, are often "sweated" at a lower temperature, to keep them from caramelizing, while sausage (right pan) needs to be cooked hot enough to trigger the Maillard reactions that give seared meats an intensely rich flavor.

Saucepans. A saucepan, roughly as wide as it is tall and with straight sides, holds two to three quarts of liquid and is used in cooking liquid foods such as sauces, small batches of soups, and hot drinks like hot chocolate. Look for a pan that has a thick base, as this will help dissipate the heat and avoid hot spots that could burn your food. Keep in mind that many of the liquids cooked in a saucepan tend to be things that can burn, so it's worth spending a bit more to purchase a pan that conducts heat better. I picked up my favorite saucepan as an "odd lot" piece from a department store set. (Be sure to snag the lid as well!) You might prefer a *saucier* pan, one that has rounded corners that are easier to get into with a whisk or a spoon.

Stockpots. A stockpot holds two or more gallons of liquid and is used in blanching vegetables, cooking pasta, and making soups. Since most applications for a stockpot involve a large amount of water, burning foods is not as much of a concern as it is with a saucepan—unless you can figure out how to burn water! The stockpot I use is one of the $20 cheap stainless steel commercial varieties. Make sure to pick up a lid as well, because commercial sellers tend to sell them separately.

Cast iron pans. You should have a good cast iron pan in your pot and pan collection. Cast iron pans are heavy, and their larger mass allows for better retention of heat. Cast iron pans can also be heated to higher temperatures than nonstick and stainless steel pans, making them ideal for searing foods such as meat. They're also handy for baking items such as cornbread or deep-dish pizza. Just remember to avoid cooking highly acidic items such as tomatoes in them, because the iron will react with acidic items.

As with frying pans, when washing cast iron, don't use soap. Instead, rinse the pan and wipe the inside to dislodge any stuck-on food, and then place the pan back on the stove. If the food is really stuck, throw in a few tablespoons of coarse salt and a spoonful or two of oil such as canola oil, and "sand" it off with a paper towel. Once your pan is clean, wipe it down with a little heat-stable oil such as canola or sunflower oil (but not extra virgin olive oil) and place on a burner set for low heat for a minute or so to thoroughly dry it out. And never let it sit in water for hours on end, because the water will ruin the finish. If you *do* end up with rust spots, don't fear. You can use a metal scrubbing brush to scrape away the rust, and then reseason the pan with a coating of oil.

Metals, Pans, and Hot Spots

What's the deal with pans made of different metals or with various combinations sandwiched together? It has to do with the differences in thermal conductivity (how quickly heat energy moves through a material) and heat capacity (how much energy it takes to heat a material, which is the same as how much energy it'll give off when cooling).

Let's start with the thermal conductivity of common metals in pans, along with a few other materials for reference.

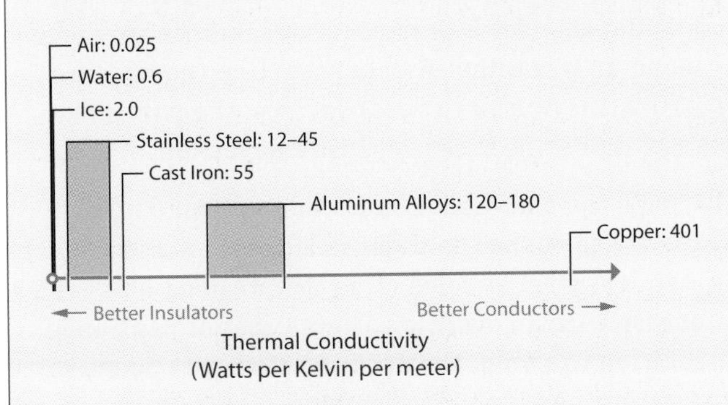

Pans made from materials with a lower thermal conductivity take longer to heat, because the thermal energy applied from the burner takes longer to transfer up and outward. In physics-speak, this is called *low thermal response time*. In cooking, pans with slow thermal conductivity (cast iron, stainless steel) are "sluggish" in response to changes in heat. Pop them on the burner, and nothing seems to happen for a while. Likewise, if you get them too hot and pull them off the burner, food in them will continue to cook for a while.

Given two pans of identical diameter, one cast iron and one aluminum, the aluminum pan will conduct the heat throughout the pan faster. Here's a picture of this, using thermal fax paper (hey, not all of us can afford a thermal imaging camera!). Since thermal fax paper turns dark where heated, dark = hot and white = cold.

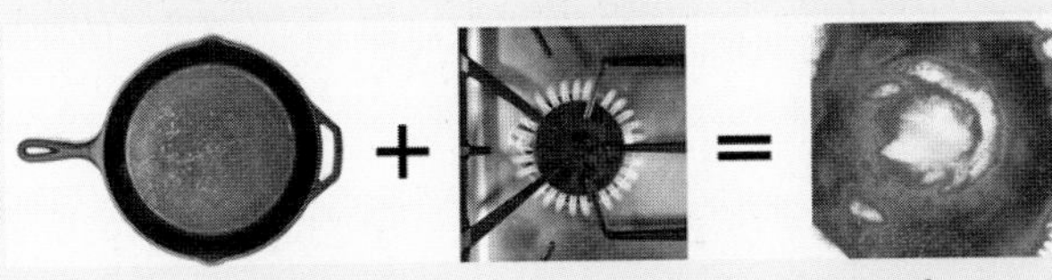

Cast iron pan on a gas burner = slower heat transfer.

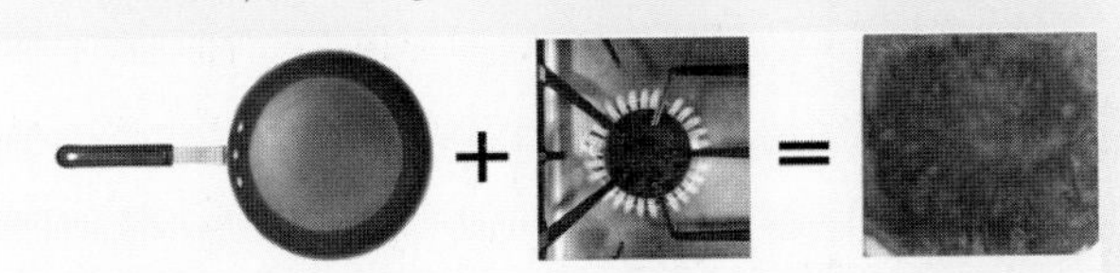

Aluminum iron pan on a gas burner = faster heat transfer.

If you're keen to try this yourself, grab a roll of thermal fax paper, heat your pan on the burner for 30 to 60 seconds, turn off the heat, and then place a square sheet of paper on top of the pan and coat it with a few cups of cold rock/kosher salt to help press the paper against the surface of the pan.

Notice that the gas burner has a wide radius and the gas jets are directed outward. Result? The *center* of the pan actually ends up being colder. The cast iron pan shows this well because the heat does not conduct through the material as quickly as it does with the aluminum pan, leading to a cold spot.

Specific heat is important, too. *Specific heat* is the thermal energy (measured in *joules*) needed to change a unit mass of material by a unit of temperature, and it differs between materials. That is, it'll take a different amount of energy to raise a

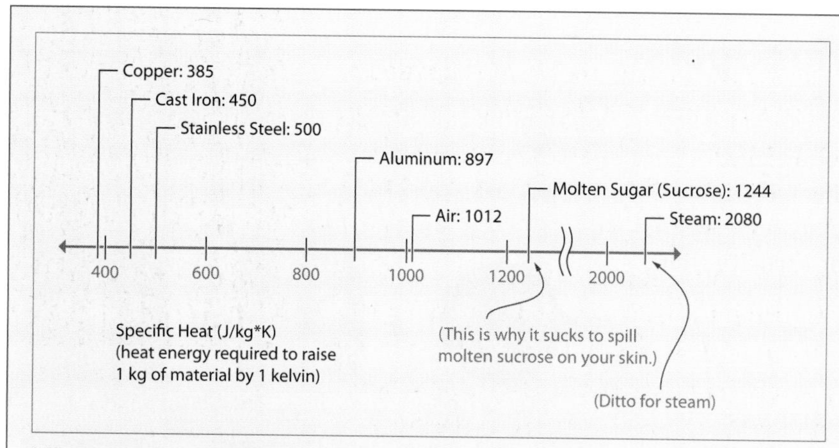

kilogram of cast iron 1°C versus a kilogram of aluminum, because of how the materials are structured at the atomic level. How do common metals in pans compare in terms of specific heat?

Cast iron has a lower specific heat than aluminum. It takes roughly twice as much energy (897 J/kg*K versus 450 J/kg*K) to heat the same amount of aluminum up to the same temperature, and because energy doesn't just disappear (first law of thermodynamics), this means that a kilogram of aluminum will actually give off *more* heat than a kilogram of cast iron as it cools (e.g., when you drop that big steak onto the pan's surface).

It's not just the thermal conductivity or specific heat of the metal that matters, though; the mass of the pan is critical. I always sear my steak in my cast iron pan. It weighs 7.7 lbs / 3.5 kg, as opposed to 3.3 lbs / 1.5 kg in the case of my aluminum pan, so it has more heat energy to give off. When searing, pick a pan that has the highest value of *specific heat * mass*, so that once it's hot, it won't drop in temperature as much when you add the food.

There are a few other factors you should consider when picking a pan. Cast iron and aluminum react with acids, so pans made of those materials shouldn't be used for simmering tomatoes or other acidic items. Nonstick pans shouldn't be heated above 500°F / 260°C. And then there are cases where the pan isn't the primary source of heat for cooking: when boiling or steaming, the water provides the heat transfer, so the material used in making the pan isn't important. Likewise, if you're using an ultra-high-BTU burner (like the 60,000-BTU burners used in wok cooking), the pan isn't a heat sink so heat capacity isn't important.

What's the deal with *cladded* metals? You know, pans with copper or aluminum cores, encased in stainless steel or some other metal? (*Clad* = to encase with a covering.) These types of pans are a solution to two goals: avoiding hot spots by evening out heat quickly (by using aluminum or copper), and using a nonreactive surface (typically stainless steel, although nonstick coatings also work) so that the food doesn't chemically react with the pan.

Finally, if you're buying a pan and can't decide between two otherwise identical choices, go for the one that has oven-safe handles. Avoid wood, and make sure the handles aren't so big that they prevent popping the pan in the oven.

Measuring cups and scales

In addition to the common items used for measuring (e.g., measuring cups and spoons), I strongly recommend purchasing a kitchen scale. If you will be following any of the recipes from this book using hydrocolloids or other food additives (see Chapter 6), it is practically required. You might not use it every day (or even every week), but there is no substitute for it when you need one.

You can pour ingredients directly into a mixing bowl by weight, skipping the need for measuring cups.

You will obtain better accuracy when measuring by weight. Dry ingredients such as flour can become compressed, so the amount of flour in "1 cup" can vary quite a bit due to the amount of pressure present when it's packed (see the sidebar Weight Versus Volume: The Case for Weight). Also, it is easier to precisely measure weight than volume. Because much of cooking is about controlling chemical reactions based on the ratio of ingredients (say, flour and water), changes in the ratio will alter your results, especially in baking. Weighing ingredients also allows you to load ingredients serially: add 390 grams of flour, hit tare; 300 grams of water, hit tare; 7 grams of salt, hit tare; 2 grams of yeast, mix, let rest for 20 hours, and you've got no-knead bread. (See the interview with Martin Lersch on page 224 in Chapter 5 for baking instructions.)

Use a high-precision scale when working with food additives.

When choosing a scale, look for the following features:

- A digital display, showing weights in grams and ounces, that has a tare function for zeroing out weight
- A flat surface on which you can place a bowl or dish (avoid scales that have built-in bowls)
- A scale that is capable of measuring up to at least 5 lbs or 2.2 kg in 0.05 oz or 1g increments

If you plan on following any "molecular gastronomy / modernist cuisine" recipes that use chemicals, you'll need to pick up a *high-precision scale* that measures in increments of 0.1 gram or finer. I use an American Weigh Scale AMW-100.

Spoons & co.

Few things symbolize cooking more than a spoon, and for good reason: stirring, tasting, adjusting the seasoning, stirring some more, and tasting again would be virtually impossible without a good spoon! I prefer the wooden variety. In an age of technology and modern plastics, there's just something comforting about a wooden spoon. Look for one that has a straight end, as opposed to a traditional spoon shape, because the straight edge is useful for scraping the inside corners and bottom of a pan to release fond. When it comes to cleaning them, I run mine through the dishwasher. True, it's bad for the wood, but I find it easier and don't mind buying a new one every few years.

Weight Versus Volume: The Case for Weight

How much of a difference does it *really* make to weigh your flour? To find out, I asked friends to measure out 1 cup of all-purpose flour and then weigh it. Ten cups later, the gram weights were in: 124, 125, 131, 133, 135, 156, 156, 158, 162, and 163. That's a whopping 31% difference between the lowest and highest measurements.

How much flour is in a cup? Depends on whether you pack it in tight (on left: 1 cup at 156 grams, then sifted) or keep it loose (on right: 1 cup at 125 grams, then sifted).

Even if you could perfectly measure the same weight with every cup, you still might end up using a different amount than what a recipe calls for. The average weight of the 10 samples above is 144 grams. The United States Department of Agriculture defines 1 cup of flour as 125 grams; Wolfram|Alpha (*http://www.wolframalpha.com*) gives 137 grams. And the side of the package of flour in my kitchen? 120 grams.

The upshot? You'll get better results by weighing ingredients, especially when baking. A cup might not be a cup, but 100 grams will always be 100 grams. Clearly, weight is the way to go.

But what about wet measurements—measurements of things that don't compress? While you're not going to see the same variability, you can still end up with a fair amount of skew just based on the accuracy of the measuring device. The following image shows what four different methods for measuring 1 cup of liquid yielded.

212 grams
Tablespoon
(16 tablespoons = 1 cup)

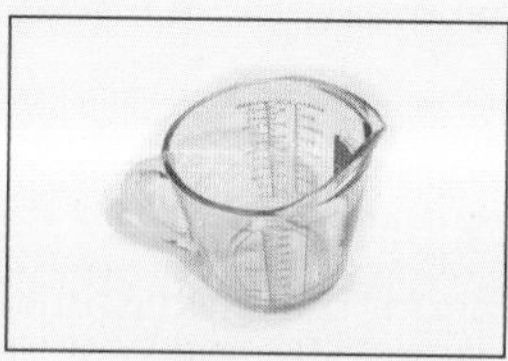

225 grams
Liquid measuring cup

232 grams
Dry measuring cup

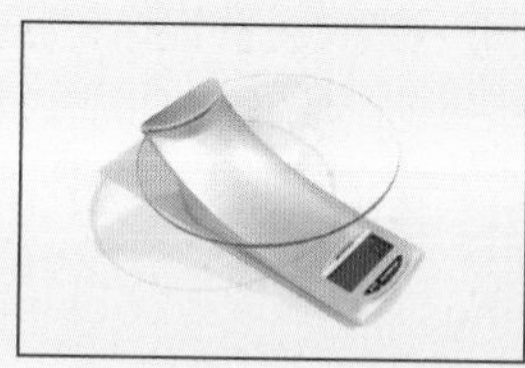

237 grams
Digital scale

Besides the ubiquitous wooden spoon, here are a few related tools that you should keep "near to hand" while cooking.

I'd rather have two or three of the commonly used tools—spoons, spatulas, whisks—than a large collection containing only one each of many specialized tools.

Silicone stirring spatulas. This type of spatula, in addition to making perfect scrambled eggs, is handy for folding egg whites into batters, scraping down the edges of bowls, and reaching into the corners of pots needing stirring. Silicone is also heat-stable up to 500°F / 260°C.

Whisks. If you're going to bake much, a whisk is essential. Go for a standard balloon whisk, not one of those funky attempts at wires with balls on the end or crazy little loopy things. Besides coming in handy when you want to whisk eggs and dressings, you should always whisk together the *dry* ingredients for baked goods to ensure that things like salt and baking powder are thoroughly blended with the flour.

Initializing the Kitchen

Kitchen shears. Essentially heavy-duty scissors, kitchen shears are useful for cutting through bones (see "Butterflied Chicken" on page 206) and are a great alternative to a knife for cutting leafy greens, both small (chives) and large (Swiss chard). If you're serving soup into bowls and want to garnish with chives, instead of using a knife and cutting board, you can hold the chives directly above the bowl and use the shears to snip them directly into the bowl: faster, and fewer dishes, too!

In addition to flipping items in pans or grabbing hot ramekins from an oven, tongs can be useful for holding on to hot foods such as just-cooked sausage while slicing them.

Tongs. Think of tongs as heatproof extensions of your fingers. They're useful not just for flipping French toast in a frying pan or chicken on the grill, but also for picking up ramekins in the oven or grabbing a cookie tray when you're out of towels. Look for spring-loaded tongs that have silicone or heatproof tips, because these can be used with nonstick coated pans. Scalloped edges are also useful, because they tend to grip things better than their straight counterparts.

Thermometers and timers

Probe thermometers are awesome because they use a thermocouple attached to a long heat-safe lead, designed so that you can stick the probe into a piece of meat and set the controller to beep when it reaches the desired temperature. Timers are handy, and if you'll be doing much baking, one will be critical. But if you expect to be doing mostly cooking, a timer is just a proxy for checking when, say, an oven roast has reached temperature, in which case why not use something that actually checks that? And when it comes to food safety, it's not possible to "see" what a hamburger cooked to 160°F / 71°C looks like, even when cut in half.

Infrared thermometers are great for taking dry temperatures, such as the surface temperature of a frying pan before you start making pancakes, or ice cream you've just made with liquid nitrogen (see page 377 in Chapter 7). The other great thing about them is that they're instant: point, click, done. You can also use them to take the temperature of liquids in a pan without having to worry about handling a hot thermometer probe or washing it after. Keep in mind, though, that stainless steel is reflective in the IR range, just like a mirror reflects visible light—you'll end up taking the temperature of your ceiling, not the pan, if you try to meter an empty stainless steel pan. Also, IR thermometers only take surface temperature, so they shouldn't be used for checking internal temperatures for food safety.

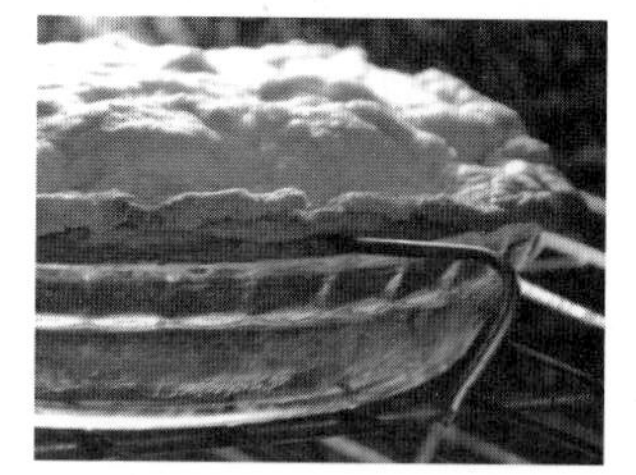

Tuck a probe thermometer into a quiche or pie to tell when the internal temperature indicates it is done. I pull my quiches out when the temperature reaches 140°F / 60°C. The egg coagulates in the range of 140–149°F / 60–65°C, and 140°F / 60°C is hot enough that the "carryover" heat will just set the egg custard without making it dry.

Finally, I'd be remiss if I didn't mention the most overlooked but useful thermometer: your hands. Learn what various temperatures feel like: hold your hand above a hot pan, and notice how far away you can be and still "feel" the heat (thermal radiation). Stick your hand in an oven set to medium heat, remember that feeling, then compare it when you're working with a hot oven. For liquids, you can generally put your hand in water at around 130°F / 55°C for a second or two, but at 140°F / 60°C it'll pretty much be a reflexive "ouch!" Just remember to use a thermometer for foods that need to be cooked to a certain temperature for food safety reasons, which we'll cover in Chapter 4.

Mixing bowls

While you can get away with using your dinner plates or soup bowls for holding some things, you'll invariably need mixing bowls for working with and storing your ingredients. I recommend two types: large metal bowls (~12 to 16" / 30 to 40 cm diameter) and small glass bowls.

For metal bowls, poke around your nearest restaurant supply store for some cheap stainless steel ones, which should cost only a few dollars apiece. These bowls are large enough to hold cookie dough, cake batter, and soup, and they have enough room for chopped leafy greens that you plan to sauté. You can also toss them in the oven at low heat to keep cooked items warm, something you can't do with plastic ones.

Small glass bowls are also very useful, especially if you're using a *mise en place* setup. Measuring out chopped ingredients into small glass bowls ahead of time will allow you to toss the ingredients together much faster during the cooking process. If you have leftovers, just wrap the bowls with plastic wrap and store in the fridge. Look for glass bowls that are all the same size and that stack well. You'll often find these bowls available at your local hardware store.

Bar towels

In addition to wiping off counters with them, you can use bar towels (typically 12″ × 18″ / 30 × 40 cm terry towels with some thickness) as potholders, under a cutting board to prevent slippage, or as a liner in a bowl to help dry washed items such as blueberries or cherries. And you can never have enough of them. I keep several dozen on hand in my kitchen.

You can use a bar towel as a potholder to handle dishes or pans that have been in the oven. Fold it in half to double the thickness, and don't use a wet towel because it'll steam up and burn you. Some people prefer oven mitts, because oven mitts are typically thicker and don't have the potential to catch on corners like a towel might.

Standard Kitchen Equipment

There is a balance between having the right tool for the right job and having too much stuff on hand. When looking at a potential kitchen tool, consider if you can do the task it's intended for with a tool you already have, and whether the new gizmo is a multitasker capable of solving more than one problem.

Storage containers

While you can use consumer-grade plastic containers, the commercial-grade polycarbonate containers used in the restaurant industry are great: they're rugged enough to last a lifetime, can handle hot liquids, and are designed for holding the larger quantities you'll be handling for group cooking. Search online for Cambro's CamSquare containers.

Ubiquitous in commercial kitchens, CamSquare containers are affordable, practically indestructible, and add a certain geek flair. You can flip the lid over and use it as an impromptu cutting board as well.

Strainers

Look for a strainer that has a metal mesh and a handle long enough to span your sink. Avoid strainers that have plastic parts; plastic isn't as strong or heat resistant and will eventually break. In addition to the normal application of straining cooked foods like pasta or washing berries, a metal strainer can double as a splatter guard when inverted above a frying pan. Depending on the types of food you are cooking, you might find a spider—a specialized spoon with a wide shallow mesh bowl and a long handle—helpful for scooping out items from pots of boiling water.

When straining out pasta from boiling water, pour ***away*** from yourself to avoid steam burns.

You can use a strainer as a splatter guard while pan-frying items such as salmon. Make sure your strainer is wire mesh and has no plastic parts.

Mixers & co.

For baking, a handheld mixer or stand mixer is pretty much indispensable. Sure, you can use a whisk or a spoon, but when it comes to creaming together butter and sugar, you'll get better results with an electric mixer that can whip microscopic air bubbles into the mix. Besides a mixer, there are a few other electric devices that are worth their counter space.

Okay, I'll level with you: I have a normal blender, too, but I use my immersion blender 10 times more often.

Immersion blender. Skip the normal blender and go for an *immersion blender*. Sometimes called a *stick blender*, the blade part of the blender is mounted on a handle and immersed into a container that holds whatever it is you want to blend. When making soup, for example, instead of transferring the soup from pot to blender for puréeing, you take the immersion blender and run it directly in the pot. Quicker to use, easier to wash.

Food processor. While not an essential, there are times when a food processor makes quick work of otherwise laborious tasks—for example, making pesto or slicing 10 pounds of onions or pulsing pie dough to incorporate flour and butter. They're expensive, though, and take up space. You might opt for a mandolin, instead, which can also be used to quickly make large piles of julienned (matchstick-cut) veggies.

Sad but true: the julienned strips you see in restaurants aren't lovingly cut by hand.

Rice cooker with slow-cook mode. I'm in love with my rice cooker. Actually, that's not true; I'm in love with the slow-cook mode of my rice cooker, and you should be, too. As we'll discuss in Chapter 4, some chemical processes in cooking require a long period of time at a relatively specific temperatures. This is why you should make room for a rice cooker with a slow-cook mode: you can safely leave it on overnight, or even for a few days, without worrying about either the utility bill or the house burning down (something that you shouldn't do with almost any other source of heat in the kitchen). This handy appliance makes an entire class of dishes (braised short ribs, duck confit, beef stew) trivially easy. You could just get a slow cooker, but a rice cooker with slow-cook mode will also come in handy for those occasions when you actually want to make rice.

Unitaskers

I know, I know...unitaskers. Some unitaskers are worth making space for, though, because of how well they perform their particular tasks.

Box grater. A simple box grater for grating vegetables, cheese, and butter (for cutting into pastry dough) can save a lot of time. Sure, you can use a food processor with a grating disk (fast but lots of cleanup) or a paring knife to cube (tedious), but there are times when it's just easiest to plop a box grater on a plate or cutting board and grate away.

Simple Beef Stew

In a pan, sear 1–2 lbs / 0.5–1 kg cubed stewing beef (it should be a cheap cut for stewing; more expensive cuts won't have as much collagen, which will affect the texture, as we'll discuss in Chapter 4). After browning the outsides of the beef, transfer the meat to the bowl of the rice cooker. Using the same pan, sauté one or two diced onions (red, yellow, white—doesn't matter). After the onions have started to caramelize, transfer them to the rice cooker. Toss in a can or two of diced tomatoes (enough to cover the beef). Add seasonings—such as oregano, thyme, or rosemary—and salt and pepper. You can add diced potatoes, canned beans, or other starches as well. I sometimes throw in a tablespoon of ketchup and port to add more flavors. Leave to slow cook for at least six hours. You can start the cooking in the morning before work and arrive home to a quick and easy dinner of beef stew.

Pepper grinder. You want a 9" Unicorn Magnum Plus. Really, that's the best pepper mill out there; never mind what it sounds like.

Garlic press. If you like a good strong kick of garlic in your food and don't mind taking a shortcut, a garlic press makes it easy to get a quick fix. By all means, if you're the type who strongly believes in always doing things the right way—a sharp knife, dicing it with precision, and reveling in the texture and nuance—then skip the garlic press. But if you're a garlic lover who, after a long day at work, just wants to cook a quick five-minute meal, a garlic press will make it easier to use your favorite ingredient. The trick is to get a garlic press with a good handle and good "teeth," so that you can pop in a clove *unpeeled* and squirt out fresh garlicky goodness. Then, make sure you pull out the just-pressed skin and wash the garlic press *right away*. With these two tricks, you can add garlic to a dish with about five seconds of work. Be advised that garlic squirted out of a garlic press will quickly oxidize, so save pressing the garlic until the moment you're ready to cook it.

Try this: cook a serving of pasta. Then, in a small sauté pan over medium heat, add a tablespoon or so of olive oil. Once the oil is hot, use a garlic press to add two or three cloves of garlic and cook until the garlic gives off a pleasant aroma. (You can "squirt" the garlic from the garlic press straight into the pan.) Toss in the cooked pasta to coat and serve. Top with Parmesan cheese and a few red pepper flakes if desired.

The Best Tool in the Kitchen?

Don't be shy about using your hands! After a good scrubbing with soap, they're just as clean as a pair of tongs and infinitely more dexterous. Tossing a salad? Putting veggies on a plate? Dropping cookie dough onto a baking sheet? Use your hands. It's faster, easier, and means one less utensil to wash.

I was making crepes at a friend's dinner party several years ago and one of the other guests was an executive chef for a prominent Boston restaurant. We were in the kitchen together chatting about our respective fields of software and cooking. He stopped me as I went to flip the crepe with a spatula, showing me how to pull it back and flip it using my hands instead. Flipping this way allows you to feel how much the crepe is sticking to the pan and avoids the awkward angling-in of a spatula against the sidewall of the frying pan. While I still usually use a spatula (my fingers aren't made of asbestos like a full-time cook's!), the idea of getting in there with my fingers has definitely made me more comfortable grabbing foods and moving them where they need to be. Just don't burn yourself, and remember to wash your hands, especially when working with raw meats.

1-2-3 Crepes

Whisk or purée until entirely mixed, about 30 seconds:

1 cup (250g) milk (preferably whole milk)
2 large (120g) eggs
⅓ cup (40g) flour (all-purpose)
Pinch of salt

Let rest for at least 30 minutes, preferably longer, so that the gluten in the flour has a chance to thicken the batter. (Stash the batter in the fridge if you're going to leave it for more than half an hour.)

Making crepes is like riding a bicycle: it takes practice before it's easy. Expect to completely screw up the first few you make (training wheels!), and keep in mind that while the batter is easy and the technique simple, the error tolerances are actually pretty tight, so don't get discouraged! Like riding a bicycle, it's far easier to go fast; going slow is hard.

Start with a nonstick frying pan over medium-high heat, heating up the pan for about 30 seconds, or until a drop of water sizzles when dropped into it. Once your pan is at temperature, plan to work quickly: butter, wipe down, pour batter in while swirling, flip, flip *again*, add fillings, plate, and repeat. Because they're fast and cheap, crepes are great for dinner parties or brunches, but you should *definitely* practice beforehand.

Butter: Grab a cold stick of butter with the wrapper partially pulled back, and using the wrapped part as a handle, spread a small amount of butter around the pan.

Wipe down: Use a paper towel to thin out the butter over the surface of the pan, wiping up almost all of it (and on repeats, any crumbs left behind from the previous crepe). The pan should look almost dry; you want a super-thin coating of butter, not noticeable streaks.

Pour: Pour in the batter while swirling the pan: pour about 1/4 cup / 60 ml of batter into a 10" / 25 cm pan, adjusting as necessary (you want enough batter just to coat the bottom evenly). While pouring in the batter with one hand, use your other hand to hold the pan in the air and swirl it so that the batter runs and spreads over the surface of the pan. If you can pour batter out of the pan after swirling, you're using too much.

If you're short on batter, you can "spot pour" a bit in to fill in the gap. This is also the point at which you should check the heat of the pan: it should be hot enough that the batter develops a lace-like quality: little holes all over the crepe as the steam tunnels up through the batter. If your crepes come out whitish, turn up the heat.

Flip: Wait until the crepe begins to brown. Don't poke, don't prod; just let it cook. Once the crepe has begun to brown around the edges, use a silicone spatula (one of those folding spatulas works well) to push down the edge all around the circumference. This will release the edge of the crepe so that it lifts off the pan. Carefully grab that little edge to flip the crepe with both hands.

Flip again: Let the crepe cook on the second side for half a minute or so, until it's cooked. The first side should come out a uniformly brown tone, so flip the crepe again before adding the fillings. This will leave the better-looking side on the outside of the finished crepe.

Add fillings: Add whatever fillings you like. You can heat and even cook the fillings by leaving the pan on the heat during this step. Or, you can move the crepe to a plate and fill it off the heat if you're using something cold (e.g., lox, cream cheese, dill). Crepes are a great vehicle for almost any filling, either savory or sweet. If a combination of ingredients works on pizza or in a pie, it'll probably work in a crepe. Try some of the following combinations:

Powdered Sugar & Lemon Juice
Jam
Granulated Sugar & Grand Marnier (Orange Liqueur)
Bananas & Chocolate Ganache or Nutella

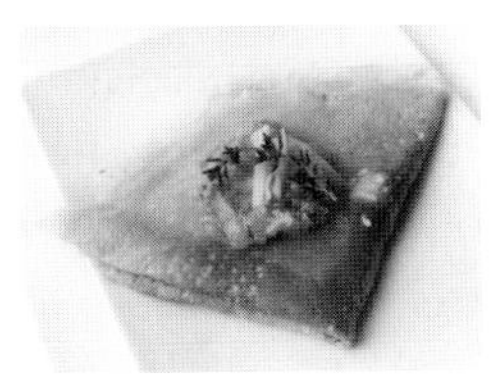

Eggs, Cheese, & Ham
Lox, Cream Cheese, & Dill
Onions, Sausage, & Cheese

Notes

- *It might not look pretty, but a rolled-up crepe with a light sweet filling is fantastic.*
- *If you have a substantial amount of filling, it's easier to fold the crepe into quarters or into a square. I put a small amount of the filling on top as well, as a reminder of what's lurking inside.*
- *When making a crepe with an egg, you can crack the raw egg directly onto the crepe after it's done cooking but while it's still in the pan. Use the back of a fork to break the yolk and scramble the egg, smoothing it over the entire surface of the crepe. Drop some cheese on top wherever the egg is setting too fast.*

Kitchen Organization

A kitchen that has been thoughtfully organized greatly helps in the process of preparing a meal. You will have a more relaxed time cooking if you are able to quickly find what you are looking for and have confidence that you have the right tool for the task at hand.

O(1) Retrieval

Julia Child's kitchen took the adage "a place for everything and everything in its place" to its logical conclusion: pots and pans were hung on pegboards that had outlines drawn around each item to ensure that they were always returned to the same location, knives were stored above countertops on magnetic bars where she could easily reach out to take one, and common cooking ingredients—oil, vermouth—were placed next to her stovetop. Her kitchen was organized around the French method *near to hand*, in which tools and common ingredients are kept out in the open and located near the cooking station where they would normally be used.

Julia Child's kitchen is part of the Smithsonian's permanent collection, including her pots and pans, which she hung on pegboards for easy access.

Store spices in a drawer to speed up the search for any given jar. For extra geek cred, sort them alphabetically (e.g., allspice on the left, wasabi on the right), so that you can use a tree-traversal search algorithm (see http://www.cookingforgeeks.com/book/spicelabels/ *for labels). If you don't have a drawer available, at least make sure to store them in a dark cupboard and not above the stove, where they would get hot.*

Ideally, every item in your kitchen should have a "home" location, to the point where you could hypothetically grab a particular spice jar or pan while blindfolded and without second thought. (This isn't hypothetical for everyone—how else would the blind cook?) This avoids the frustration of digging through a dozen jars to find the one you're looking for. In practice, this isn't always worth the work, but try to keep your kitchen organized enough to be able to select what you're looking for with a minimum of shuffling.

Instead of keeping spice containers in a cupboard, where they get stacked *N* deep (invariably resulting in endless digging for a container that turns out to be right in front), see if you have a drawer where you can see them from the top down. If they're too tall for you to close the drawer, check to see if there is a way to modify the drawer to give you more clearance. In my kitchen, the cabinet had a nonstructural 1.5" wooden slat at the front that, once removed, allowed for storing the bottles upright. I slapped labels on the tops of all my jars to make it easier to find things. (Why is it that a solid third of all spices seem to start with the letter C? Cinnamon, Cardamom, Cumin, Caraway, Cloves...)

For pull-out drawers or fridge doors with a top-down view, labeling the top gives a quick way to find an item.

Hanging up pots, pans, and strainers not only ensures you have a convenient "home" location for each item, but also frees up the cabinet space that they would otherwise occupy. In my kitchen, I created a hanging system using supplies from the hardware store: *S* hooks and a steel *L* beam with holes every few inches (an outside corner support for drywall, made in steel, not aluminum!).

Functional Grouping

Consider storing your everyday kitchen tools near the food items with which they are most commonly used. This approach cuts down on the number of trips between cupboards and counters. That is, instead of having a drawer for storing measuring spoons, measuring cups, small mixing bowls, garlic presses, etc., store those items next to the foods with which they are commonly used:

- Measuring spoons and mortar and pestle with spices
- Garlic press with garlic
- Measuring cups with bulk foods
- Small mixing bowls (8 oz) with oils, vinegars, small bulk goods
- Teapots with tea; coffee beans with the French press / coffee pot

Uniform Storage Containers

There are several benefits to using food-grade storage containers for bulk items such as flours, sugars, salts, beans, rices, cereals, grains, pastas, lentils, chocolate chips, cocoa powder, etc. Using standard-sized containers makes optimal use of space, and using plastic containers for storage keeps pantry moths in check. Pantry moths (weevils) can enter your kitchen as free riders in packaged dry goods such as grains or flours. If you're concerned, freeze newly purchased bags of rice, beans, flour, etc. for a week before transferring their contents to storage containers.

Storing dry bulk goods in standardized containers is a more efficient use of space and prevents spills from torn paper bags. If you have the cupboard space, consider getting wide-mouthed containers for flour and sugar that are big enough for you to scoop directly from.

> Yes, there are bugs in dry goods like flour and cereal. Bugs happen. Take their presence as a sign that the food you are buying is nutritious.

I store my bulk items in food-grade PVC containers, roughly 3″ × 3″ × 12″ / 7 cm × 7 cm × 30 cm, that I purchased online from U.S. Plastic Corp. (*http://www.usplastic.com*, search for "PVC clear canister with lid"). Look for a product that has a screw-on lid and meets FDA standards, that has clear sides so that you can clearly see the food inside, and that has a narrow enough opening that you can easily pour from the container into dry measuring cups without spillage. (For flour, you might want to use one of the larger Cambro storage containers.)

Having a hard time getting stuff to pour out of the container? Try rocking the container back and forth or rolling it in your hands to tumble out things like flour in a controlled manner.

If you have a particular food product that you buy regularly that comes in a suitable container (*mmm, licorice!*), you might be able to reuse the empty containers and skip the expense of buying new ones. As with spices, I label the tops of the containers and store them so that I can view the labels at a glance. This way, they can be stored sideways in a cabinet for a front view or in a pull-out drawer for top-down access.

Counter Layout

Should you have the luxury of designing your own kitchen, there is one rule that can make a profound difference: design your space so that you have three distinct countertop or work surfaces, each of which has at least 4 feet / 1.2 meters of usable space. Think of it like swap space: without enough space for raw ingredients about to be cooked (first counter), plates for cooked food (second counter), and dirty dishes (third counter), your cooking can crash mid-process as you try to figure out where to stack that dirty pan. This isn't to say the three counter sections will always be used for those three functions, but as a rule of thumb, having three work surfaces of sufficient length (and depth!) seems to make cooking easier.

The 3 × 4 counter rule is a slight variation on the "Cooking Layout" design pattern from Christopher Alexander et al.'s *A Pattern Language: Towns, Building, Construction* (Oxford; see p. 853). It's a great book that examines the common design patterns present in good architecture and urban development.

If your current kitchen setup violates the three-counter, four-feet rule, see if you can come up with a clever way to extend a countertop or create a work surface. If you have the space, the easiest option is to buy a "kitchen island" on wheels, which you can move around as needed and also use to store common tools. If you don't have the space for a floating island, see if there's a spot where you can mount a cutting board onto a wall, hinged in such a way that you can latch it up out of the way while not cooking. Or, you might be able to extend a countertop over an unused space. (Ikea sells excellent and cheap wooden kitchen countertops.)

Kitchen Layout Tips

Most commercial kitchens are optimized to turn out meals as efficiently as possible. What tips can you borrow from the commercial world and apply in your home kitchen?

Cabinet doors. Restaurants don't use them because they slow down access. If you cook often enough that dust isn't an issue, see if removing a few strategic cupboard doors and going to open shelves might work. If you're tight on storage, consider getting a Metro Cart or similar freestanding wire shelving.

Hanging pots. Yeah, hanging pots, pans, and strainers can look a little showy. But it's also handy: they're faster to find and easier to get to. And again, if you're tight on space, hanging up your pots and pans will free up the cupboard space that they would have otherwise taken up. If you're on a budget, look for a steel bar and some *S* hooks. For a couple of dollars you might be able to rig up a serviceable solution.

Counter space. Running out of space can be more than just frustrating; it can lead to kitchen lock-up. The kitchen I had in college was miniscule. I once resorted to putting a warm pot that I was finished with on a rug near the kitchen, having run out of counter space for dirty dishes, only to discover that the carpet was synthetic nylon, followed shortly by the discovery that synthetic nylon melts at a rather low temperature. If you're short on counter space, see if you can rig up a removeable cutting board between two counters.

Cleanability. Consider ease of cleaning in your setup. Commercial kitchens are usually designed to be scrubbed down: white tile, drains, stainless steel. While you're probably not going to go that far, keeping the countertop free of various containers, jars, coffee grinders, etc. makes wiping down the space easier.

I once had a studio apartment with two feet of counter space in a tiny galley-style kitchen. I was able to add another work surface by building a "temporary" counter that spanned the galley space: I screwed a 2″ × 4″ board to the wall opposite the sink and found a cutting board large enough to span from wall to sink. Two dowel pins kept the board from moving. Whenever I needed the counter space, I could just pick up the board and drop it into place. It was simple, cheap, and easy—and well worth the two hours of time it took to put it in place.

If hacking your kitchen space isn't for you, you might still be able to reclaim some space through judicious relocation of kitchen appliances from counter to cupboard (do you really use that bread maker every day?). Spending a few hours creatively reorganizing your counter setup will avoid a lot of potential headaches down the road.

Kitchen Pruning

Keeping your countertops and cupboards junk-free is just as important as having enough storage space for all of your kitchen accoutrements. By pruning out unused or uncommonly used items, you'll find it easier to locate the everyday items.

Start by inventorying the gadgets you have in your drawers. Anything that you haven't used for more than a year should be foisted off on others. If you're not sure you can part with something or have emotional attachments (*but that's the mango slicer from our honeymoon!*), find another home for it, outside the kitchen. Duplicate items (three bread knives?!) and rarely used gadgets should be moved out of the kitchen and, if never recalled, recycled. When you're unsure, err on the side of relocating stuff away from the kitchen. Remember, you can always pull it back into your kitchen if you need it!

Some of your kitchen tools will be seasonal. If you're tight on space, large roasting pans for Thanksgiving turkeys and egg decoration supplies for Easter might be better off stored in a garage or closet.

Broken cheese graters, chipped glasses, cracked dishes—anything that can cause injury should it break while in use—should be fixed or replaced. (Dull knives count. Keep those knives sharp!) Should something break while you're cooking and leave bits of glass or ceramic in your food, toss the whole dish out and order pizza. (*Mmm, pizza: cheaper than a visit to the emergency room.*)

Parkinson's Law: Work expands so as to fill the time available for its completion.

Potter's Corollary to Parkinson's Law: Kitchen stuff expands so as to fill every last shelf and drawer.

Kitchen stuff expands to fill all available space, and then some. Any time you introduce a new chop-o-whiza-matic to your kitchen, try to remove something that takes up a similar amount of space. If your kitchen is already crammed full of stuff and you find the idea of a marathon pruning session overwhelming, try doing your clean-out one cupboard at a time. Still too overwhelming? Remove just one thing a day, no matter how small or big, until you reach a Zen state of tranquility. Kitchen pruning is much easier as an ongoing habit than as an annual ritual.

Adam Ried on Equipment and Recipes

PHOTO BY ANDREW BARANOWSKI

Adam Ried writes the Boston Globe Magazine's *cooking column and appears as the kitchen equipment specialist on the PBS series* America's Test Kitchen. *His personal website is at* http://www.adamried.com.

How did you end up writing for the *Globe* and working at *America's Test Kitchen*?

I didn't intend to be involved in food for a living. I went to school for architecture. I was quick to realize that a) I never should have been admitted to architecture school, and b) even though I was admitted, it would be a grave mistake for me to pursue it, because to quote Barbie, "Math is hard."

So I was doing marketing for architecture firms. I spent a whole lot of time cruising cookbooks, making dinners, and having friends over, but the light bulb hadn't quite gone off. I would come in every Monday morning after a weekend of cooking, and regale my officemates with the various things that I had tried, and how they worked, and what I wanted to change. One day, someone just looked at me and said, "What are you doing here? Why don't you just go to cooking school?" I mean talk about feeling like a doofus. It had never gelled for me, even though my sister had been to cooking school, and my whole family cooks. I promptly quit my job and went to the Boston University Culinary Certificate Program.

At one point, I was in the office of the director, and there was another woman waiting in the office to speak with her. The woman and I struck up a conversation. She had done the program a year or two before me. She was an editor at *Cook's Illustrated,* which I read, but again, doofus moment, it had never registered that it was just down the street in Brookline. I started talking to her about what her job was and how she liked it. Then and there, I decided that I wanted to write about food instead of actually cooking it.

I was on the poor woman like white on rice and just kept after her for a freelance assignment here and there. That finally snowballed into a real job at *Cook's Illustrated.* This was in the early 1990s. I remember being in school thinking, "Oh, God, I don't want to work on the line in a restaurant. That's too hard. I'm too old. I don't like the heat. What am I going to do?" It's one of those incredibly irritating right place, right time stories that you never want to hear when you're on the other end of it.

From the perspective of cooking in the kitchen, what has turned out to matter more than you expected?

This sounds a little geeky, but the thing that I didn't realize going into it, especially because I don't have a scientific mind, was that understanding some of the science behind cooking is important. Leavening is still an uphill battle for me to understand. All these recipes rely primarily on baking powder, but sometimes include a little baking soda. Really understanding the acid neutralization in baking soda as an ingredient and what ingredients are acidic is not something that they really teach you in cooking school.

What turned out to be less important?

Not to shoot myself in the foot here, but kitchen tools. You really don't need every conceivable tool to cook well.

What would you consider to be the few basic tools a kitchen needs?

Certainly a chef's knife. A serrated knife is also really useful. A good, heavy aluminum core sauté pan is important. You can do a million different things in it: sauté obviously, braise, shallow fry, roast, bake... A good strainer, measuring cups, and spoons are useful. I love bowls that have the measurements on them so you can get the volume as you are mixing. I have an immersion blender that I use a ton. I would not want to go anywhere without my immersion blender. I use the food processor quite a bit. I have a standing mixer, but I could probably get away with a hand mixer for most of what I do. Those are some of the basics.

What's your overall approach when you look at a piece of kitchen gear?

I do my best to dump all preconceptions. Because I have had years of experience in the area and exposure to the various tools and talked to various experts, I automatically know what I'm looking for. But I have to try and let go of that stuff and do the test as objectively as possible, because I may be surprised.

I remember testing grill pans that had ridges in the bottom, which are supposed to create the visual effect of real grill marks. I'm a big cast iron pan man. I like cast iron, and one of the pans in the line-up was a cast iron grill pan. Even doing my best to drop the preconceptions, I still thought, "It's going to be fabulous." In fact, it did heat up reasonably evenly, and it retained its heat. It made good grill marks. But I was surprised by the fact that it was a pain to clean because of the shape and placement of the ridges. Gunk would collect between them. I try not to use detergent and abrasives on cast iron, because I want to care for the seasoning. If I have really stuck-on gunk, I get in there with coarse salt and stiff brush, and there just wasn't enough room for the salt to really do its thing. After cleaning it twice, I swore I would never use it again.

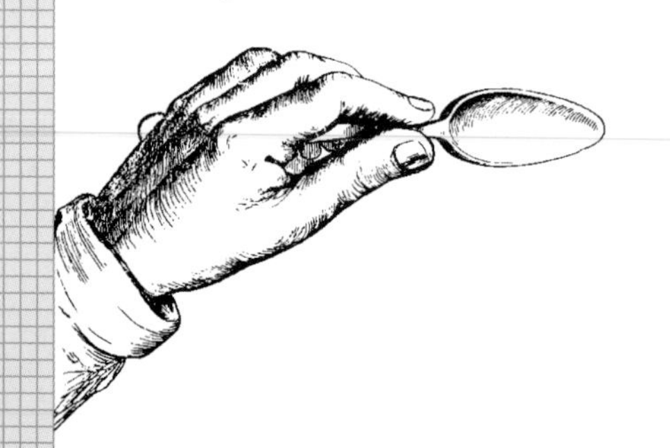

What's your process for going from a first version, or concept, to a final recipe for a *Boston Globe* article?

I've never really shaken the cook's process, so I probably research and test more than I have to. For instance, I'm currently working on fruitcake for a Christmas holiday column. I start by looking online. I have a whole bunch of cookbooks at home, and I also make liberal use of all the libraries in our area. So I'll look at as many fruitcake recipes as I can, say 40 or 50, or whatever is practical given my deadline. I will make a little chart for myself, just a quick handwritten thing, of the types and variables in a fruitcake recipe. Then I overlay my own food sensibilities.

For example, what color scheme I want, what ratio of batter to fruit and nuts I want, what shapes and so forth. I will do what I call "cobbling together" a recipe. I'll give it a try. I convene my tasters and we taste it and analyze it. There's no such thing as a casual, thoughtless meal in this house. I want feedback on pretty much every bite that everyone puts in his or her mouth. Then I'll go back and make it a second time. If I'm really, really, lucky, I can nail it on the second try. More often than not, I will make it a third time. It's a constant process of critique and analysis.

Are there cases where you just get stuck and can't figure out why it's not working?

I'm really lucky to have worked in the food world for long enough that I know a lot of people, much smarter than I am, who I can always call with questions. Actually, for one of my first columns for the *Globe*, I was doing a thing on mangos and I wanted to do mango bread. I was trying to get the leavening right. There was some molasses in there, and some puréed mangos, and this question of baking powder and baking soda came up. I ended up calling a million different bakers to help me understand the role of the baking powder and how it affected the browning.

I've been known to scrap recipes if they don't work the way I want them to after the third or fourth try, or if it doesn't taste as good as I want it to. But I don't remember being so stuck in a problem that I wasn't able to work it out without the help of many smart people.

Has there ever been a case where you've published a recipe, and in hindsight, said "oops," or where the reaction was unexpected?

Oh, God, yes. It's really difficult to please all of the people all of the time. I remember publishing one recipe early on and when I went back and looked at it a couple of years later, I thought, "What the hell was I thinking? That is just as convoluted as can be."

Have any of your recipes caught you off guard by how well liked they were?

There was a lemony quinoa pilaf and asparagus with shrimp scampi recipe that I did. I had discussed quinoa off and on with my editor for a while, because I really like it. Now it's in pretty much any supermarket, but at the point I was writing this recipe it was new to me. People loved it. I got so much positive response from readers on that one.

Lemony Quinoa and Asparagus with Shrimp Scampi

¼ cup (50g) olive oil

3 tablespoons (40g) butter

1 medium (100g) onion, finely chopped

1½ cups (280g) quinoa, rinsed

Salt and black pepper

½ pound (225g) asparagus, ends snapped off and cut into 1.5" / 4 cm lengths

1½ teaspoons (2g) lemon zest (about 1 lemon's worth)

¼ cup (60g) lemon juice (about 1 lemon's worth)

2 pounds (900g) large shrimp, peeled, deveined (if desired), rinsed, and dried

4 cloves (12g) garlic, minced

½ cup (125g) dry white wine

Cayenne pepper

¼ cup (15g) minced fresh parsley

Adjust the oven rack to the center position, place an ovenproof serving dish on the rack, and heat the oven to 200°F / 95°C. In a large nonstick sauté pan set over medium heat, heat 2 tablespoons of oil and 1 tablespoon of butter. Add the onion and cook until soft, about five minutes. Add the quinoa and cook, stirring constantly, until it smells toasty, about four minutes. Add 2¾ cups / 650g of water and 1 teaspoon of salt, increase the heat to high, and bring to a boil. Reduce the heat to low, cover, and simmer until the quinoa is just tender, about 12 minutes. Off heat, sprinkle the asparagus over the quinoa, replace the cover, and set the pan aside until the quinoa has absorbed all of the liquid and the asparagus is tender, about 12 minutes. Add the lemon zest and juice, season with black pepper and additional salt, if desired, and stir. Transfer the quinoa to the warmed serving dish, spread it out to make a bed, and place it in the oven to keep warm.

Wipe out the sauté pan with a paper towel, add 1 tablespoon of oil, and set it over high heat. When the oil just begins to smoke, add half of the shrimp and cook them, without moving, until they begin to turn opaque, about a minute. Quickly turn the shrimp and cook them until fully opaque, about 45 seconds longer, and transfer them to a bowl. Add the remaining 1 tablespoon of oil to the pan, and repeat the process to cook the remaining shrimp. Add the remaining 2 tablespoons of butter to the pan, place over medium-low heat, and when the butter has melted, add the garlic and cook, stirring constantly, until fragrant, about 45 seconds. Add the wine and a pinch of cayenne and stir to blend. Return the shrimp and any accumulated juices to the pan, add the parsley, season with salt to taste, and stir to combine. Remove the serving dish from the oven, pour the shrimp and its juices over the bed of quinoa, and serve at once.

RECIPE USED BY PERMISSION OF ADAM RIED;
ORIGINALLY RAN IN THE MAY 18, 2008 *BOSTON GLOBE MAGAZINE*

Giving Kitchen Tools As Gifts

Don't.

Or, at least not without talking to the lucky recipient. It's virtually impossible to predict someone else's needs and tastes in kitchen tools, and for all the reasons just discussed, saddling them with the wrong tool might be worse than giving them nothing at all. The one exception is if your recipient is just embarking on her culinary adventures, in which case the bare minimum essentials are *probably* okay: chef's knife, scalloped paring knife, wooden cutting board, frying pan, a stack of bar towels, and a gift certificate to the local grocery store. If the recipient has a sense of humor, get 'em a fire extinguisher, too.

3

Choosing Your Inputs: Flavors and Ingredients

YOU OPEN YOUR FRIDGE AND SEE PICKLES, STRAWBERRIES, AND TORTILLAS. What do you do?

You might answer: create a pickle/strawberry wrap. Or if you're less adventurous, you might say: order a pizza. But somewhere between just throwing it together and ordering takeout is another option. If you're reading this book, you're hopefully opening another door and taking the path toward answering one of life's deeper questions: *how do I know what goes together?*

The answer is, as with so many things, "it depends." It depends on how the flavors combine, how those combinations line up with your past experiences, and how the tastes and smells stimulate the regions of your brain responsible for generating and satisfying cravings.

The secret to achieving that blissful sensation of *yummy* in your cooking is to pick good inputs: ingredients that carry good flavor, generate pleasure, and make your mouth water. The single most important variable in predicting the outcome of your culinary attempts is choosing the right ingredients. I'll say it again, because this is probably the second most important sentence in the book: picking the right ingredients for your dish is the biggest predictor of its success.

And here is the most important sentence of this book: *the secret to a good meal is having fun making it and enjoying the entire experience!*

True, you need some skill to manipulate those inputs once they land in the frying pan—don't burn the dinner! But no amount of skill can correct for bad inputs. Cooking and engineering definitely share the maxim *garbage in, garbage out* (GIGO). This chapter covers what you need to know in order to avoid the "garbage in" condition while cooking.

The easiest way to turn a bunch of ingredients into something that tastes great is to buy good ingredients, pick a great recipe, and execute it faithfully. But as the type of geek who likes to be creative, I don't always want to follow a recipe slavishly. I want to understand how to improvise on one or create my own—how to write my own "code" in the kitchen to create something new.

Great chefs can imagine the taste of a combination of ingredients without picking up a spoon. Chef Grant Achatz, of Alinea fame, went through a bout with tongue cancer during which he wasn't able to taste anything, but he was still able to conceive flavor combinations, coming up with what some consider the best food in the nation. As you cook, take time to imagine how the dish you're working on will eventually taste, and check yourself by comparing that imagined taste against the real taste of the final product. Ultimately, knowing which inputs go together is based on having had a wide range of experience putting things together and having taken note of what works and what fails.

If you can't yet imagine what a pickle/strawberry wrap would taste like—or if you can but still want to explore some new ideas on flavor—use the methods for combining ingredients described in this chapter as ways of building up your experiential memory and stirring up new ideas:

Adapt and Experiment Method
: Learn how to recognize the basic tastes and how to adjust them by starting with something you already know how to make, even if it's spaghetti with store-bought pasta sauce.

Regional/Traditional Method
: Use ingredients and recipes specific to a particular place to understand the traditional ways flavors and tastes are combined. A truly new dish in cooking is rare; almost everything "new" can be traced back to tradition.

Seasonal Method
: Limit yourself to produce and meats that are in season. With this approach the number of potential combinations goes down, while the higher quality of the in-season foods results in better flavors.

Analytical Method
: Use a "geek" approach: look at chemicals in foods or at the co-occurrence of ingredients in recipes, and build heuristics that try to predict what flavors will work together.

Because combining ingredients is about how the flavors meld, we'll start with a primer on the physiology of smell and taste, including some experiments to help shed light on how the olfactory and gustatory senses work. Then we'll tackle each of these methods in turn.

Smell + Taste = Flavor

Taste is the set of sensations picked up by taste buds on the tongue (gustatory sense), while smell is the set of sensations detected by the nose (olfactory sense). Even though much of what we commonly think of as taste is really smell, our perception of flavor is actually the result of the combination of these two senses.

When you take a sip of a chocolate milkshake, the flavor you experience is a combination of tastes picked up by your tongue (sweet, a tiny bit salty) combined with smells detected by your nose (chocolate, dairy, a little vanilla, and maybe a hint of egg). Our brains trick us into thinking that the sensation is a single input, located somewhere around the mouth, but in reality the "sense" of flavor is happening up in the grey matter. In addition to the taste and smell, our brains also factor in other data picked up by our mouths, such as chemical irritation (think hot peppers) and texture, but this data plays only a minor role in how we sense most flavors.

The most important variable for good flavor is the quality of the individual ingredients you use. If the strawberries smell so amazing that they make your mouth water, they're probably good. If the fish looks appealing, doesn't feel slimy, and smells "clean," you're good to go. But if an avocado has no real smell and feels like it would be better suited for a game of mini-football, there's little chance that guacamole made from it will be particularly appealing. And if the meat is a week past its use-by date and is home to bacteria that have evolved to be smart enough to say, "well, hello there" when you open the package? Definitely not good.

For a tomato-based dish to taste good, the tomatoes in it should taste and smell like tomatoes. Just because the grocery store has a sign that reads "tomato" next to a pile of red things that look like tomatoes but don't smell like much doesn't make them automatically worthy of a place on your dinner plate. Though they just might not be ripe yet, more likely than not they're a variety that will never be truly flavorful. While serviceable in a sandwich, many of the current mass-produced versions rarely bring the *pow* or *bam* that is the hallmark of great food.

This isn't to say that mass-produced tomatoes can't be flavorful. It's just that the most important variables for taste—primarily genetics, but also growing environment and handling—haven't been given very much attention in recent years.

It can be discouraging, especially for someone new to cooking, to spend the time, money, and energy trying something new only to have a disappointing outcome. Starting with good inputs gives you better odds of getting a good output. You're better off substituting something else that does pack a wallop of flavor than using a low-quality version of a specified ingredient. If you're shopping for a green hardy leaf like kale but what you find looks like it has seen better days, keep looking. Maybe the store has a pile of beautiful collard greens. Would that work? Give it a try.

When it comes to detecting quality, your nose is a great tool. Fruits should smell fragrant, fish should have little or no smell, and meats should smell mild and perhaps a little gamey, but never bad. Smelling things isn't foolproof—some cheeses are supposed to smell like sweaty gym socks and there are some foodborne illness-causing bacteria that have no odor—so you should still use common sense. Still, your sense of smell remains the best way to find good flavor as well as root out whatever evil might be lurking inside.

One caveat on fruit needing to smell fragrant when you purchase it: this doesn't necessarily pertain to types of fruit that will continue to ripen after picking (see the "Storage Tips for Perishable Foods" sidebar on page 45 in Chapter 2 for details). But between giving a peach a squeeze or a sniff, what you smell is going to tell you a lot more about the ripeness of the peach.

Taste (Gustatory Sense)

Our tongues act as chemical detectors: receptor cells of the taste buds directly interact with chemicals and ions broken down by our saliva from food. Once triggered, the receptor cells send corresponding messages to our brains, which assemble the collective set of signals and compile the data into a taste and its relative strength.

The basic tastes in western cuisine that Leucippus (or more likely one of his grad students, Democritus) first described 2,400 years ago are salty, sweet, sour, and bitter. Taste researchers are beginning to discover that Leucippus and Democritus described only part of the picture, though. It turns out that our tongues are able to sense a few secondary tastes as well. About a hundred years ago, Dr. Kikunae Ikeda identified a fifth taste, which he named *umami* (sometimes called *savory* in English) and described as having a "meaty" flavor. Umami is triggered by receptors on the tongue sensing the amino acids glutamate and aspartate in foods such as broths, hard aged cheeses like Parmesan, mushrooms, meats, and MSG. Recent research suggests that we might also have additional receptors for chemicals such as fatty acids and some metals.

Our taste buds also detect and report oral irritation caused by chemicals such as ethyl alcohol and capsaicin, the compound that makes hot peppers hot. Try tasting a small pinch of cinnamon and then some cayenne pepper while keeping your nose plugged. Notice the sandy, flavorless sensation caused by the cinnamon as compared to the sandy, flavorless, burning sensation caused by the cayenne pepper. Capsaicin literally irritates the cells, which is why it's used in pepper sprays like Mace and in some antifouling paints used by the boating industry. (Zebra mussels don't like the cellular irritation either.)

Cellular irritation isn't limited to the "hot" reaction generated by compounds like capsaicin. Pungent reactions are triggered by other compounds, too. Szechuan (also known as Sichuan) peppers, used in Asian cooking, and Melegueta peppers, used in Africa, cause a mild pungent and numbing sensation. Another plant, *Acmella oleracea*, produces *Szechuan buttons*, edible flowers that are high in the compound spilanthol. Spilanthol causes a tingling reaction often compared to licking the terminals on a nine-volt battery.

Spilanthol, the active ingredient in Szechuan buttons (which are also known as sansho buttons or electric buttons), triggers receptors that cause a numbing, tingling sensation. The "buttons" are actually the flowers of the Acmella oleracea *plant.*

Regardless of how many types of receptors there are on the tongue or the mechanisms by which taste sensations are triggered, the approach in cooking is the same: try to balance the various tastes (e.g., not too salty, not too sweet).

Whether you find a set of flavors to be enjoyable or how you prefer tastes to be balanced depends in large part on how your brain is wired and trained to respond to basic tastes. If you're like many geeks I know, you might have an affinity for coffee with lots of sugar and milk or find a certain candy bar loaded with caramel and nuts and covered in chocolate irresistible. But *why* are these things delicious? Because our bodies find fats, sugars, and salts to be highly desirable, perhaps due to their scarcity in the wild and the relative ease with which we can process them for their nutrients.

Besides basic physiology, your cultural upbringing will affect where you find balance in tastes. That is, what one culture finds ideally balanced won't necessarily be the same for another culture. Americans generally prefer foods to taste sweeter than our European counterparts. Umami is a key taste in Japanese cuisine but has historically been given less formal consideration in the European tradition (although this is starting to change). Keep this in mind when you cook for others: what you find just right might be different than someone else's idea of perfection.

Taste (Flavor) Test

Identifying foods by flavor is harder than it sounds. Here are two experiments you might enjoy. The first experiment uses both taste and smell and requires a small amount of advance prep work. The second experiment uses just smell and is easier to set up (but not quite as rewarding).

Experiment #1: Taste and smell

This exercise uses common foods from your grocery store, items that might not be part of your day-to-day diet but are still generally familiar to many. Dice or purée the items to remove any visual clues about their normal size and texture. You might be surprised at the degree of difficulty in identifying some of them! It's surprising to discover how much "knowing" what a food item is—seeing the cilantro leaf or being told it's a hazelnut chocolate cupcake—allows us to sense the flavors and tastes we expect from it.

This exercise is best done in a group, because the experience can be surprising and the resulting conversation really educational. I find that a group size of around six to eight participants works best. Have tasters write down their guesses individually, and then start a conversation about the experience when everyone's finished. The person preparing the exercise will, unfortunately, not be able to participate.

In a set of small bowls, divide the following food items, setting with spoons or toothpicks as appropriate:

White turnip, cooked and diced

Cooked polenta, diced (some stores carry packaged cooked polenta that can be easily sliced)

Hazelnuts ground to the size of coarse sand

Cilantro paste (look in the frozen food section; or buy fresh cilantro and use a mortar and pestle to make a paste)

Tamarind paste or tamarind concentrate

Oreo cookies, ground (both cookie and filling; will result in a black powder)

Almond butter (or any nut butter other than peanut butter)

Caraway seeds

Jicama root, diced

Puréed blackberry

Notes

- *If you're setting this up for more than a few people, use ice cube trays instead of small bowls so that you can put a tray down in the center of a table with six to eight participants around each tray.*
- *Jicama root and tamarind paste are a bit obscure, but they serve as fun challenges for tasters familiar with common flavors. If your local grocery store doesn't carry these, they can be found in almost any Asian grocery store.*
- *Try to keep the diced items all of a consistent size, around ⅓" or 1 cm.*

Hazelnuts or filberts? They're actually different things—the outer husk of a filbert is longer than that of a hazelnut—but either is fine.

Experiment #2: Smell

If you'd rather avoid food prep work, you can do a smell test instead. Place the following items into paper cups, one per cup, and cover the cups with gauze or cheesecloth to prevent peeking (you can use a rubber band around the perimeter to hold the gauze in place; blindfolds work too for small groups):

Almond extract

Baby powder

Chocolate chips

Coffee beans

Cologne or perfume (sprayed directly into the cup or onto a tissue)

Garlic, crushed

Glass cleaner

Grass, chopped up

Lemon, sliced into wedges

Maple syrup (*real* maple syrup, not that "Pancake Syrup" stuff)

Orange peels

Soy sauce

Tea leaves

Vanilla extract

Wood shavings (e.g., saw dust, pencil shavings)

Label each cup with a number, and have test takers write down their guesses on a sheet of paper.

Notes

- *You might find that some people are much better at detecting odors than others. Just as with taste, any given smell can be detected at various strengths. Some people are very sensitive to smells; others have a harder time detecting odors (a condition known as* hyposmia*). Like eyesight and hearing, our sense of smell begins to deteriorate sometime in our thirties and starts to fall off faster once we reach our sixties. It's a slow decline, and unlike hearing and eyesight, is hard to notice as it changes, but the loss does impact our enjoyment of food to some degree.*
- *If you are interested in taking a "real" smell test, researchers at the University of Pennsylvania have developed a well-validated scratch-and-sniff test called UPSIT that you can mail order. Search online for "University of Pennsylvania Smell Identification Test."*

Lingering Tastes: Carryover and Adaptation

What you eat does leave a taste in your mouth. Try the following experiment.

You'll need sugar, a slice of lemon or some lemon juice, and a glass of water. Take a sip of the water (tastes like, well...water). Suck on the slice of lemon, or if you're using lemon juice, take a spoonful and let it sit on your tongue. Take another swig of water; it should taste sweet. Now, take a bit of sugar on a spoon and let it coat your tongue for at least 10 seconds. If you try the water again now, you may find that it tastes sour. Taste researchers call this phenomenon *carryover and adaptation*.

When planning what to serve together, you should consider how the tastes in a group of dishes and drinks will interact. When serving one course after another, the tastes from the first course can linger, which is where a palate cleanser course comes in during multicourse meals. A common traditional palate cleanser is a sparkling beverage (carbonated water or sparkling wine), although some studies suggest crackers are more effective. Maybe that bread basket on the table isn't just about filling you up!

Smell (Olfactory Sense)

While the sensation of taste is limited to a few basic (and important) sensations, smell is a cornucopia of data. We're wired to detect somewhere around 1,000 distinct compounds and are able to discern somewhere over 10,000 odors. Like taste, our sense of smell *(olfaction)* is based on sensory cells (*chemoreceptors*) being "turned on" by chemical compounds. In smell, these compounds are called *odorants*.

In the case of olfaction, the receptor cells are located in the olfactory epithelium in the nasal cavity and respond to volatile chemicals—that is, compounds that evaporate and can be suspended in air such that they pass through the nasal cavity where the chemoreceptors have a chance to detect them. Our sense of smell is much more acute than our sense of taste; for some compounds, our nose can detect odorants on the order of one part per trillion.

Trained whiskey tasters can differentiate among around 100,000 odors.

There are a few different theories as to how the chemoreceptors responsible for detecting smell work, from the appealingly simple ("the receptors feel out the shape of the odor molecule") to more complex chemical models. The more recent models suggest that an odorant can bind to a number of different types of chemoreceptors and a chemoreceptor can accept a number of different types of odorants. That is, any given odor triggers a number of different receptors, and your brain applies something akin to a fuzzy pattern-matching algorithm to recall the closest prior memory. Regardless of the details, the common theme of the various models suggests that we smell based on some set of attributes such as the shape, size, and configuration of the odor molecules.

This more complex model—in which a single odorant needs to be picked up by multiple receptors—also suggests an explanation for why some items smell odd when you receive only a weak, partial whiff. To use a music analogy, it's like not hearing the entire set of notes that make up a chord: our brains can't correctly match the sensation and might find a different prior memory closer to the partial "chord" and misidentify the smell.

It also appears that we smell in stereo: just as our ears hear separately, we use our left and right nostrils independently. Researchers at UC Berkeley have found that with one nostril plugged up, we have a much harder time tracking scents, due to lack of "inter-nostril communication."

While you might think of smell as being only what you sense when leaning forward and using your nose to take a whiff of a rose, that's only half the picture. Odors also travel from food in your mouth into the nasal cavity through the shared airway passage: you're smelling the food that you're "tasting."

When cooking, keep in mind that you can smell only volatile compounds in a dish. You can make nonvolatile compounds volatile by adding alcohol (e.g., wine in sauces), which raises the vapor pressure and lowers the surface tension of the compounds, making it that much more likely that they will evaporate and pass by your chemoreceptors.

Chemists call this *cosolvency*. In this case, the ethanol molecule takes the place of the water molecules normally attached to the compounds, resulting in a lighter molecule, which then has a higher chance of evaporating.

Temperature also plays an important role in olfaction. We have a harder time smelling cold foods because temperature partially determines a substance's volatility.

Your sense of taste is affected by temperature, too. Researchers have found that the intensity of primary tastes varies with the temperature both of the food itself and of the tongue. The ideal temperature is 95°F / 35°C, the approximate temperature of the top of the tongue. Colder foods result in tastes having lower perceived strength, especially for sugars. It's been suggested that red wines are best served at room temperature to help convey their odors, while white wines are better served

Here's a simple experiment that shows the difference between taste and smell. You'll need a test subject, two spoons, a grater, an apple, and a potato. Without the subject seeing, grate some of the apple (without skin) onto one spoon, and grate some of the potato (again, without skin) onto the other spoon. Have the subject pinch her nose shut, and give her both spoons to taste. Make sure she keeps that nose pinched the entire time! This prevents the air carrying the odorants from circulating up into the nasal cavity. After she's tried both spoons, instruct her to stop pinching her nose, and note what sensations occur. If you want to do this with a large group, flavored jelly beans work, too.

Jim Clarke on Beverage Pairings

PHOTO USED BY PERMISSION OF JIM CLARKE

Jim Clarke is a wine writer whose work has appeared in the New York Times, *the* San Francisco Chronicle, Imbibe, *and* Foreign Policy, *as well as on Forbes.com and StarChefs.com. He is also the wine director at Megu in New York City. Above, Jim stands in front of wood-fired brewing kettles at the Brasserie Caracole in Belgium.*

How do you pair beverages with food?

What you're looking for is the structure of the beverage and the dish. For example, if you have sweetness in a dish, you're going to want a similar level of sweetness in the wine. If you don't, the wine will taste relatively flat compared to the dish, or if the wine is too sweet, the dish will be less expressive.

Another example would be acidity. Salads can be very difficult because you have acidities in the vinaigrette dressing, so you need a wine that is going to have that acidity as well, or the wine is going to seem flat or even a little bit bitter next to the dressing. If you want high acidity, you can buy an Austrian Grüner Veltliner, a New Zealand Sauvignon Blanc, a New York Dry Riesling, as a couple of examples. Sancerre are 99% of the time going to be dry and acidic.

If you're not into learning about different wines in and of themselves, then when you're buying the wine you should have a retailer who you can talk to. Tell them what the protein is as well as the sauce or preparation. You want to say, "Grouper with a red wine sauce as a reduction: should I do that with a red wine or a white wine?" If you know the dish well and you say, "There's a really buttery sauce," then your retailer will probably direct you to a California Chardonnay.

The convention I'm familiar with is to look at the type of wine. It sounds like this is not a bad approach?

It's not, especially when you talk about Old World wines. A lot of times Old World wine drinkers wouldn't differentiate. They would say, "I'm making *coq au vin,* so I need a burgundy." There are good ones and there are bad ones, but as far as a pairing goes that's a good pairing. So it can be fairly generic. There is really a broad range of wines within a given category that are going to work pretty well with any given dish.

What are the key variables in wine pairings?

Acidity, tannins if you're going with a red wine, body or alcohol level, and sweetness are really the four main things. Flavor is kind of a bonus. The aromas aren't so important to the pairing as the other elements. High alcohol, say California Chardonnay, can be pretty overwhelming for a delicate fish dish. On the other hand, with lobster it's often fantastic because of lobster's rich flavor.

Look where a wine comes from; that really tells you quite a bit. In 95% of cases, if it's somewhere warm, the wine probably has fuller body. If it's fuller-bodied, then it has lower acidity, because those two have an inverse relationship. Sugar in the grapes eventually becomes alcohol in the wine. Unripe fruit is tart; that's acidity. As it becomes ripe, it comes into balance. If it's overripe, it actually tastes kind of plain because the acidity has dropped out. Take a place like California, where there are these beautiful warm vintages. Ask me for a light-bodied California white and I've really got to dig on my list. There are a few little isolated areas that do it, but it's really not what they do best. On the other hand, Austrian whites are generally from a cool-climate growing area; those are usually pretty good in acidity, light to medium bodied, and lower alcohol.

Why do so many people talk about flavor and aroma instead?

People find it much more poetic to talk about the aromas and the flavors. From a practical point of view, if I'm a sommelier on the floor, I'll be very careful about using the word "acidity" with guests. I'll use all sorts of euphemisms like "crisp" or "fresh," because people don't think about acidity in the context of food. "I don't want to drink acid! That sounds terrible!" The fact of the matter is that every beverage has acid in some balance with sugar. Coca Cola technically has more acid than any wine you can think of.

It seems like the way that somebody inside the industry talks about wine is not the same language that's used publicly. Why is that?

If I'm talking to someone in the industry and they say a modern Rioja, I know exactly what they mean, so we skip a lot of stuff. When I'm training servers or people new to the restaurant industry who need to learn about wine, I talk about that cold climate/hot climate thing and then—and this is a little bit harder, because wine making is changing—about Old World versus New World. New World has more fruit expression; Old World has more earth and spice sort of things. It's still generally true, but there are always exceptions. If someone tells me it's a classic Napa Cab, I know it's full bodied because of the warm climate and has more fruit expression because it's New World. So when we talk, we only have to say how it's different from that model, whereas when I talk to the consumer, I can't assume that they have that understanding already.

Are there similar kinds of variables that you use to describe beer pairings?

Alcohol is still a factor, as is sweetness—acidity not so much except in some unusual beers like lambics. You don't have tannins but you have hops. Those play into how you pair, just like acidity and those other things in wine. One of my favorite classics is oysters and dry stout.

Are there particular things that you would avoid in pairings?

Don't go overboard trying to get the right pairing, especially if you have a style of wine or beer that you really don't like. People will say, "I'm trying to eat fish because it's healthier, but I love red wine." Well, don't let that stop you. I drink all sorts of wine depending on the occasion or how I feel, but if I'm not in the mood for a big red and I'm having a steak, I'll find something else. You need to match with your own tastes along with matching the food. Certainly, as a sommelier, the first thing I'm trying to find out is not what the guest is having, but what they like. Pairing is to give you more pleasure.

Any tips for a consumer speaking with a sommelier?

Well, certainly telling sommeliers what you like. Also, a lot of people feel like they need to dance around price. There is an easy way to do this if you're entertaining guests and you don't want to make a big show of not spending a lot of money. When I'm talking to a guest I will have the list open right there so they can run their finger along the name of the wine to the price. They tell me what they're interested in spending. If you wanted to say, "We're looking for something around $100," that's fine, too, but this is kind of a genteel way of doing it. This will save a lot of the feeling like we're sparring with each other because now I know what you're looking for, both in style and in price range, and we'll find a wine for you.

Is there an exercise that one could do to better understand how to do wine pairings?

Get four glasses, one with lemon juice and water, one with very overbrewed ice tea, one with some sugar dissolved in water, and one with half vodka, half water. Then get a few dishes or ingredients and taste each of them with the different elements—tasting with lemon juice, the sugar water, the tannins in the tea, and then alcohol. You can see what each individual element is doing to, say, a piece of cheese. What is each doing to this piece of asparagus? You'll see what those different elements are and how they affect the food. We're not talking about wine flavors at all, just the four elements that occur in wine that are really important to the pairing process.

Smelling Chemicals

Our noses are veritable chemical detectors on par with modern lab equipment. Our sense of smell is capable of distinguishing the difference a single carbon atom makes (e.g., octane versus nonane) and sniffing out compounds all the way down to the level of 0.00002 parts per billion (for one compound in grapefruit). That said, factors such as age, hormonal levels, and exposure mean that some of us notice smells at lower thresholds than others.

Octane

Nonane

Not all compounds can be smelled. Size, shape, and something called chirality all determine whether a molecule is smellable or not. *Chirality* has to do with whether or not a molecule and its mirror version (the pair is known as *enantiomers*) are identical. Your left and right hands, for instance, are chiral because they are not identical, even though they have the same fundamental shape. Carvone is a classic example in chemistry: the compound D-carvone smells of caraway, while R-carvone smells of spearmint.

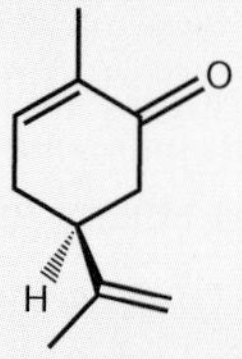

D-Carvone

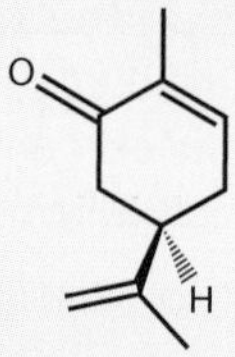

R-Carvone

Some chemical structures have distinct smells, and the families of compounds that contain those structures generally end up smelling similar. Esters (compounds with the general formula of R-CO-OR') are classically thought of as having fruity aromas. Amines smell stinky and rotting, like week-old raw fish, with cadaverine and putrescine being two of the more well-known odors. And aldehydes (organic compounds that have a carbon atom both double-bonded to an oxygen atom and bonded to a hydrogen atom) tend to smell green or plant-like.

While smelling an aldehyde won't bring the entire smell of, say, green ivy, it's similar enough that industry can use aldehyes as artificial odorants to trick our brains into thinking we're smelling the real thing. Artificial flavorings are used in products from laundry detergent to candies, because they cost less and in some cases are chemically more stable than the original scents. Artificial vanilla extract, for example, generally contains just vanillin, which has the molecular formula $C_8H_8O_3$, which happens to be the most common chemical in vanilla. Although the artificial stuff is missing all the other compounds from vanilla, we still find it to be enjoyable.

Many other herbs and spices are also composed of just a few key chemicals, making artificial extracts of them relatively close to the real thing. Fruits, however, have hundreds of compounds that are involved in creating their aromas. Even adding the dozen or so most common chemicals for an artificial strawberry flavoring, for example, leaves 200+ volatile compounds missing in the "odor spectrum." This is why artificial fruit flavors tend to taste, well, artificial (chocolate, too).

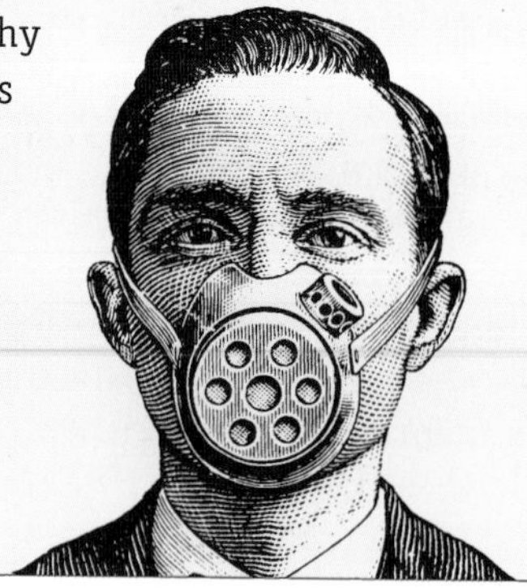

Here are a few examples of compounds and their smells. A number of these compounds can be purchased online if you happen to have an account with an industrial supplier, such as sigma.com. *(Make sure you acquire food or medical grade versions!) Flavored jelly beans and scratch-and-sniff stickers are just a few products that rely on these compounds, so if you don't just happen to have an account with an industrial supplier, try popping open a package of jelly beans and seeing if you can identify some of the odors with these compounds.*

Name	Description	Comment
2,4-dithiapentane	Black truffle	Black truffle oils commonly use this in lieu of oil from real truffles.
Isoamyl acetate	Banana	Creating artificial banana extract is one of the classic chem lab projects, to the annoyance of teachers in adjoining rooms. It's also the pheromone that honey bees use to signal attacking, so don't take overly ripe bananas on a picnic during the height of bee season.
Benzaldehyde	Almond	Primary component of bitter almond oil.
Diacetyl	Like butter	Used in microwave popcorn and Jelly Belly's "Buttered Popcorn" flavor, in large doses it causes a lung disease called "popcorn lung."
Furaneol	Strawberry	Also occurs in pineapple, tomatoes, and buckwheat.
Hexanal	Generic fruity flavor, "tutti frutti" (like pink bubble gum)	Search Sigma.com for 115606-2ml. Used in Jelly Belly's "Tutti Frutti" flavor.
Hexyl acetate	Golden Delicious apple	Search Sigma.com for 25539-1ml.
Maple lactone	Like maple syrup	Search Sigma.com for 178500-10g.
1-p-menthene-8-thiol and nootkatone	Grapefruit	Grapefruit has at least 126 volatile compounds, but these two seem to be the primary ones. Jelly Belly's "Grapefruit" flavor likely includes this compound.

chilled to moderate the levels of volatile compounds and sweetness. This would make sense—by chilling white wine, it'll be less likely to overpower the milder meals that they customarily accompany, such as fish.

Then there's the effect of the temperature of the tongue itself. For example, when drinking a cold soda, as you consume more and more of it, your tongue will begin to cool down. And as your tongue cools, you should perceive the soda as being less sweet. There's a reason why warm soda is gross: it tastes sweeter, cloyingly so, than when it's cold. What does all this mean when you're in the kitchen? Keep the impact of temperature on your senses of smell and taste in mind when making dishes that will be served cold. You'll find the frozen versions of things like ice cream and sorbet to be weaker tasting and smelling than their warmer, liquid versions, so adjust the mixtures accordingly.

As an example, try making the following pear sorbet. Note the difference in sweetness between the warm liquid and the final sorbet. Yes, you could just buy a container of sorbet and let some of it melt, but where's the fun in that?

DIY Lego Ice Cream Maker

Don't have an ice cream maker, but have a pile of Lego bricks? Make your own ice cream maker! Ice cream is made from a base (traditionally, milk or cream with flavorings added) that's agitated as it freezes. Stirring the base as it sets prevents the ice crystals that form from solidifying into one large ice cube.

Of course, the fun with Lego is in figuring out how to build things with it. To make an ice cream maker, grab a Lego Technic kit and an XL motor and snap away. Once you have your motorized stand and agitator put together, mix up your base, transfer it to a large yogurt container, and prechill it by putting the container in your freezer until the base just begins to freeze, about 30 to 60 minutes. Once the base is cold (but not frozen!), slide the container into your Lego rig, place it back in the freezer, and flip the switch. (Dangle the battery outside, because cold environments slow down the chemical reactions that generate energy.) Check on your ice cream every 10 minutes or so, until it begins to set. You'll probably need to stop the motor before the ice cream completely sets, lest the torque tear your Lego creation apart.

For a video of a Lego ice cream maker in action, see *http://www.cookingforgeeks.com/book/legoicecream/*.

Pear Sorbet

In a pan, create a simple syrup by bringing to a boil:

½ cup (120g) water

¼ cup (50g) sugar

Once the simple syrup has reached a boil, remove from heat and add:

15 oz (425g = 1 can) pears (if fresh, peel and core them)

1 teaspoon (5g) lemon juice

Purée with an immersion blender, food processor, or standard blender, being careful not to overfill and thus overflow the container. Transfer to a sorbet maker and churn until set. If you don't have a sorbet maker, you can make sorbet's sister dish, granita, by freezing the mix in a 9" × 13" / 23 cm × 33 cm glass pan, using a spoon to stir up the mixture as it sets. Or, see Chapter 7 to learn about using dry ice or liquid nitrogen to make ice cream.

Notes

- *The lemon juice helps reduce the sweetness brought about by the sugar. The sugar is added not just for taste, but also to lower the freezing point of the liquid (salt does the same thing). Adding a small quantity of alcohol will further help prevent the sorbet from setting into a solid block. Ice cream and sorbets have a fascinating physical structure: as the liquid begins to freeze, the remaining unfrozen liquid becomes more concentrated in sugar, and as a result, the freezing point of the unfrozen portions drops. Harold McGee's* On Food and Cooking *(Scribner) has an excellent explanation of this process for the curious reader.*
- *You can make a more concentrated simple syrup and then dilute it (after letting it cool) with champagne, pear brandy, or ginger brandy. The alcohol is a solvent and will help carry the smells. Alternatively, try adding a pinch of ginger powder, cardamom, or cinnamon either in the sorbet liquid or as a garnish.*

Gail Vance Civille on Taste and Smell

PHOTO USED BY PERMISSION OF GAIL VANCE CIVILLE.

Gail Vance Civille is a self-described "taste and smell geek" who started out working as a sensory professional at General Foods' technical center and is now president and owner of Sensory Spectrum, Inc., in New Providence, New Jersey.

How does somebody who is trained to think about flavor, taste, and sensation perceive these things differently than the layperson?

The big difference between a trained taster and an untrained taster is not that your nose or your palate gets better, but that your brain gets better at sorting things out. You train your brain to pay attention to the sensations that you are getting and the words that are associated with them.

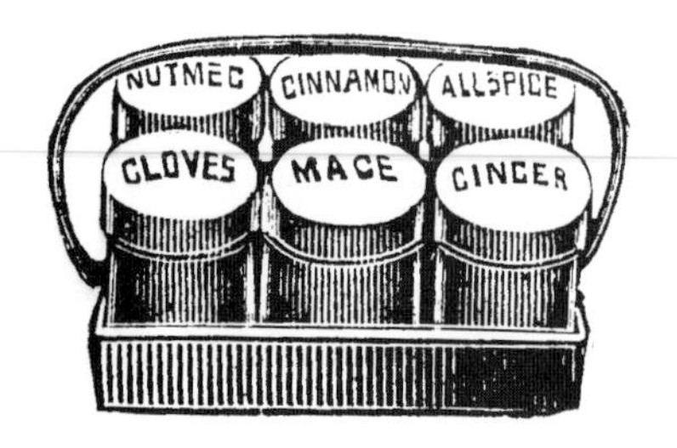

It sounds like a lot of it is actually about the ability to recall things that you've experienced before. Are there things that one can do to help get one's brain organized?

You can go to your spice and herb cupboard and sort and smell the contents. For example, allspice will smell very much like cloves. That's because the allspice berry has clove oil or eugenol in it. You'll say, "Oh, wow, this allspice smells very much like clove." So the next time you see them, you might say, "Clove, oh but wait, it could be allspice."

So, in cooking, is this how an experienced chef understands how to do substitutions and to match things together?

Right. I try to encourage people to experiment and learn these things so that they know, for example, that if you run out of oregano you should substitute thyme and not basil. Oregano and thyme are chemically similar and have a similar sensory impression. You have to be around them and play with them in order to know that.

With herbs and spices, how do you do that?

First you learn them. You take them out, you smell them, and you go, "Ah, okay, that's rosemary." Then you smell something else and you go, "Okay, that's oregano," and so on. Next you close your eyes and put your hand out, pick up a bottle, and smell it and see if you can name what it is. Another exercise to do is to see if you can sort these different things into piles of like things. You will sort the oregano with the thyme and, believe it or not, the sage with the rosemary, because they both have eucalyptol in them, which is the same chemical and, therefore, they have some of the same flavor profile.

What about lining up spices and foods, for example apples and cinnamon?

You put cinnamon with an apple because the apple has a woody component, a woody part of the flavor like the stem and the seeds. And the cinnamon has a wood component, and that woody component of the cinnamon sits over the not-so-pleasant woodiness of the apple, and gives it a sweet cinnamon character. That's what shows. Similarly, in tomatoes you add garlic or onion to cover over the skunkiness of the tomatoes, and in the same way basil and oregano sit on top of the part of the tomato that's kind of musty and viney. Together they create something that shows you the best part of the tomato and hides some of the less lovely parts of the tomato. That's why chefs put certain things together. They go, they blend, they merge and meld, and actually create something that's unique and different and better than the sum of the parts.

It takes a while to get at that level, because you have to really feel confident as both a cook and getting off the recipe. Please, get off the recipe. Let's get people off these recipes and into thinking about what tastes good. Taste it and go, "Oh, I see what's missing. There's something missing here in the whole structure of the food. Let me think about how I'm going to add that." I can cook something and think to myself, there's something missing in the middle,

I have some top notes and I have maybe beef and it's browned and it has really heavy bottom notes. I think of flavor like a triangle. Well, then I need to add oregano or something like that. I don't need lemon, which is another top note, and I don't need brown caramelized anything else because that's in the bottom. You taste it, and you think about how you are going to add that.

How does somebody tasting something answer the question, "Hey, if I wanted to do this at home, what should I do?"

I can sit in some of the best restaurants in the world, and not have a clue what's in there. I can't taste them apart, it's so tight. So it's not just a matter of experience; it's also a matter of the experience of the chef. If you have a classically trained French or Italian chef, they can create something where I will be scratching my head, going "Beats me, I can't tell what's in here," because it's so tight, it's so blended, that I can't see the pieces. I only see the whole.

Now this does not happen with a lot of Asian foods, because they are designed to be spiky and pop. That's why Chinese food doesn't taste like French and Italian food. Did you ever notice that? Asian foods have green onions, garlic, soy, and ginger, and they're supposed to pop, pop, pop. But the next day they're all blending together and this isn't quite so interesting.

This almost suggests that if one is starting out to cook, that one approach is to go out and eat Asian food and try to identify the flavors?

Oh, definitely. That's a very good place to start, and Chinese is a better place to start than most. I've had some Asian people in classes that I've taught get very insulted when I talk about this, and I'm like, no, no, no, that's the way it's *supposed* to be. That's the way Asian food is; it's spiky and interesting and popping, and that's not the way classic European food, especially southern European food, is.

In the case of classic European dishes, let's say you're out eating eggplant Parmesan, and it's just fantastic. How do you go about trying to figure out how to make that?

I would start identifying what I am capable of identifying. So you say, "Okay, I get tomato, and I get the eggplant, but the eggplant seems like it's fried in something interesting, and not exactly just peanut oil or olive oil. I wonder what that is?" Then I would ask the waiter, "This is very interesting. It's different from the way I normally see eggplant Parmesan. Is there something special about the oil or the way that the sous chef fries the eggplant that makes this so special?" If you ask something specific, you are more likely to get an answer from the kitchen than if you say, "Can you give me the recipe?" That is not likely to get you an answer.

When thinking about the description of tastes and smells, it seems like the vocabulary around how we describe the taste is almost as important.

It's the way that we communicate our experience. If you said "fresh" or "it tasted homemade," you could mean many things. These are more nebulous terms than, say, "You could taste the fried eggplant coming through all of the sauce and all of the cheese." This is very, very specific, and in fact, "fresh" in this case is freshly fried eggplant. I once had a similar situation with ratatouille in a restaurant. I asked the waiter, "Could you tell me please if this ratatouille was just made?" The waiter said, "Yes, he makes it just ahead and he doesn't put all the pieces together until just before we serve dinner." When people say "homemade," they usually mean that it tastes not sophisticated and refined, but that it tastes like it had been made by a good home cook, so it's more rustic, but very, very well put together.

Is there a certain advantage that the home chef has because he is assembling the ingredients so close to the time that the meal is being eaten?

Oh, there's no question that depending upon the nature of the food itself, there are some things that actually benefit from sitting long in the pot. Most home chefs, either intuitively or cognitively, have a good understanding of what goes with what, and how long you have to wait for it to reach its peak.

You had said a few minutes ago, "We need to get off the recipe." Can you elaborate?

When I cook, I will look at seven or so different recipes. The first time I made sauerbraten, I made it from at least five recipes. You pick things from each based on what you think looks good, and what the flavor might be like. I think the idea of experimenting in the classic sense of experimenting is fine. Geeks should be all about experimenting. What's the worst thing that's going to happen? It won't taste so great. It won't be poison, and it won't be yucky; it just may not be perfect, but that's okay. I think when you do that, it gives you a lot more freedom to make many more things because you're not tied to the ingredient list. The recipe is, as far as I'm concerned, a place to start but not the be all, end all.

Tastes: Bitter, Salty, Sour, Sweet, Umami, Others

You'll have an easier time seasoning dishes if you understand the five primary tastes the tongue can detect, as well as how it responds to "other" things (for example, the chemicals that give hot peppers their kick, carbonated drinks their effervescence, and peppermint candies their cooling sensation).

When cooking, regardless of the recipe and technique, you *always* want to adjust and correct the primary tastes in a dish. There is just too much variability in any given product for a recipe to accurately prescribe how much of a taste modifier is necessary to achieve a balanced taste for most dishes: one apple might be sweeter than another, in which case you'll need to adjust the amount of sugar in your applesauce, and today's batch of fish might be slightly fresher than last week's, changing the amount of lemon juice you'll want. Because taste preferences vary among individuals, you can sometimes solve the balance problems by letting the diners adjust the taste themselves. This is why fish is so often served with a slice of lemon, why we have salt on the table (don't take offense at someone "disagreeing" with your "perfectly seasoned" entrée), and why tea and coffee are served with sugar on the side. Still, you can't serve a dish with every possible taste modifier, and you should adjust the seasonings so that it's generally pleasing.

Bitter

Bitter is the only taste that takes some learning to like. Some primitive part of our brain seems to reject bitter tastes by default, probably because many toxic plants taste bitter. This same primitive mechanism is why bitter foods are unappealing to kids: they haven't learned to tolerate, let alone enjoy, the sensation of bitterness. Dandelion greens, rhubarb, and uncooked artichoke leaves all contain bitter oils that cause them to taste bitter; not surprisingly, I couldn't stand those things as a kid.

Adding salt can neutralize bitterness, which is why a pinch of salt in a salad that contains bitter items such as dandelion greens helps balance the flavor. Sugar can also be used to mask bitterness. Try grilling or broiling Belgian endive lightly sprinkled with sugar. Quarter the endive down the center to get four identical wedges and place them on a baking sheet or oven-safe pan. Sprinkle with a small amount of sugar. You can also drizzle a small amount of melted butter or olive oil on top. Transfer the tray to a grill or place it under a broiler for a minute or two, until the endive becomes slightly soft and the edges of the leaves begin to turn brown. Serve with blue cheese or use the endive as a vegetable accompaniment to stronger-flavored fish.

Try this simple "bitter taste test" to demonstrate how salt interacts with bitter tastes. Modern tonic water (a much weaker version of the traditional medicinal drink of quinine and carbonated water that was then spiked with gin to make it palatable) uses quinine as a bittering agent and is easy to get at the grocery store. Pour tonic water into two drinking glasses. In one, add enough salt to neutralize the taste. Compare the taste of the tonic water in the two cups.

Bitterness seems to lend itself exceedingly well to drinks: unsweetened chocolate, raw coffee, tea, hops (used in making beer), and kola nuts (kola as in cola as in soft drinks) are all bitter. And many before-meal aperitifs are bitter, from the classic Campari to the simple parsley-dipped-in-salt-water customary during Passover. Conventional wisdom states that bitter foods increase the body's production of bile and digestive enzymes, helping in digestion. The food science literature doesn't seem to support the conventional wisdom, though.

Salty

Salt (sodium chloride) makes foods taste better by selectively filtering out the taste of bitterness, resulting in the other primary tastes and flavors coming through more strongly. The addition of a small quantity of salt (not too much!) enhances other foods, bringing a "fullness" to foods that might otherwise have what is described as a "flat" flavor. This is why so many sweet dishes—cookies, chocolate cake, even hot chocolate—call for a pinch of salt. How much salt is in a pinch? Enough that it amps up the food's flavor, but not so much that the salt becomes a distinct flavor in itself. A "pinch" isn't an exact measurement—traditionally, it's literally the amount of salt you can pinch between your thumb and index finger—but if you need to start somewhere, try using ¼ teaspoon or 1.5 grams.

In larger quantities, salt acts as an ingredient as much as a flavor enhancer. Mussels liberally sprinkled with salt, bagels topped with coarse salt, salty lassi (an Indian yogurt drink), even chocolate ice cream or brownies with sea salt sprinkled on top all taste inherently different without the salt. When using salt as a topping, use a coarse, flaky variety, not rock/kosher salt or table salt. (I happen to use Maldon sea salt flakes.)

In larger quantities, salt brings a distinct taste to a dish. Try cooking mussels sprinkled with copious amounts of sea salt. Place a cast iron pan over high heat until the pan is screaming hot, and drop in the mussels. After two to three minutes, they'll have opened up and cooked; sprinkle with salt. You can optionally add in diced shallots or crushed garlic, cooking another minute or so until done. Serve with forks and a small bowl of melted butter for dipping the mussels. You should rinse the mussels before cooking, discarding any that have broken shells or that aren't closed tightly.

Differences in Taste and Supertasting

Imagine you're slaving over the stove, cooking dinner with your girlfriend or boyfriend, and you get into a heated discussion about the dish needing more salt. To you it's not salty enough, while to her (or him), it's already way too salty. What's going on? Why can't you *ever* agree on the seasoning?

As it turns out, some of us really do taste things differently. Just like variations in eye color, there are variations in our taste buds. What one person senses and perceives can differ from what another person experiences. In terms of taste, there are a number of known differences, one of the most prominent being *supertasting*.

Supertasting was accidentally discovered in the 1930s when a DuPont chemist, Arthur Fox, unwittingly spilled phenylthiocarbamide (PTC) powder. He didn't notice, but a colleague complained of a bitter sensation from the dust kicked up in the air. Curious, Fox started testing on friends and family (this was clearly in the days before internal review boards) and found that about one in four couldn't discern any bitterness.

More recent research by Dr. Linda Bartoshuk has shown that those of us of who can sense PTC can be broken down into two groups: a *supertaster* group that detects these compounds as unbearably bitter (~25% of the general population of European heritage) and a second group of *medium tasters* who find the compounds bitter, but not overwhelmingly so (50%).

If you're looking at the percentages and thinking "Mendelian trait?," you're right: you're a supertaster if you've inherited both dominant alleles from your parents. As with other Mendelian traits, the percentage breakdowns do differ by ethnicity and gender. For example, white females have a 35% chance of being supertasters, while white males have only a 10% chance. Asians, Sub-Saharan Africans, and indigenous Americans have a much higher chance of being supertasters.

If you're wondering if you're a supertaster, there are a couple of ways to tell.

Method #1: PTC or PROP test strips

The best way to tell if you're a supertaster is to see if you can taste the chemical directly. Two chemical compounds are commonly used to test for taste differences: phenylthiocarbamide (PTC) and 6-*n*-propylthiouracil (PROP). You'll need to order paper strips impregnated with either chemical (search online for "supertaster test paper" or see *http://www.cookingforgeeks.com/book/supertaster/* for up-to-date sources).

Place the test strip on your tongue and let it rest there for 10 seconds. You'll know if you're a supertaster if you experience an extremely bitter taste. Supertasters will generally yank the piece of paper out of their mouths really fast. Medium tasters (individuals with only one dominant allele) will sense a mild but tolerable bitter taste, and *nontasters* will enjoy the pleasant sensation of, well, wet paper.

Method #2: Taste bud count

If you don't have test strips, you'll have to stick your tongue out (all in the name of science, of course). Because supertasters generally have more taste buds on their tongues than medium tasters, the low-tech (and low-accuracy, unfortunately) way of checking to see if you're a supertaster is to count the fungiform papillae, which contain taste buds and are correlated to the number of taste buds you have.

You'll need blue food coloring, a cotton swab or spoon, and a sheet of binder paper (i.e., three-hole punched paper that has a 5⁄16" / 8mm-diameter hole).

Place a drop of the food coloring on the cotton swab, and then stain your tongue with it. Place the paper on top of your tongue such that you or a partner can see the tongue through one of the holes. Choose the area that is densest with spots, usually the front portion of the tongue. Count the number of pink dots visible (fungiform papillae aren't stained by the food coloring). If you count more than 30 papillae, you're probably a supertaster. Normal tasters tend to have between 15 and 30 papillae, while nontasters have fewer than 15, on average. These numbers are only broad generalizations, so it's hard to say for sure which group you fall into based on the counts.

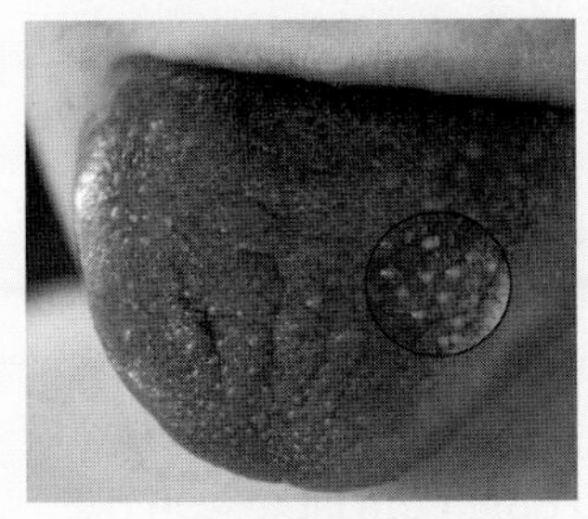

Counting the number of fungiform papillae visible in a three-hole-punch-sized area of the tongue takes a bit of dexterity and good lighting. Look for the densest area, the location of which varies among people. Count the lighter dots in the circle. This image shows approximately 12.

Being a supertaster or a nontaster isn't necessarily good or bad. Supertasters might find some foods—especially dark-green leafy vegetables such as kale, cabbage, broccoli, and Brussels sprouts—to be overly bitter, because of phenylthiourea-like compounds that their tongues can sense. Supertasters generally also find astringent, acidic, and spicy foods to be stronger, due to the higher number of taste buds and thus larger number of cells experiencing oral irritation. Researchers have found that in addition to bitter tastes (tested using quinine), supertasters also experience sweet (sucrose), sour (citric acid), and salty (sodium chloride) tastes as being more intense. Nicotine is more bitter to supertasters, and sure enough, supertasters are less likely to smoke. Caffeine also tastes more bitter, and researchers have found that supertasters are more likely to add milk/cream or sugar to coffee and tea.

Keep in mind that supertasting is just one of many factors that impact our sense of taste and our food habits. Physiological factors and disease can affect our sense of taste, as can our experiences. Stress leads to an increase in the hormone cortisol, which, among other things, dampens the stimuli strength of taste buds. Our environment can also impact our taste buds. For example, drier conditions change the amount of saliva in the mouth, resulting in a decrease of taste sensitivity.

As we touched on earlier, temperature also impacts taste sensation, just as it impacts our sense of smell: foods served warmer (by some accounts, above 86°F / 30°C) will be detected as stronger by the taste buds than colder dishes, due to the heat sensitivity of at least one of the receptors (TRPM5) responsible for taste. Foods served below body temperature won't register as warm, so if you want a dish—say, a spinach and bacon salad—to taste stronger, serve it on the warmer side (but below body temp). If you want a dish to carry milder tastes—e.g., to moderate the bitterness of beer or sweetness of ice cream—serve it colder.

Finally, if you're a cilantro hater—if it tastes like dish soap and you can't stand it—you're not alone; even Julia Child hated cilantro. While there's no known scientific mechanism or genetic marker for determining this reaction, preliminary research based on differences between identical and fraternal twins does suggest that a distaste for cilantro is genetic.

In some recipes, salt is used for its chemical properties, such as the osmosis of cellular fluids for food preservation. We'll cover more uses for salt in Chapter 7.

As described in the sidebar "Differences in Taste and Supertasting," there are known genetic differences in the way people taste some bitter compounds. Because salt masks bitterness, those of us who taste things like broccoli, Brussels sprouts, and kale as being bitter tend to add more salt to compensate.

Sour

Sour tastes are caused by acids in foods. The sensation of sourness is detected by part of the taste bud (ion channels) interacting with the hydrogen ions in the acids. Quite literally, your sour taste buds are a primitive chemical pH tester. In cooking, lemon juice and vinegar are commonly used to make dishes more sour, sometimes for effect but more often to bring balance. When cooking, taste the food and think about the balance of both saltiness and sourness, adding an ingredient such as vinegar to "brighten up" the flavors.

In Latin American and Asian cuisines, tamarind paste is often used to adjust the sourness of a dish.

From an evolutionary perspective, we appear to have evolved to taste sourness as one method of determining spoilage, because a number of acids are produced by bacteria during the breakdown of food. This isn't to say that sourness in food is always due to bacterial breakdown or that the fermentation caused by bacterial breakdown necessarily results in bad food. Lemon juice is sour due to citric acid, and yogurt (pH of 3.8–4.2) acquires a sour taste because of the lactic acid created by the bacteria breaking down the lactose in the milk (pH of 6.0–6.8).

To get a better understanding of how bacteria make the taste of a food more sour, try making your own yogurt using the recipe on page 102, tasting the liquid before and after fermentation.

Sweet

We're hardwired to like sweet foods—no surprise here. Sweet tastes signal quickly digestible calories (and thus fast energy), which would have been more important in the days when picking up the groceries also involved picking up a spear.

Our desire for sweetness changes over our lifespan. Researchers have found that our preference for sweetness decreases as we mature. A child's preference for sweet things is biologically related to the physical process of bone growth. (Quick, kids, run and tell your parents that your sweet tooth is because of *biology*!) And the infamous sweet tooth isn't unique to American kids either; this finding holds up in other cultures.

Sugar is good at simultaneously promoting other flavors while masking sour and bitter tastes. Take ginger, which has a strong, pungent, and slightly sour taste. With a bit of sugar, it becomes enjoyable on its own; sugared and dipped in chocolate, it becomes irresistible. Try making a simple ginger-flavored syrup (recipe at right).

Simple Ginger Syrup

In a pot, bring to a boil and then simmer on low heat:

2 cups (470g) water

½ cup (100g) sugar

6 oz (65g) ginger (raw), finely chopped or minced

Simmer for 30 minutes, let cool, and then strain the mix into a bottle or container, discarding the strained-out ginger pieces.

Besides adding it to club soda for a simple ginger soda, try using this syrup on top of pancakes or waffles or in mixed drinks (ginger mojitos!). You can also add a vanilla bean, split lengthwise, to the mix while boiling to impart a richer flavor. And if the idea of chocolate-covered candied ginger is still bouncing around your head, take a look at the "Sugar" section of Chapter 6 and use ginger instead of citrus rind.

Umami (a.k.a. Savory)

Umami (a Japanese word that roughly translates to "savory") generates a meaty, broth-like, lip-smacking sensation typically triggered by some amino acids and nucleotides (glutamate is the poster child; inosinate, guanylate, and aspartate are also not uncommon). Glutamate is naturally present in a number of foods, especially mushrooms. To an average American palate, umami is more subtle than the four primary tastes. It tends to amplify our other senses of taste. For example, in dishes with salt, umami "brings out" the salty taste, meaning that you can cut the amount of salt in a dish by adding umami-tasting ingredients.

If you're unable to imagine the taste of umami, make a simple broth by rehydrating a tablespoon of dried shiitake mushrooms in 1 cup (240g) of boiling water. Let the mushrooms steep for at least 15 minutes, and then remove them and save them for something else (*mmm*, stir-fry). Taste the liquid; it will have a high glutamate content dissolved out from the mushrooms. (If this is too much work for you, I suppose you could just snag a container of MSG from your local Asian grocery store and dissolve a small amount in a glass of water.)

Why we've evolved to have taste sensors for umami isn't fully clear. Sweetness and saltiness are both associated with positive attributes of food (quick energy in the case of sweets and an element essential for regulating blood pressure in the case of saltiness), while sourness and bitterness indicate potential danger. Perhaps umami is a more subtle indicator of protein content, as a way of ensuring we ingest enough amino acids to maintain muscle function. Regardless, umami is worth understanding for the hedonistic value alone. MSG (monosodium glutamate) is to umami what sugar is to sweetness: as a chemical, it's relatively odorless (still full of taste!), but it triggers the umami receptors on the tongue. MSG has gotten a bit of a bad rap in the United States, but so have salt and sugar at various times.

Yogurt

In a pan (or, preferably, a double-boiler), gently heat:

1 cup (240ml) milk (any type other than lactose-free)

Bring the milk up to 200°F / 93.3°C and hold at that temperature for 10 minutes using a candy or IR thermometer. Do not boil, because that will affect the yogurt's flavor.

After 10 minutes at temperature, transfer the milk to an open thermos, and wait until it cools to 115°F / 46°C.

Add and stir to combine:

1 tablespoon (14g) yogurt

Screw the lid onto the thermos and incubate for four hours. Transfer the liquid to a storage container and put it in the fridge immediately.

Notes

- *The small addition of yogurt acts as a* starter *because it contains the proper types of bacteria for "good" yogurt. Make sure you use yogurt that states it has "active cultures."*
- *This recipe sterilizes the milk (pasteurized milk can still have a low level of bacteria) and keeps the incubation period to four hours to reduce the chance of growth of foodborne illness-related bacteria. Longer incubation times lead to a stronger, more developed flavor. As with anything you eat, keep in mind that if it tastes bad, smells off, or looks up at you and cracks a joke, you probably shouldn't eat it. (The inverse is, of course, not true: just because something smells fine doesn't mean it's necessarily safe.)*
- *Try adding honey to the hot milk to take the sour edge off the finished product (sweet helps mask sour). For "Greek-style" thick yogurt, place the yogurt in a strainer over a bowl and let it drip-strain overnight in the fridge. For additional yogurt-making tips, see* http://extension.missouri.edu/publications/DisplayPub.aspx?P=GH1183.

You can make an impromptu "bain-marie" (double-boiler) by placing a metal bowl in a sauté pan and wedging a spoon under one side of the bowl. This allows the water to circulate between the frying pan and bowl.

There are plenty of natural sources of glutamate. Many traditional Japanese dishes call for *dashi*, a stock made from ingredients high in natural glutamate such as kombu seaweed (2.2% glutamate by weight). Making dashi is super easy: in a pot, place 3 cups (700g) cold water and a 6"/ 15 cm strip of kombu (dried kelp), and let rest for 10 minutes. Bring to a boil slowly on low heat. Remove the kombu just before the water begins to boil and add 10g of bonito flakes (flakes of dried and smoked bonito fish). Bring to a boil, remove from heat, and strain out the bonito flakes. This liquid is dashi. To make miso soup, add miso paste, diced tofu, and (optionally) sliced green onions, nori, or *wakame* (an edible seaweed).

Glutamate occurs naturally in many other foods—for example, beef (0.1%) and cabbage (0.1%). And if you're like most geeks and pizza makes your mouth water, it might be because of the glutamate in the ingredients: Parmesan cheese (1.2%), tomatoes (0.14%), and mushrooms (0.07%).

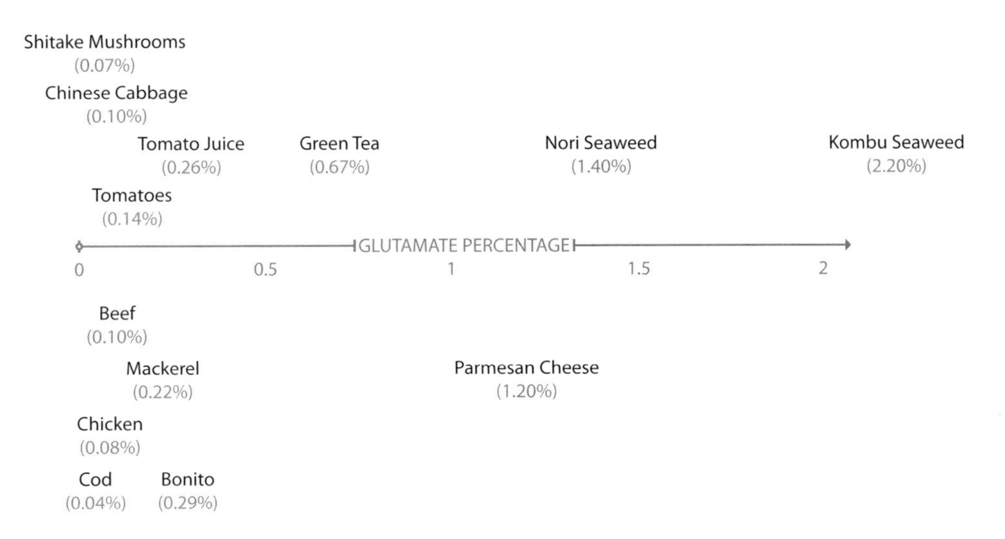

Glutamate content of common ingredients.

Others

In addition to the primary sensations of taste, our taste buds also respond to oral irritation brought about by hot peppers (typically from the chemical capsaicin), cooling sensations (typically menthol from plants like peppermint), and carbonation. The reaction to hot peppers is governed by a neurotransmitter called *substance P* (P is for *pain*; go figure). In one of nature's more subtle moves, substance P can be depleted slowly and takes time—many days, possibly weeks—to replenish, meaning that if you eat hot foods often, you literally build up a tolerance for hotter and hotter foods as your ability to detect their presence goes down. Because of this, asking someone else if a dish is spicy won't always tell you if it's safe to jump in. Carbonation in soft drinks also irritates the taste buds, but in a different way that stimulates the somatosensory system. Carbonation also interacts with an enzyme (carbonic anhydrase 4) to trigger our sour taste receptors, but for now it's unclear as to why it doesn't actually taste sour to us.

Our mouths also capture data for a few chemical families present in some foods, along with noticing texture and "mouthfeel." Some of the sensations picked up by our mouths include pungency, astringency, and cooling. Pungency is commonly described as being like some strong, stinky French cheeses: a sharp, caustic quality. Astringency results when certain compounds literally bind to taste receptors and causes a drying, puckering reaction. Astringent foods include persimmon, some teas, and lower-quality pomegranate juices (the bark and pulp are astringent). Cooling is the easiest to understand: the chemical menthol, which occurs naturally in mint oils from plants such as peppermint, triggers the same nerve pathways as cold. Menthol is commonly used in chewing gum and mint candies.

Different cultures give different weights to some of the sensations listed here. Ayurvedic practices on the Indian subcontinent include food recommendations as part of their prescriptions, defining six types of taste: sweet, sour, salty, warm (like "hot" but not the same kind of kick), bitter, and astringent. No umami, but two additional variables: warm and astringent. Thai cooking also defines hot as a primary taste. For most European cuisines, these additional variables are of lesser importance, possibly due to genetic differences in taste receptors related to supertasting between Europeans and Asians.

Taste Aversions

Your reaction to a particular taste is based in part on your prior experiences with similar flavors. Have prior exposures been pleasant, or revolting? Taste aversions—a strong dislike for a food, but not one based on an innate biological preference—typically stem from prior bad experiences with food. Sometimes only a single exposure that results in foodborne illness (and usually an unpleasant night near the bathroom) is all that it takes for your brain to create the negative association.

The food that triggers the illness is correctly identified only part of the time. Typically, the blame is pinned on the most unfamiliar thing in a meal (this is known as sauce béarnaise syndrome). Sometimes the illness isn't even food-related, but a negative association is still learned and becomes tied to the suspected culprit. This type of conditioned taste aversion is known as "the Garcia Effect." As further proof that we're at the mercy of our subconscious, consider this: even when we know we've misidentified the cause of an illness ("It couldn't be Tim's mayonnaise salad, because everyone else had it and they're fine!"), an incorrectly associated food aversion will still stick.

One of the cleverest examinations of taste aversion was done by Carl Gustavson as a grad student stuck at the ABD (all but dissertation) point of his PhD. Reasoning that taste aversion could be artificially induced, he trained free-ranging coyotes to avoid sheep by leaving (nonlethally) poisoned chunks of lamb around for the coyotes to eat. They soon learned that the meat made them ill, and thus "learned" to avoid the sheep. I don't recommend this method for kicking a junk food habit or keeping your coworkers from stealing unmarked food from the company fridge, as tempting as it might be.

Combinations of Tastes and Smells

Most dishes involve a combination of ingredients that contain at least two different primary tastes, because the combination brings balance and adds depth and complexity. Whether the dish is a French classic or a simple item of produce, the taste will be simple ("one note") unless it's paired with at least one other.

To alter the flavor of fresh fruit, you can sprinkle it with sugar (try this on strawberries) or salt (on grapefruit), wet it with lime juice (papaya, watermelon, peaches with honey), or combine it with an ingredient from another taste family (sweet watermelon and salty feta cheese). If you can find fresh papaya, try slicing it and sprinkling a bit of cayenne pepper and salt on top of the pieces for a salty/sweet/hot combination. Try replacing the papaya with other tropical fruits and the cayenne pepper with other hot items. Guava and chili pepper? Mango salad with jalapeños and cilantro? Strawberries and black pepper?

Black pepper has no capsaicin (the chemical that gives cayenne pepper and jalapeños their heat) but is still pungent due to another chemical, piperine.

For another twist, try mixing foods high in fats with hot ingredients. They should pair well with ingredients that contain capsaicin, because capsaicin is fat soluble. Experiment with avocado and sriracha sauce, commonly known as *rooster sauce* for the drawing on the bottle of one popular brand.

Rooster sauce (sriracha sauce—Thai hot sauce), it has been said, can improve the taste of *any* dorm food, but beware, it's spicy. As one friend quipped to me, it'll hit you like a freight train and then leave like a freight train.

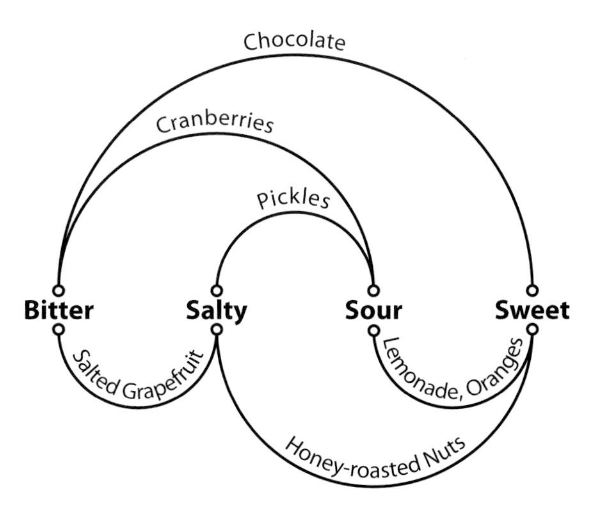

For you visual thinkers, here's a diagram of the combinations of the four basic flavors, with a few foods labeled for each combination. Ask yourself: what other foods have these combinations? When cooking, think about which tastes your dish emphasizes and in which direction you want it to go.

Many foods are combinations of three or more primary tastes. Ketchup, for example, is surprisingly complex, with tastes of umami (tomatoes), sourness (vinegar), sweetness (sugar), and saltiness (salt).

Taste combinations are equally important in drinks. The hallmark of a well-mixed cocktail is the balance between bitter (bitters) and sweet (sugar). Likewise, unless you've learned to enjoy bitterness, coffee and tea (slightly bitter) are commonly combined with sweeteners (milk, sugar, honey) or acidifiers (lemon juice, orange juice) to balance out the tastes.

In some cases, the combination of different primary tastes is achieved by serving two separate components together, pairing one dish with a second on the basis that the two will complement each other. In Indian food, for example, the salty sweetness of a yogurt lassi balances out the spicy hotness of curries. Consider the following combinations of primary flavors. With the exception of bitter/salty, every pair of primary tastes is a common combination.

Combination	Single-ingredient example	Combination example
Salty + sour	Pickles Preserved lemon peel	Salad dressings
Salty + sweet	Seaweed (slightly sweet via mannitol)	Watermelon and feta cheese Banana with sharp cheddar cheese Cantaloupe and prosciutto Chocolate-covered pretzels
Sour + sweet	Oranges	Lemon juice and sugar (e.g., lemonade) Grilled corn with lime juice
Bitter + sour	Cranberries Grapefruit (sour via citric acid; bitter via naringin)	Negroni (cocktail with gin, vermouth, Campari)
Bitter + sweet	Bitter parsley Granny Smith apples	Bittersweet chocolate Coffee/tea with sugar/honey
Bitter + salty	(N/A)	Sautéed kale with salt Mustard greens with bacon

Watermelon and Feta Cheese Salad

If it's summertime and you're able to get good watermelon, try this simple salad to experience the contrast in flavors between the salt in feta cheese and the sweetness of watermelon.

In a bowl, toss to coat:

2 cups (300g) watermelon, cubed or scooped

½ cup (120g) feta cheese, cut into small pieces

¼ cup (40g) red onion, sliced super thin, soaked and drained

1 tablespoon (14g) olive oil (extra virgin because it imparts flavor)

½ teaspoon (3g) balsamic vinegar

Notes

- *Try using a teaspoon or two of lime juice, instead of vinegar, as the source of acidity. Alternatively, play with the tastes by adding black olives (salty), mint leaves (cooling), or red pepper flakes (hot), thinking about how each variation pushes the tastes in different directions.*
- *Always soak onions that will be served raw. When cut, an enzyme (allinase) reacts with sulfoxides from the onion's cells to produce sulfenic acid, which stabilizes into a sulfuric gas (syn-propanethial-S-oxide) that can react with water to produce sulfuric acid. This is why we cry when cutting onions: the sulfuric gas interacts with the water in our eyes (the lacrimal fluid) to generate sulfuric acid, which triggers our eyes to tear up to flush the sulfuric acid. Because sulfides are water soluble, soaking the cut onions removes most of the undesirable odors. You can soak them in water, or try vinegar to impart a bit of additional flavor. Also, cutting onions in a wet environment provides liquid for the sulfur compounds to dissolve into. Try pulling off the onion skin under water and then cutting with a wet blade on a rinsed-but-not-dried cutting board. Another method to reduce tearing is to chill the onion, because this makes the cell structures firmer and reduces the amount of intracellular fluid available for the allinases to react with.*
- *If you're lazy, skip cubing the watermelon and feta and instead serve a slice of watermelon alongside a slice of feta, and alternate back and forth. You can also make appetizers by skewering a cube of watermelon and a cube of feta with a toothpick.*

There is some evidence that suggests our taste receptors can interact with the capsaicin in hot peppers (for you bio geeks, via TRPV1) and possibly menthol in mint (via TRPM8), but these interactions are not yet well understood in the science domain, let alone the culinary world.

Dicing a watermelon is easier and faster than using a melon baller. Using a knife, make a series of parallel slices in one direction, and then repeat for the other two axes.

Virginia Utermohlen on Taste Sensitivity

Virginia Utermohlen is an associate professor of nutritional sciences at Cornell University, where she studies individual differences in taste and smell sensitivity and how those differences relate to our personality and ability to perform.

Do different people taste things differently?

Yes, there are genetic variations all over the place. We have differences in the taste buds and the trigeminal nerve, plus people differ in their saliva, which influences taste. What I taste and what you taste in the same food is going to be different.

One difference is sensitivity to 6-n-propylthiouracil (PROP). People are genetically capable or incapable of tasting it. Certain populations such as the Sub-Saharan Africans tend to be very sensitive to it, and then there are populations in Europe where insensitivity is very common. The British are famous for simple flavor systems and they tend to be fairly insensitive tasters, whereas if you look at the flavors in Asian, Sub-Saharan African, and indigenous American cuisines, those people are highly sensitive tasters and their foods have complex flavor profiles.

The trigeminal nerve senses hot and cold, pain, texture, and to some extent sweet, independently of the taste bud cells. Compounds like menthol bind to the same receptor that changes shape in response to cold temperatures, so the brain interprets what menthol is doing as cooling, even though it's not colder than anything else around. Similarly with capsaicin: it binds with the receptor that changes conformation with warm or hot temperatures so the brain says, "This is hot!" It really isn't hot, but the brain interprets it as that. There are differences [in trigeminal sensitivity] from one individual to the next. Some people will put an Altoid on their tongue for the first time and think, "This isn't bad!" Another person puts one on their tongue and thinks it's got to get out of there as soon as it can. Once those trigeminal nerves get overactivated, it's painful. Another thing that the trigeminal nerve senses is pungency. French cheeses are very pungent.

Is there something about the French that causes them to like pungency in cheese?

In my rather small sample, the people of French descent tend to be more sensitive to the cooling sensation from mint. So on the average my guess is that they would be more trigeminally sensitive.

If the French tend to be more trigeminally sensitive, it seems like they would be more sensitive to the pungency in cheeses and thus not like them very much.

This is something important and interesting: if you are sensitive to something, it can be either adverse or pleasant. Now pungency, in my opinion, is not aversive. Some people think it is. I personally like chocolate that's quite bitter. Just because something has a quality doesn't mean it's aversive or pleasant. That varies from person to person.

Something you said makes me wonder about beer and wine, and low trigeminal sensitivity versus high trigeminal sensitivity.

The carbonation and pungency of beer give it a trigeminal kick. If you are sensitive to the pungent, the trigeminal side of things, but not sensitive to the bitterness of beer, you're going to like beer a lot better than I do. I can't stand it!

So could one take a PROP test strip and a menthol candy and between those two things figure out which a person likes more, beer or wine?

That's possible. I've never done that experiment before.

Dr. Utermohlen had previously explained to me that individuals who define "reasonable" as "logical" are generally less trigeminally sensitive and those who define it as "justifiable, fair" tend to be more trigeminally sensitive. Note that trigeminal sensitivity is a separate phenomenon from PROP sensitivity.

We talked about how people define the word "reasonable" in a previous conversation. Why would somebody who is more trigeminally sensitive define reasonable as fair?

Well, here is my hypothesis—not that I necessarily have any proof of this—but taste and smell go to the orbitofrontal cortex, which is the part of the brain that is critical in evaluating whatever you experience. That's its job: to *evaluate* whether something is good or not.

When you *reason* something through, it's another way of getting at whether something is good or not. Logical reasoning makes use primarily of another part of the brain called the dorso-lateral prefrontal cortex. That part of the brain gets no input whatsoever from taste and smell.

Which way you decide to think something through will depend on which way you decide how something has value. It depends on whether you go to the that-smells-fishy-to-me sort of evaluation versus a logical evaluation.

Does this mean that when people talk about making decisions from their gut versus logical decisions, they're making decisions with their orbito-frontal cortex versus their dorso-lateral prefrontal cortex?

Yeah. I think so. I really do.

What about geeks?

In my experience, geeks are very mixed in their sensitivities. In our data, the more mathematically oriented computer scientists tend to be on average nontasters. The computer scientists who are more interested in the purpose of a program tend to be more sensitive tasters. One group of geeks will be creative in ways that are highly logical and scientific. That other crowd, however, will be interested in emotion and expression and look at programming in a holistic way. That crowd will "get it" when you ask them if sunsets or the crackle and flame of a wood fire spark their imagination.

Wait, what's this?

Some of the questions that we ask have to do with a phenomenon called *absorption*, the capacity to become completely immersed in a sensory experience. On the average, from the data we have, people who are moved by a sunset, or for whom the crackle and flames of a wood fire spark the imagination or may produce visual images are highly trigeminally sensitive.

I believe that people who have a high capacity for absorption should imagine what a dish would be like, and then work toward it by maybe adding a pinch of this or that, tasting as they go. They should really spend time experimenting at what the differences in tastes are like, and not be religiously bound to a recipe.

And the other type?

The other type might stick to a recipe, because they are probably going to have better success if they follow A, follow B, follow C, follow D, and the thing will come out. It requires less guesswork, and so is less dependent on a person's sensitivity.

Trigeminal Sensitivity Experiment

Researchers, of course, go about their work in a controlled, reproducible way. Scientists in Germany looked at one way of measuring differences in trigeminal sensitivity by using strips of filter paper coated with various levels of capsaicin and asking subjects if they could perceive any sensation (such as burning, prickling, stinging) when tasting.

For the "home scientist" (or the just plain curious), there's an easier experiment you can do to get a rough sense of how sensitive you are to trigeminal stimulation. Menthol, the compound in mint that gives it its cooling sensation, is the primary flavor in candies such as Altoids and Peppermint Lifesavers. First, get a fresh peppermint candy. No, the one you recently discovered between the couch cushions from who-knows-when won't work: menthol is a volatile compound and evaporates away from the mint over time.

Pop the fresh mint in your mouth, clamp down, and breathe through your nose for half a minute or so, giving your saliva a chance to soften up and break down the candy, then chomp down on it without opening your mouth. If the cooling sensation you have is a really strong, *whooo that's strong*, then you're likely to be very trigeminally sensitive. If you hardly notice anything, then you are likely to be mildly sensitive. Most people, however, find that the cooling effect lies between these two extremes. Then, just for fun, breathe through your mouth. You should notice the cooling effect become even stronger.

"Flavor Tripping" with Miracle Berries

Try tasting chocolate, blackberries, apples, strawberries, lemons, and blue cheeses while "under the influence" of miraculin.

Our taste buds are chemical detectors full of receptor cells waiting for a chemical to come along that "fits" to trigger them. You can think of it a bit like a lock waiting for the right key to fit before it opens. But what if there were a way to pick that lock?

Miraculin and curculin are two proteins that do exactly that. They bind to sweet receptors and trigger them when acidic compounds wander along, thus causing foods that would normally taste sour (due to the acids) to taste sweet.

The miracle fruit plant produces a small red berry, aptly named the "miracle berry," which contains a large concentration of miraculin. Chewing the berry flesh for a few minutes is enough to "dose" yourself with enough miraculin that chomping down on a lemon will give the taste of lemonade.

You can order the berries online, but they are perishable. Dried tablets derived from the berry are also available. (For sources, see *http://www.cookingforgeeks.com/book/miraculin/*.) Once you have the berries or tablets in hand, invite a bunch of your friends over, munch on them, and serve up some sour foods. Grapefruit works amazingly well; try slices of lime and lemon as well.

The "flavor tripping" isn't limited to sour foods. I've had one friend swear that the roast beef sandwich he was eating was made with a honey-glazed variety and other friends try Worcestershire sauce and compare it to sashimi. Try foods such as salsas, tomatoes, apple cider vinegar, radishes, parsley, stout beers, Tabasco, and cheeses. Keep in mind that miraculin makes sour foods *taste* sweet but doesn't actually alter their pH, so don't pig out on lemons, lest you give yourself a bad case of heartburn.

Other compounds, such as lactisole, do the opposite of miraculin and suppress the sensation of sweetness, but without affecting our perception of saltiness, sourness, or bitterness. The food industry uses these types of compounds to alter the taste of things like jams to reduce the taste of sugar and bring out the fruit. Lactisole is used at around a 0.1% to 1% concentration by weight; search for Domino Sugar's "Super Envision" (it's listed on food labels as part of the general category "artificial flavors"). We'll cover more of these types of food additives in Chapter 6.

Adapt and Experiment Method

You'll have an easier time cooking as you learn about more flavors and the ingredients that provide them. Take time to notice the odors in the foods you are eating, taking note of smells that you don't recognize. Next time you're eating out, order a dish you're not familiar with and try to guess its ingredients. If you're eating with a friend who doesn't mind sharing, play a guessing game with your dinner companion to see if you can identify the tastes and flavors in both dishes.

If you're stumped, don't be shy to ask the staff. I remember being served a roasted red pepper soup and being completely stumped as to what provided the body (thickness) of the soup. Five minutes later, I found myself sitting across from the chef, who had brought me the kitchen's working copy of the recipe and told me the real secret of the recipe (Armenian sweet red pepper paste). I learned not just about a new type of flavor that day, but also about a new technique (toasted French bread puréed into the soup—an old, old trick to thicken soups) and the location of a great Armenian grocery store.

Another way to learn new flavors is to play "culinary mystery ingredient." Next time you're at the grocery store, buy one thing you've never cooked with before. For "intermediate players," pick up something you're familiar with but have no idea how to cook. And if you've progressed to the "advanced" level, choose something that you don't recognize at all. You'd be surprised at how many foods can be unfamiliar in their ingredient form, but once cooked into a meal are familiar, maybe even downright commonplace. Yucca roots? Try making yucca fries. Lemongrass, kaffir lime leaves? Try making tom yum soup. With the thousands of items available in your average American grocery store, you should be able to find something new to inspire you.

If you're just now learning your way around the kitchen and aren't yet familiar with that many recipes, think about the ingredients that go into dishes you like. If you like peanut butter and jelly sandwiches (and what self-respecting geek wouldn't?), it's not too far a leap to imagine a grilled chicken skewer coated with a sweet jelly and sprinkled with toasted peanuts. Or take another geek favorite: pizza. Maybe you like it topped with artichokes, feta, dried tomatoes, and anchovies. You could experiment by taking those toppings and adapting them for use in a pasta dish or as toppings on bread as an appetizer. Serving this to guests? Coat the bread with olive oil, toast it, and you've got bruschetta.

If items A, B, and C go together in one dish; and another dish uses B, C, and D; then A might work in the second dish as well. Transitive relationships aren't guaranteed to work, but they're a great place to start. Say you like guacamole and know that it commonly contains avocado, garlic, onion, lime juice, and cilantro. When tossing together a salad that has similar ingredients—say, tomato, avocado slices, and onion—it's reasonable to guess that some coarsely chopped cilantro will work well in it, and maybe even some crushed garlic in a vinegar/oil dressing.

Ingredients that commonly show up with chicken (left) and chocolate (right). This screenshot is from a visualization I did that displays related ingredients by building a co-occurrence map of ingredients in recipes—in essence, generating a network graph of the ingredients. See http://www.cookingforgeeks.com/book/foodgraph/ *for an interactive version.*

Like tastes in music, tastes in food aren't completely transitive. At the extreme, what one culture values often proves to be jarring to another culture. Traditional Chinese music uses a pentatonic scale (five notes per octave); European music uses a heptatonic (seven-note) scale. This is why some Chinese music sounds just plain odd to western listeners. The same is true in flavors: combinations used in one culture are invariably different from those used in another. When the difference is too great, the tastes lack appeal. Take cheese: plenty of European and American foods use it (who would the French be without cheese? how could lasagna be *lasagna* without it?), but the Chinese barely recognize it as a food.

Gourmet Magazine ran a good article (August, 2005) about three acclaimed Sichuanese chefs from China eating at one of the absolute top restaurants in the United States. They didn't "get" the food at all. The flavors just didn't strike chords in the ranges for which the chefs had formed any appreciation. This isn't to say that the flavors that excite you will be unstimulating for the friends that you cook for, unless you're cooking for a bunch of world-famous Chinese chefs. But there will be some differences between your tastes and experiences and those of your guests, so don't be surprised if a combination of flavors tastes great to you and merely okay to your eating companions.

Another method for finding ingredients to experiment with is to run a simple online search with a list of ingredients you already have in the dish. If you're making an improvised stew (i.e., following no recipe) and have tomatoes, onions, and lamb but aren't sure what foods and spices might round out the flavors, run the existing ingredients in an online search and see what the Internet says. Just scanning the titles of pages found can be enough (in this case, coriander, potatoes, and chili powder).

When experimenting, you might find it easier to test a new flavor in a small portion set aside in a separate bowl. Cooking has no "undo," so if you're not sure that the chili power will work, put a few spoons of the stew in a bowl along with a pinch of the spices and taste that. That way, if it turns out yucky, you still have an unadulterated pot of stew to try something else with.

Similarity is also a good gauge of compatibility. If a recipe calls for A, but B is extremely similar, try using B instead and see if it works. Kale and chard are both hardy green leaves that can be substituted for each other in many dishes. Likewise, Provolone and mozzarella cheese both have mild flavor and share similar melting properties, so using one in place of the other in foods like omelets makes sense. I'm not suggesting that like foods are always interchangeable. They each have their distinct flavors, and if you attempt to recreate a traditional dish with substitutions, you won't faithfully reproduce the original. But if your goal is to make an enjoyable dish, experimenting with similar ingredients is a great way to see where things line up and where they diverge.

One easy place to start experimenting, especially if you're at the pasta sauce and spaghetti phase (ah, those were the days!), is to take whatever it is you're doing now and start making minor changes. If one of your standard dinners is store-bought pasta sauce over spaghetti (and there's nothing wrong with this; it's easy, yummy, and satisfying), toss in some additional vegetables next time. Take a nonstick frying pan, put in a tablespoon or two of olive oil, chop up an onion, and sauté the onion for a while, until it tastes good. Think about the taste. Does it need salt? Onion and olive oil by itself, while good, is a little flat. So, toss in a pinch of salt. Now, take the jar of pasta sauce, heat it up in that same frying pan, and toss in your spaghetti, just as before. "Spruced-up pasta sauce" is easy and tastes better than the original (unless you burned the onions, in which case, try again!).

Spruced-up Pasta Sauce

Here's an example of how to extend pasta sauce, courtesy of one of my readers: Try mixing in a can of quality tuna and chopped olives. Microwave a few ounces of pancetta until crispy, and then crumble it into the sauce.

Next, begin to expand your flavor repertoire. The next time you make the dish, try something other than an onion. At the grocery store, take a look at other veggies and ponder whether they might go in that pasta sauce equally well. Zucchini? Squash? Mushrooms? They all sound good to me. One sure-fire way of guessing is by peeking at the ingredient list on your jar of pasta sauce. If it includes mushrooms and oregano, you know adding more mushrooms won't cause an unpleasant flavor. The key is to try varying these things one at a time, starting slowly and building up various bits of information one variable at a time. It's just like writing and then debugging code: instead of making a whole bunch of changes, try one thing at a time. As you get more comfortable, you can work for longer intervals before testing your logic.

If you fold back the wrapper on a stick of butter, you can butter the pan with it directly. It's not as elegant as slicing off a pat and dropping it into the pan, but it does save fetching and washing another utensil. Make sure to fold the paper back over the end when storing, so the butter is wrapped when you put it back in the fridge. As an experiment, figure out if "crowding" the mushrooms matters. Conventional wisdom says that overcrowding the pan will cause the mushrooms not to taste as good, but is this really true?

If you're already a master of the "spruced-up pasta sauce method," try experimenting with what the French call *mother sauces*. The French chef Marie-Antonin Carême began this classification scheme in the 17th century with a handful of sauces from which virtually all traditional French sauces are derived. Both Béchamel and velouté sauces are fast to make and can carry a lot of flavors.

Beyond picking harmonious flavors, you should also consider how they blend together (the "loudness" of the notes). Good cooking is very much about balance. One of the most important corrections you can make is to adjust the balance of tastes and smells to where you think the dish tastes ideal. Lean in and smell the dish. Take a taste. Then ask yourself: what would make this dish better? Does it need more salt? If it's dull or flat, would adding a sour note (lemon or lime juice) add some brightness? Is it too heavy on one of the tastes? Some pasta sauces, for example, can benefit from a splash of balsamic vinegar (sour) to help balance the sweetness brought by tomatoes. Chocolate chip cookies can be improved with a bit of salt to help balance the sweetness from sugar and slight bitterness from cocoa.

If you're following a recipe closely, approach your tweaking cautiously, being sure not to add too much of a taste modifier such as salt or lemon juice, because it is nearly impossible to remove it. Add only a small amount at the beginning, and adjust the taste at the end of cooking by adding a little more at a time until it tastes balanced.

If you do add too much seasoning, you can make minor adjustments of other taste sensations to partially mask overzealous seasoning. If, however, you've added too much of something, your best bet is to dilute the dish to reduce the seasoning's concentration. Contrary to folk wisdom, adding potato does nothing to reduce saltiness (how could it? evaporation?), but might "work" for the same reasons that adding more broth does (dilution). You're better off just removing some of the too-salty liquid and adding more unsalted liquid. If your dish is too sweet, you can adjust it by adding spicy ingredients.

Capsaicin is sugar and fat soluble, but not water soluble. This is why drinking milk or consuming something fatty tames the mouth, while drinking water doesn't help. If a dish is too hot, add either something with sugars or fats to reduce the heat.

Think about what tastes are missing, and add one of the following seasonings to adjust the taste.

To increase:	Bitter	Sour	Umami	Sweet	Salty
Add…	Cocktail bitters Quinine powder Tonic water Cocoa	Lemon juice Vinegar Verjuice (verjus) or dry wine	Soy sauce Fish sauce Mushrooms Kombu (seaweed) MSG	Sugar Honey Sweet wines (sherry, Madeira) Mirin	Salt Umami-based ingredients (because they amplify salt perception)

As a general tenet when seasoning a dish, start slowly. You can always add more. You can partially mask some tastes by increasing other tastes. If you do end up with one taste being too dominant, try one of the following adjustments.

To counteract:	Bitter	Sour	Umami	Sweet	Salty
Add…	Increase saltiness or sweetness	Increase sweetness to mask	None known; try dilution	Increase sourness or heat (e.g., cayenne pepper) to mask	Increase sweetness (low concentrations) to mask

Béchamel Sauce (White Sauce)

In a pan, melt 1 tablespoon / 14g butter over medium heat. Stir in 1 tablespoon / 8g flour and continue stirring, making sure to combine the flour and butter thoroughly, cooking for several minutes until the mixture begins to turn a blond to light brown color (this butter-flour combination is called a *roux*). Add 1 cup / 256g milk, increase the heat to medium-high, and stir continuously until the mixture has thickened.

Traditional additions include salt, pepper, and nutmeg. Try adding dried thyme, or preheating the milk with bay leaves. If you're anti-butter, you can use a half butter/half oil mixture.

This sauce can be "subclassed" into other sauces. After making the roux and adding the milk, try the following instances:

Mornay sauce (a.k.a. cheese sauce)

Béchamel sauce with equal parts of Gruyère and Parmesan added. If you're not a stickler for tradition, almost any cheese that melts well will work.

Bayou sauce

Béchamel sauce that's had the roux cooked until it reaches a dark brown color. Commonly used in Louisiana-style Cajun cooking, in which onions, garlic, and Creole seasonings are also added.

Mustard sauce

Béchamel sauce with mustard seed or a spoonful of mustard (try one with whole seeds in it). Mustard sauce can be further subclassed with the addition of cheddar cheese and Worcestershire sauce. Or, try sautéing some diced onions in the butter while making the roux and adding mustard at the end for a mustard-onion sauce.

Velouté Sauce

Start like you're making Béchamel: create a blond roux by melting 1 tablespoon / 14g of butter in a pan over low heat. Stir in 1 tablespoon / 8g of flour and wait for the flour to cook, but not so much that it browns (hence the term *blond roux*). Add 1 cup / 256g of chicken stock or other light stock (one that uses raw bones instead of roasted bones) and cook until thickened.

You can make derivative sauces by adding various ingredients. Here are a few suggestions. Note the absence of specific measurements; use this as an exercise to take a guess and adjust the flavors to suit what you like:

Albufera sauce

Lemon juice, egg yolk, cream (try on chicken or asparagus)

Bercy sauce

Shallots, white wine, lemon juice, parsley (try on fish)

Poulette sauce

Mushrooms, parsley, lemon juice (try on chicken)

Aurora sauce

Tomato purée; roughly 1 part tomato to 4 parts velouté, plus butter to taste (try on ravioli)

Hungarian sauce

Onion (diced and sautéed), paprika, white wine (try on meats)

Venetian sauce

Tarragon, shallots, chervil (try on mild fishes)

Grilled Fish with Bayou Sauce or Mustard Sauce

Make either a bayou sauce or mustard sauce per previous instructions, playing with the amount of onions and seasoning.

Select a mild fish, such as cod or halibut. Season with a light amount of salt and pepper and transfer to a heated grill. Cook for about 5 minutes, flip, and cook until flaky, about another 5 minutes. Place on serving plate and spoon sauce on top. (Try serving this with simple steamed vegetables!)

Simple Ravioli Sauce

Here's a quick experiment: try making two batches of Aurora sauce, one with 1 cup / 240 ml of light cream and the second with 1 cup / 240 ml of chicken stock. Add ¼ cup / 60 ml of either tomato sauce or puréed tomatoes to each and stir to combine. (If you're cutting corners and cooking for yourself, you can use ketchup instead of tomato purée.)

What do the two sauces remind you of? Try using the cream-based sauce (sometimes called a pink sauce) on top of ravioli. And the stock-based sauce? Toss in some carrots, celery, beans, and pasta, and you've got the beginning of a minestrone soup.

Mac 'n Cheese

Start with a double batch of Béchamel sauce. Add and slowly stir until melted:

1 cup (100g) grated mozzarella

1 cup (100g) grated cheddar cheese

In a separate pot, bring salted water to a rolling boil and cook 2 cups / 250g pasta. Use a small pasta, such as elbow, fusilli, or penne—something that the sauce can cling to. Test for doneness by tasting a piece of the pasta. When ready, strain and transfer to pan with cheese sauce. Stir to combine.

You can stop here for a basic Mac 'n Cheese, or spruce it up by mixing in:

¼ cup (60g) sautéed onions

2 slices (15g) bacon, cooked and chopped into pieces

A pinch of cayenne pepper

Transfer to a baking pan or individual bowls, sprinkle with bread crumbs and cheese, and broil under medium heat for 2 to 3 minutes, until bread and cheese begin to brown.

Notes

- *You can add more milk to the cheese sauce to make it thinner.*
- *To make your own bread crumbs, you can drop a slice of day-old or toasted bread into a food processor or blender and pulse it. Or, use a knife and chop it up into small pieces.*

Regional/Traditional Method

Say your Aunt Suzie sends you a jar of her famous (or is it infamous?) homemade quince jelly. What to do with it? Someone suggests that you try it with Manchego cheese and crackers and, sure enough, the combination is delicious. But why? One potential explanation can be found in the history of the ingredients: they come from the same geographic region and its corresponding cuisine.

This method for thinking about flavor combinations is expressed in the idiom "if it grows together, it goes together" and encompasses everything from a loose interpretation of what the French call *le goût de terroir* ("taste of the earth") to what an American gourmand would term "regional cooking" for broad styles of cooking. In addition to the limitation of ingredients based on what can be grown in any given area, regional cooking also involves the culture and tradition of a region. Back to Aunt Suzie's jelly: Manchego cheese and quince jelly both have long histories in Spain, so the pairing is likely rooted in history.

Given an ingredient, you can look at how that ingredient has been used historically in a particular culture to find inspiration. (Think of it as historical crowdsourcing.) If nothing else, limiting yourself to ingredients that would traditionally be used together can help bring a certain uniformity to your dish, and serve as a fun challenge, too. And you can extend this idea to wines to accompany your dishes, from the traditional (say, a French rosé with Niçoise salad) to modern (Aussie Shiraz with barbeque).

Another way of looking at historical combinations is to look at old cookbooks. A number of older cookbooks are now in the public domain and accessible via the Internet Archive (*http://www.archive.org*), Project Gutenberg (*http://www.gutenberg.org*), and Google Books (*http://books.google.com*). Try searching Google Books for "Boston Cooking-School Cook Book"; for waffles, see page 80 (page 112 in the downloadable PDF). If nothing else, seeing how much—and, really, how little!—has changed can be great fun. And then there are classic gems, foods that have simply fallen to the sidelines of history for no discernable reason.

Sally Lunn

Mix one pint of flour, two teaspoonfuls baking powder, one-half teaspoonful salt, yolks of two eggs well beaten, one-half cup milk, one-half cup butter melted, whites of two eggs beaten stiff. Bake in muffin pans or drop loaf fifteen to twenty minutes. If for tea, add two tablespoonfuls sugar to flour.

—*from* The Community Cook Book *on Project Gutenberg* http://gutenberg.org

The older the recipe, the harder it can be. One reason is that language has changed. A lot. Take this example (also taken from Project Gutenberg) for apple pie from *The Forme of Cury*, published around 1390 A.D.:

> *Tak gode Applys and gode Spycis and Figys and reysons and Perys and wan they are wel brayed coloure wyth Safron wel and do yt in a cofyn and do yt forth to bake well.*

Almost as bad as a condensed tweet, this translates to: "Take good apples and good spices and figs and raisins and pears and when they are well crushed, color well with saffron and put in a coffin (pie pastry) and take it to bake." (The "coffin"—little basket—is an ancestor to modern-day pie pastry and would not have been edible at that point in time.) Still, as a starting point for an experiment, the idea of making a mash of apples and pears, some dried fruit, spices, and saffron suggests not just a recipe for pie filling, but also a festive apple sauce for Thanksgiving.

Old recipes aren't always so concise. Take Maistre Chiquart's recipe for *parma torte* in *Du Fait de Cuisine*, 1420 A.D. He starts with "take 3 or 4 pigs, and if the affair should be larger than I can conceive, add another, and from the pigs take off the heads and thighs, and..." He goes on for four pages, adding 300 pigeons and 200 chicks ("if the affair is at a time when you can't find chicks, then 100 capons"); calling for both familiar spices like sage, parsley, and marjoram, and unfamiliar ones such as hyssop and "grains of paradise"; and ending with instructions to place a pastry version of the house coat of arms on top of the pie crust and decorate the top with a "check-board pattern of gold leaf" (diamond-studded iPhone cases have *nothing* on this guy).

Needless to say, you'll likely need to do some scaling and adaptation of older recipes—again, part of the fun and experimentation! For parma tortes, I worked out my own adaptation. I later found that Eleanor and Terence Scully's *Early French Cookery: Sources, History, Original Recipes and Modern Adaptations* (University of Michigan Press) includes a nice adaptation. You can peek at it on Google Books; search for "parma torte."

Modernized version of parma torte, without the gold leaf, from Du Fait de Cuisine, *by Maistre Chiquart—France, 1420 A.D.*

Besides studying older recipes, you can look at traditional recipes from particular regions to see how ingredients are normally combined. Different cultures have different "flavor families," ingredients that are thought of as having an affinity for one another. Rosemary, garlic, and lemon are pleasing together—hence, traditional dishes like chicken marinated in those ingredients. It can take time to build up a familiarity with flavor families, but taking note of what ingredients show up together on menus, bottles of salad dressings, or in seasoning packets is a good shortcut.

	Common ingredients	Served with...
Chinese	Bean sprouts, chilies , garlic, ginger, hoisin sauce, mushrooms, sesame oil , soy, sugar	Rice
French	Butter, butter, and more butter, garlic, parsley, tarragon, wine	Bread
Greek	Garlic, lemon, oregano, parsley, pine nuts, yogurt	Orzo (pasta)
Indian	Cardamom seed, cayenne, coriander, cumin, ghee, ginger, mustard seed, turmeric, yogurt	Rice or potatoes
Italian	Anchovies, balsamic vinegar, basil, citrus zest, fennel, garlic, lemon juice, mint, oregano, red pepper flakes, rosemary	Risotto or pasta
Japanese	Ginger, mirin, mushrooms, scallions, soy	Rice
Latin American	Chilies, cilantro, citrus, cumin, ginger, lime, rum	Rice
Southeast Asian	Cayenne, coconut, fish sauce, kaffir lime leaves, lemon grass, lime, Thai pepper	Rice or noodles

Common ingredients used in chicken dishes by a few common cuisines. (Note that not all of these ingredients would be used simultaneously.)

The ingredients used to bring balance to a dish will vary by region. For example, the Greeks use lemon juice in *horta* to moderate the bitterness of the dark leafy greens like dandelion greens, mustard greens, and broccoli rabe, while the Italian equivalent uses balsamic vinegar.

Set some of the marinade aside before adding the meat for use as dipping sauce.

With even a short list of culturally specific ingredients as inspiration, you can create simple marinades and dipping sauces without too much work. Pick a few ingredients, mix them in a bowl, and toss in tofu or meat such as chicken tenderloins or steak. Allow the tofu or meat to marinate in the fridge for anywhere from 30 minutes to a few hours, and then grill away.

When creating your own marinade, if you're not sure about the quantities, give it a guess. This is a great way to build up that experiential memory of what works and what doesn't.

Simple Greek-Style Marinade

In a bowl, mix:

¼ cup (60g) yogurt

1 tablespoon (15g) lemon juice (about ½ lemon's worth)

1 teaspoon (2g) oregano

½ teaspoon (3g) salt

Zest of 1 lemon, minced finely

Simple Japanese-Style Marinade

In a bowl, mix:

¼ cup (70g) low-sodium soy sauce (regular soy sauce will be too salty)

2 tablespoons (10g) minced ginger

3 tablespoons (20g) minced scallions (also known as green onions), about 2 stalks

2 tablespoons (40g) honey

	Bitter	Salty	Sour	Sweet	Umami	Hot
Chinese	Chinese broccoli Bitter melon	Soy sauce Oyster sauce	Rice vinegar Plum sauce (sweet and sour)	Plum sauce (sweet and sour) Jujubes (small red dates) Hoisin sauce	Dried mushrooms Oyster sauce	Mustard Szechwan peppers Ginger root
French	Frisée Radish Endive Olives	Olives Capers	Red wine vinegar Lemon juice	Sugar	Tomato Mushrooms	Dijon mustard Black, white, and green peppercorns
Greek	Dandelion greens Mustard greens Broccoli rabe	Feta cheese	Lemon	Honey	Tomato	Black pepper Garlic
Indian	Asafetida Fenugreek Bitter melon	Kala namak (black salt, which is NaCl and Na_2S)	Lemon Lime Amchur (ground dried green mangoes) Tamarind	Sugar Jaggery (unrefined palm sugar)	Tomato	Black pepper Chilies, cayenne pepper Black mustard seed Garlic Ginger Cloves
Italian	Broccoli rabe Olives Artichoke Radicchio	Prosciutto Cheese (pecorino or parmigiano-reggiano) Capers or anchovies (commonly packed in salt)	Balsamic vinegar Lemon	Sugar Caramelized veggies Raisins / dried fruits	Tomato Parmesan cheese	Garlic Black pepper Italian hot long chilies Cherry peppers
Japanese	Tea	Soy sauce Miso Seaweed	Rice vinegar	Mirin	Shitake mushrooms Miso Dashi	Wasabi Chiles
Latin American	Chocolate (unsweetened) Beer	Cheeses Olives	Tamarind Lime	Sugar cane	Tomato	Jalapeño and other hot peppers
Southeast Asian	Dried tangerine peel Pomelo (citrus fruit)	Fish sauce Dried shrimp paste	Tamarind Kaffir limes	Coconut milk	Fermented bean paste	Bird chili Thai chili in sauces and pastes

Examples of ingredients used by different cultures to balance out flavors. Use this chart as an inspiration to try out new combinations and take note of how the various flavors change your perceptions.

Rice, Wheat, Grains ≅ Congee, Cream of Wheat, Porridge

A billion people eat congee daily, but you're unlikely to find it on many restaurant menus in the United States, for the same reason that "porridge" and "gruel" don't appear very often: it's a dish meant to stretch the filling power of a few cheap ingredients as much as possible. (Think Oliver Twist: "Please, sir, I want some more.") That doesn't mean it can't be delicious and nutritious; it just means that unless your cultural background includes it, you might not know it. For some, it's the equivalent of chicken noodle soup: something nourishing to turn to when sick or looking for comfort.

Since everybody has to eat, every culture has something like congee based on the staple crop that grows regionally. Different regions of the world support growing different crops: wheat in the United States, grains such as oats in Europe, and rice in much of Asia. Wheat becomes cream of wheat, oats become porridge, and rice becomes congee.

Congee can be "subclassed" into several different versions, depending upon the culture. The Chinese call it *zhou* (runny rice porridge with eggs, fish paste, tofu, and soy sauce); in India, it's called *ganji* (rice "soup" that has flavorings such as coconut milk, curry, ginger, and cumin seeds added to it). When cooked in sweet milk with cardamom and topped with pistachio or almonds, you have the dessert version.

If you want a further challenge, try fusion cooking: blending the ingredients and flavors of two regions together. Why not try porridge with traditional congee toppings, or congee with porridge toppings? Or, pick two random locations (the tried-and-true semi-random method: dartboard and map of the world; if you hit water, go for fish), and create a meal blending the flavors from the different cultures, or using one culture's ingredients with another culture's techniques. Italian and Mexican? Try taco pizza: pizza with cheese, tomatoes, salsa, beans, and cilantro on top. Vietnamese and Classic American? How about a Vietnamese hamburger, seasoning the meat with fish sauce, lemongrass, and red pepper flakes, and adding cucumber and bean sprouts to the bun? Japanese and Classic European? Go for miso ice cream; it's salty and sweet, and delicious!

Fusion cooking often results from the mixing of two cultures via immigration. There are plenty of fusion-like dishes that have come out of cultures situated where two different regions meet or two different cultures mingle: Mediterranean (North African + Southern European), Southeast Asian (Asian + European colonialism), and Caribbean (African + Western European), for example. Israeli markets carry ingredients from the surrounding western regions of North Africa (especially Moroccan) and Eastern Europe; their cuisine is influenced by the traditions of both areas. Modern Vietnamese food was heavily impacted by French occupation in the 19th century. The United States is perhaps the most diverse example of fusion cooking; with so many different cultures mingling, you might not even

Rice Congee

Cook for at least several hours in a slow cooker, or in a pot set over a very low flame:

4 cups (1kg) water or stock

½ cup (100g) rice, *unwashed* (so that the starches remain in the congee)

½ teaspoon (3g) salt

When you're ready to eat, heat the rice to near boiling to finish cooking. The long, low-heat cooking will have broken down the starches; boiling the liquid will cause them to gelatinize and quickly thicken. I have a rice cooker that has a slow-cook mode, so I switch it from slow-cook mode to rice mode, which is hotter and will take the rice up to near boiling. If you are doing this in a pot on the stovetop, set the pot over medium heat, periodically stirring and checking it while working on the rest of these instructions so that it does not burn on the bottom.

While the rice is cooking, prepare a number of toppings. I enjoy:

Tofu, cut into small cubes and browned on all sides

Scallions, chopped into small pieces

Garlic, sliced into thin discs and toasted on each side to make "garlic chips"

Sriracha sauce

Soy sauce

Toasted almond slices

You can serve this family-style, with the toppings in small bowls where your guests can help themselves (or not, in the case of sriracha sauce), or you can portion the toppings out more formally: a tablespoon or two of tofu, a few teaspoons of scallions, a sprinkling of garlic chips, and a dash of sriracha and soy sauces. Quantity is not particularly important, except for the hot and salty sauces.

Notes

- *This isn't a fancy or precise dish, and there's no right or wrong set of toppings or quantities. (Millions of cooks can't be getting this wrong.) A simple rice congee is a great place to try different combinations of ingredients!*
- *To toast the garlic, use a sharp knife to slice a few cloves (or more, if you're a garlicphile) into thin discs. Place a frying pan on a burner set to medium-high heat, but do not add oil. Arrange the garlic wafers in a single, thin layer. Toast one side until medium brown, about two to three minutes, and then flip (try using tongs) to toast the second side.*
- *Try cracking an egg into the congee at the end of cooking, either in the pot (and then mix it in), or in the individual bowls (you might need to pop the congee into the microwave for a minute if it isn't hot enough to fully cook the egg). Adding an egg will alter the texture and give the dish a much richer taste.*
- *Try substituting other salty ingredients for the soy sauce and hot ingredients for the sriracha sauce, using the flavor-by-culture table presented earlier.*

think of using the term "fusion" to describe our cuisine, but it is. Just think of African-influenced Southern cooking, the French and African backgrounds in Cajun food, and the impact of Mexican cuisine on Tex-Mex.

One of the keys to a successful blend of two culinary traditions is to choose recipes for which the ingredients are readily available. Indian cuisine has translated extremely well to the United States, in large part because the ingredients commonly used either are already present here (onions, lentils, peppers) or ship and store well (cumin, paprika, curry powders). Much Egyptian food, on the other hand, relies on goat meat, which is extremely uncommon in the American grocery store. One great way to find inspiration is to visit local ethnic markets and stores. They tend to be small storefronts with "weird" smells from the different produce and spices, and are typically located in obscure neighborhoods, so ask around to discover where they're hidden.

Tomato Basil Mozzarella Salad

Tomato, basil, and mozzarella are a classic Italian combination, and a good example of "what grows together goes together." This recipe is all about the freshness of the ingredients, so you'll need to wait until the height of summer for the ingredients to be in season.

If you are adventurous, try making your own cheese. See Mozzarella on page 288 in Chapter 6.

Toss in a bowl and serve:

1 cup (180g) sliced tomatoes, about 2 medium ones

1 cup (15g) fresh basil leaves, from about 3 or 4 stems

½ cup (100g) mozzarella

1 tablespoon (15g) olive oil

Salt and pepper to taste

Notes

- *The ratio of basil to cheese to tomato is really up to you. Hold back some of each ingredient, take a look at the resulting salad, and toss in more of whatever you think will make it better. The only thing to be careful with is the salt; once there's too much in there, it's hard to fix. How to slice the tomatoes and cheese is also up to you. Try thick slices of tomato and cheese, alternating in layers on the plate and served with a fork and knife. Or, slice the tomato and cheese into bite-sized pieces to be served with just a fork.*
- *Try making this twice, once with conventional tomatoes and a second time with heirloom tomatoes, to see the difference made by the quality of ingredients.*

Xeni Jardin on Local Food

PHOTO OF XENI JARDIN USED BY PERMISSION OF JEN COLLINS

Xeni Jardin is a coeditor of Boing Boing (http://www.boingboing.net).

Could you tell me a little bit about yourself and food?

I've been fascinated with cooking as long as I've been fascinated with creating and exploring technology, if not longer. To me, the two worlds aren't mutually exclusive. On the contrary, they feed each other. Just recently, one of our coeditors on the blog, Lisa Katayama, was in Nepal, and over the weekend she posted a single sentence: "I could eat dal bhat every day of my life." Dal bhat is basically rice and stewed lentils. It's what you eat at almost every meal in Nepal. I traveled to the region myself. I was remembering how good the simple food of that Himalayan country was. So I said, "You know what? I'm going to make some dal bhat right here in Los Angeles." I had some split yellow peas in the cupboard and dug out some different spices. I didn't know exactly how to make it so I started Googling. I do this a lot. I'll spend half an hour poking around at different recipes. I end up kind of improvising something in the end usually based on my own cooking experience and different little bits of the recipes I find.

What is it like exploring food both through the Internet and through traveling and seeing the traditional ways food is prepared in other countries?

I was in a Mayan village with some people that I work with, a nonprofit. It was Christmas, and in Guatemala, tamales and Christmas go together. In this particular village, the women have a particular way that they prepare the Christmas tamales. They use locally grown white corn. I followed them around and took notes and, with their permission, filmed the preparation and watched every step. This woman was toasting sesame seeds over a wood fire, and then grinding them in a stone grinder. Another was making the spicy sauce. Other women in another part of the room boiled prepared corn into a mash. This particular preparation was kind of runny and soft and white, a lot like the grits that I grew up with in the South. I sat in the middle of this assembly line of women all wearing their brightly colored woven blouses as they glopped a big dollop of that soft white corn into big green leaves from the corn plant. Then they added a little bit of meat and sauce, and then they tied them all up and steamed them.

A few days later we drove back into Guatemala City, where we stayed with a nonindigenous family. The house was just packed to the gills with tamales that were purchased from local vendors. When a guy is on the street walking around with a bag of tamales, the answer is always yes. They have a million different kinds of tamales that are prepared for Christmas in that country alone. I remember sitting at some Christmas celebrations there, too. We're sitting there at the table and there's sheets of all these different kinds of tamales. One of the Guatemalan people at the table said, "What the hell is in this? Cherries?" That's the kind of sweet and savory Guatemalan tamale you can get for Christmas.

I can't get enough of that in the same way that I frequently fall down these Internet search rabbit holes when I'm just chilling out thinking about what kind of yummy, healthy food I'm going to prepare for my family and friends. I love exploring food in traditional cultures as a part of the reporting that I do, and I love exploring food back here. It's a fairly important part of my life.

Seasonal Method

Cooking "within season" means using only those ingredients that have good, fresh flavor and are ripe. Restricting yourself to ingredients that are in season in your region is a great way of creating constant challenges and exposing yourself to new ingredients. And because in-season ingredients tend to be of higher quality and pack more of a flavor punch, it's that much easier to make the resulting dishes taste good. Next time you're at the grocery store, take note of what new fruits and vegetables have arrived and what is in dwindling supply.

Of course, not every ingredient in a dish is a "seasonal" ingredient. Cellar onions, storage apples, and pantry goods such as rice, flour, and beans are year-round staples and fair game. What is off-limits with this approach are those foods that are outside their growing season for where you live. Put another way, don't try making grilled peaches in February. Even if you can get a peach in February, it won't have the same flavor as a mid-summer peach, so it will invariably taste flat. Even if those peaches shipped from Chile taste okay, they won't be as good as the local in-season peaches, because they have to be of a variety that favors shipping durability and disease resistance over taste and texture. (Unless you happen to live in Chile.)

One of the perks of using in-season ingredients, besides the quality, is that the abundance of the in-season produce generally means lower prices, too, as the supply-and-demand curves change. Grocery stores have to figure out how to sell all those zucchinis when they come up for harvest, and running specials is one of the standard ways of moving product. The same challenge applies even more if you're growing your own fruits and vegetables, because a home garden can produce an abundance over a short period of time. If you figure out what to do with the 100 pounds of zucchini that all come ready in late summer, I know plenty of people who would like to hear it!

In-season, local foods have the advantage of typically being fresher than their conventional counterparts, which is especially important for flavor in highly perishable foods such as heirloom tomatoes and fresh seafood. Local isn't always better, though. For example, if you live in a northern climate, you might find that produce such as radishes from traditional farms located where it gets hotter might taste better.

One suggestion for all of that zucchini: make "fake pasta" using a matchstick blade on a mandolin, then pan sear the resulting pieces in olive oil.

Caterina Fake's Roasted Potatoes

*Caterina Fake cofounded the photo-sharing site Flickr (*http://www.flickr.com*) and Hunch (*http://hunch.com*), a decision-making site. She also serves on the board of directors for both Creative Commons (*http://creativecommons.org*) and Etsy (*http://www.etsy.com*).*

Tell me about what role food and cooking play in your life.
I never considered myself to be much of a cook, but at some point I realized that I was a much better cook than I had thought I was. It was a revelation to me. I had mostly been cooking family recipes that had been handed down; I didn't refer much to cookbooks. I've been told by guests to my house that I was a better cook than I thought I was and it took me a while to come around to actually believing that. It came very gradually over time, adding one recipe a year maybe, but over time you get many, many recipes.

Why did you think that you weren't a good cook?
Just because I cooked such simple things, using such simple ingredients. I would get waxy potatoes and cut them up and put olive oil on them with rosemary and garlic and salt and pepper and grill them and for whatever reason I didn't think that that qualified. My mother would occasionally do these incredible James Beard and Julia Child recipes, these elaborate recipes that took hours and hours of preparation and special ingredient shopping and that was what qualified as good cooking. It was haute cuisine, whereas the kind of cooking that I always did was just put something yummy on the table for dinner without a lot of effort.

I think so often it's the very simple dishes that actually can be some of the most special, most significant meals of our lives. What created this interest in learning to cook?
Necessity. I loved having people over and I loved cooking for people. When I was working as a web developer in San Francisco, dinner parties were really my favorite ways of socializing. I would have people over and as a result, I had to learn how to cook. When I lived in New York I actually didn't cook very much because there are so many restaurants and New York kitchens are so tiny. It's much more of a dinner party culture here in San Francisco.

Any parting thoughts that you would like to share?
I'm very lucky to live in California, where fresh food is plentiful. I started receiving a box of food from the Eat Well Farms CSA. I learned a bunch of new recipes and ways of cooking from having had that basket of vegetables delivered to my doorstep with a bunch of things that I normally would avoid at the grocery store. I didn't know what to do with fennel. I didn't know what to do with sweet potatoes. I didn't know what to do with parsnips. It got me out of my comfort zone and taught me some new ways to think about food. One of my favorite cookbooks is *The Art of Simple Food* by Alice Waters (Clarkson Potter). It would not look down upon my potato recipe. Those are the exact kinds of recipes that it includes. It's really very intuitive cooking; it gives you a very short list of very simple ingredients from which you can cook a great many things.

Caterina Fake's Roasted Potatoes with Garlic and Rosemary

Cut in half some waxy potatoes, such as yellow potatoes like Yukon Gold, or red potatoes. Blanch the potatoes in a pot of boiling water for a minute or two. Strain the potatoes out and spread them on a cookie sheet. Mix with a generous amount of olive oil and diced-up garlic and rosemary. Season with salt and pepper to taste.

Place the cookie sheet on the top shelf of an oven set to broil. Once the top side is golden brown and crispy on the outside, turn them over, and broil the other side until done.

These are great for breakfast or you can have them as a side dish or snack.

Environmentally Sound Inputs

First they say farm-raised salmon is better, then it's wild salmon. Or the blogs light up with posts about how many metric tons of greenhouse gases or gallons of water are associated with producing an average cheeseburger. Then there's the whole "local food" movement wanting to help with reducing our collective carbon footprint. What's an environmentally conscious geek to do?

It depends. How much are you willing to give up?

Let's start with the good news, with the greenest of the green: your veggies. Locally grown veggies combined with a minimum of transportation and sold unpackaged are about as good as you can get for the environment, and they're about as good as you can get for yourself. Look for a farmers' market in the summer (or if you're lucky enough to live in California, year-round). Farmers' markets are a great way to really understand where your food is coming from. Plus, your local economy will thank you.

You can also subscribe to a CSA (community-supported agriculture) share, where every week or two you receive a box of local and seasonal produce. It's a great way to challenge yourself in the kitchen, because invariably something unfamiliar will show up in your CSA share, or you'll find yourself with 10 pounds of spinach and be looking for something new to do with it. Regardless of anything else, your mother was right: eat your veggies. (On a personal note unrelated to the environment, I believe the typical American diet doesn't include enough veggies. Eat more veggies! More!)

In the other corner, there are red meats like corn-fed beef. It's environmentally expensive to produce: the cow has to eat, and if fed corn (instead of grass), the corn has to be grown, harvested, and processed. All this results in a higher carbon footprint per pound of slaughtered meat than that of smaller animals like chickens. Then there's the fuel expended in transportation, along with the environmental impact of the packaging. By some estimates, producing a pound of red meat creates, on average, four times the greenhouse-gas emissions as a pound of poultry or fish. See Weber and Matthews' "Food-Miles and the Relative Climate Impacts of Food Choices in the United States" (*http://pubs.acs.org/doi/abs/10.1021/es702969f*).

Regardless of where you fall on the spectrum between local-shopping vegan and delighting in a ginormous bacon-wrapped slab of corn-fed beef, limiting consumption in general is the best method for helping the environment. Choose foods that have lower impact on the environment, and be mindful of wasted food. See the caveat below, but current data suggests the total impact on the environment of consuming fish is less than the impact of eating chicken and turkey, which likewise is more sustainable than pork, which is in turn better for the environment than beef.

This isn't to say all red meats are bad. If that steak you're cutting into came from a locally raised, grass-fed cow, she might actually be playing a positive role in the environment by converting the energy stored in grass into fertilizer (i.e., manure) for other organisms to use. See Michael Pollan's classic *The Omnivore's Dilemma* (Penguin) for more on this topic, but as a general rule, the more legs it has, the "less good" it is for the environment. (By this logic, centipedes are pure evil...)

By most analyses, cutting the amount of red meat you consume is the easiest way to make a positive impact on the environment. One friend of mine follows the "no-buy" policy: he happily eats it, but won't buy it. I've heard of others following variants of Mark Bittman's "vegan before 6" diet: limiting consumption of meats during the day but pigging out at dinner.

With respect to fish, whether farm raised or wild-caught is better depends upon the species of fish, so there's no good general rule. There are issues with both types: some methods of farm fishing generate pollutants or allow fish to escape and commingle with wild species, while wild-caught contributes to the depletion of the ocean's stocks, and the impact of a global collapse in the fisheries from overfishing is a very real threat to the food supply to hundreds of millions. For a good article on fisheries and the global impact of over-fishing, see the *New Republic*'s "Aquacalypse Now" (*http://www.tnr.com/article/environment-energy/aquacalypse-now*).

The biggest contribution you can make—at least on the dinner plate—is to avoid wild-caught seafood of species that are overfished. The Monterey Bay Aquarium runs a great service called "Seafood WATCH" that provides a list of "best," "okay," and "avoid" species, updated frequently and broken out by geographic region of the country. Search online for "seafood watch" or visit *http://www.montereybayaquarium.org/cr/seafoodwatch.aspx.*

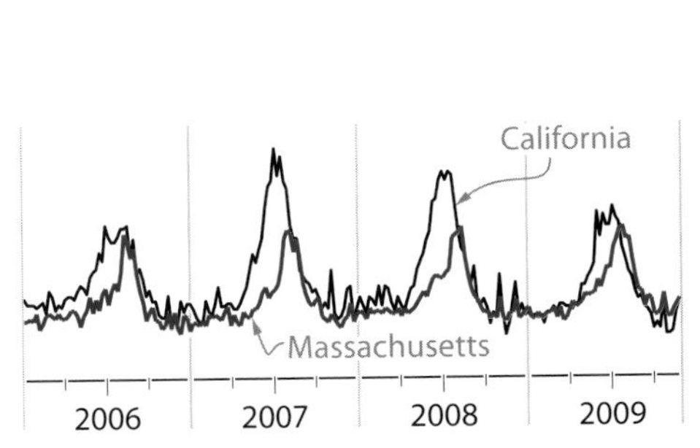

Google Searches for "Peach"
(Users in California vs. Massachusetts)

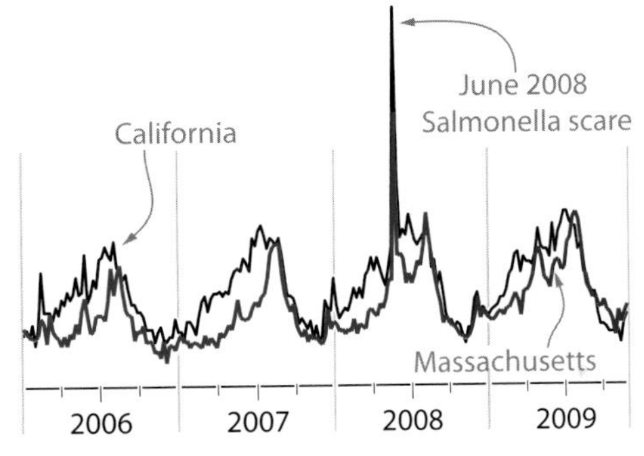

Google Searches for "Tomato"
(Users in California vs. Massachusetts)

Data from Google Trends showing search volumes for the terms "peach" (left) and "tomato" (right) for California users and Massachusetts users. The growing season in Massachusetts starts later and is much shorter than in California. There's a tight correlation of this with Google's search volumes for those terms.

If it's the dead of winter and there's a foot of snow on the ground (incidentally, *not* the best time to eat out at restaurants specializing in local, organic fare), finding produce with "good flavor" can be a real challenge. You will have to work harder to produce flavors on par with those in summer meals. Working with the seasons means adapting the menu. There's a reason why classic French winter dishes like cassoulet (traditionally made with beans and slow-cooked meats, but that description does *not* do this amazing dish justice—I make mine with duck confit, bacon, sausage, and beans, then slow roast it overnight) and coq au vin (stewed chicken in wine) use cellar vegetables such as onions, carrots, turnips, and potatoes and slow, long-cooking simmers to tenderize tougher cuts of meat. I can't imagine eating cassoulet mid-summer, let alone venting the heat generated from keeping the oven on for that long. Yet in the dead of winter, nothing's better.

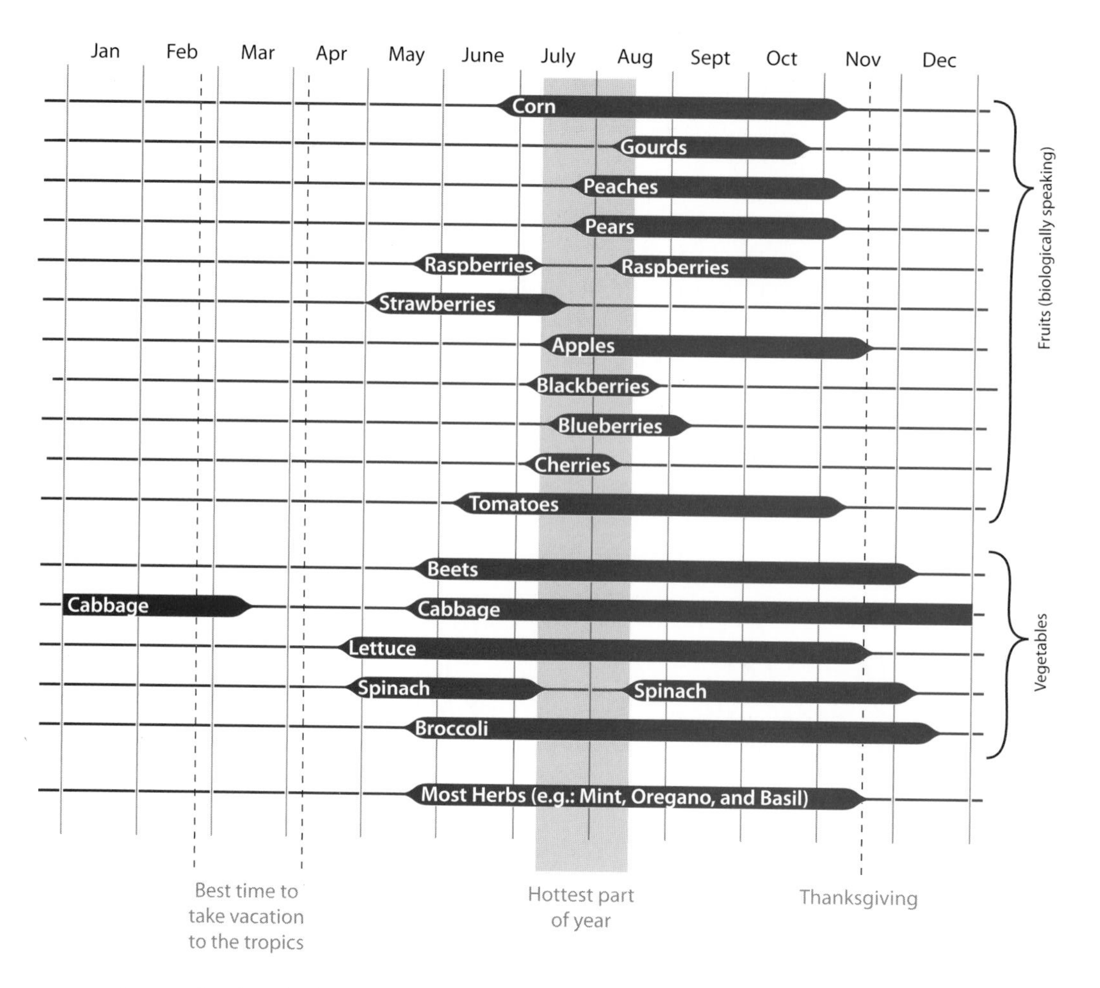

Seasonality chart for fruits and vegetables in New England. Fruits have a shorter season than vegetables, and only a few vegetables survive past the first frost. Some plants can't tolerate the hottest part of the year; others do best during those times. If you live in the Bay Area or New York, see http://www.localfoodswheel.com *for a nifty "what's in season" wheel chart.*

Consider the following three soups: gazpacho, butternut squash, and white bean and garlic. The ingredients used in gazpacho and butternut squash soup are seasonal, so they tend to be made in the summer and autumn, respectively (of course, modern agricultural practices have greatly extended the availability of seasonal ingredients, and your climate might be more temperate than the sources of these traditions). White bean and garlic soup, on the

Gazpacho (Summer)

Purée, using an immersion blender or food processor:

2 large (500g) tomatoes, peeled, with seeds removed

Transfer the puréed tomato to a large bowl. Add:

1 (150g) cucumber, peeled and seeded

1 cob (125g) corn, grilled or broiled and cut off the cob

1 (100g) sweet red bell pepper, grilled or broiled

1/2 small (30g) red onion, thinly sliced, soaked in water, and drained

2 tablespoons (20g) olive oil

2 cloves (6g) garlic, minced or pressed through a garlic press

1 teaspoon (4g) white wine vinegar or champagne vinegar

1/2 teaspoon (1g) salt

Stir to combine. Adjust salt to taste and add ground black pepper as desired.

Notes

- *The weights in this recipe are for the prepared ingredients (i.e., after removing seeds, trimming stems, or soaking).*
- *If you prefer a smooth gazpacho, purée all of the ingredients at the end. Or, add a portion of the veggies, purée, and then add the remainder to achieve a partly smooth, partly chunky texture. It's all about your preference!*
- *Gazpacho is one of those dishes that is really about the fresh ingredients that you have on hand. There's no mechanical or chemical reason for these quantities to be written as they are, so add more of this, less of that, whatever you like to suit your tastes. Try expanding this recipe to include other ingredients, such as hot peppers or fresh herbs.*
- *Grilling or broiling the corn and bell peppers adds a smoked flavor to the soup, due to the chemical reactions that take place at higher heat, as we'll discuss later in this book. You might find you prefer a "raw" version of this soup. Or, if you really like the smoky flavor, try adding some "liquid smoke" to amp it up.*

Whenever you see a recipe calling for a grilled vegetable, you should default to rubbing it with a light coating of olive oil before grilling it, because this will prevent the vegetable from drying out while cooking.

How to Peel a Tomato

I have a friend whose boyfriend tried to make her a surprise dinner involving tomato soup, but he didn't know how to peel tomatoes. She came home to find her guy frantically trying to use a vegetable peeler on the tomatoes to no avail...

To peel tomatoes, drop them in boiling water for a few seconds and then pull them out with tongs or a mesh spider, and then just pull the skins off. You can cut an "x" shape into the skin before blanching, although I find the skin on some varieties of tomatoes will pull back regardless, as long as the water is at a full rolling boil. Experiment to see if it makes a difference!

Butternut Squash Soup (Fall)

Purée in a food processor or with an immersion blender:

2 cups (660g) butternut squash, peeled, cubed, and roasted (about 1 medium squash)

2 cups (470g) chicken, turkey, or vegetable stock

1 small (130g) yellow onion, diced and sautéed

½ teaspoon (1g) salt (adjust to taste)

Notes

- *As with the gazpacho recipe, the weights are for the prepared ingredients and only rough suggestions. So, prepare each item individually. For example, for the squash, peel it, then coat it with olive oil, sprinkle it with salt, and roast it in the oven at a temperature around 400–425°F / 200–220°C until it begins to brown. When you go to purée the ingredients, hold back some of the squash and some of the stock, taste the purée, and see which you think it needs. Want it thicker? Add more squash. Thinner? Add more stock.*
- *This soup by itself is very basic. Garnish with whatever else you have on hand that you think might go well, such as garlic croutons and bacon. Or top with a small dab of cream, some toasted walnuts, and dried cranberries to give it a feeling of Thanksgiving. How about a teaspoon of maple syrup, a few thin slices of beef, and some fresh oregano? Chives, sour cream, and cheddar cheese? Why not! Instead of purchasing items to follow a recipe exactly, try using leftover ingredients from other meals to complement the squash soup.*
- *If you're in a rush, you can "jump-start" the squash by microwaving it first. Peel and quarter the squash, using a spoon to scoop out the seeds. Then, cube it into 1–2" / 3–5 cm pieces, drop it into a glass baking pan that's both oven and microwave safe, and nuke it for four to five minutes to partially heat the mass. Remove from microwave, coat the squash with olive oil and a light sprinkling of salt, and roast it in a preheated oven until done, about 20 to 30 minutes. If you're not in a rush, you can skip the peeling step entirely: cut the squash in half, scoop out the seeds, add oil and salt, roast it for about an hour (until the flesh is soft), and use a spoon to scoop it out.*

To cut thick gourds such as squash and pumpkins, use a large chef's knife and a mallet. First, slice off a thin piece of the gourd so that it lies flat and doesn't roll, then gently tap the knife blade through the gourd.

White Bean and Garlic Soup (Winter)

In a bowl, soak for several hours or overnight:

2 cups dry white beans, such as cannellini beans

After soaking overnight, drain the beans, place them in a pot, and fill it with water (try adding a few bay leaves or a sprig of rosemary). Bring to a boil and simmer for at least 15 minutes. Strain out the water and put the beans back in a pot (if using an immersion blender) or in the bowl of a food processor.

Add to the pot or bowl with beans and then purée until blended:

2 cups (500g) chicken or vegetable stock
1 medium (150g) yellow onion, diced and sautéed
3 slices (50g) French bread, coated in olive oil and toasted on both sides
½ head (25g) garlic, peeled, crushed, and sautéed or roasted
Salt and pepper, to taste

Notes

- *Don't skip soaking and boiling the beans. Really. One type of protein present in beans—phytohaemagglutinin—causes extreme intestinal distress. The beans need to be boiled to denature this protein; cooking them in a slow cooker or sous vide setup (see Chapter 7) will not denature the protein and actually makes things worse. If you're in a rush, use canned white beans; they'll have already been cooked.*
- *Variations: try blending some fresh oregano into the soup. Toss some bacon chunks on top or grate on some Parmesan cheese as well. As with many soups, how chunky versus how creamy to blend the soup is a personal preference.*
- *Make sure to toast the French bread to a nice golden brown. This will add the complex flavors from caramelization and Maillard reactions in the sugars and proteins from the bread. You can pour olive oil onto a flat plate and dip the bread in to coat it.*

Choosing Your Inputs: Flavors and Ingredients

other hand, uses pantry goods that can be had at any time of year. Thus, it is traditionally thought of as a winter soup, because it's one of the few dishes that can be made that time of year.

If you're not much of a soup person, try making savory sorbets using seasonal ingredients. A summer sorbet with tomatoes and tarragon? Yum. Carrot ice cream? Why not? And for winter, while unusual, bacon ice cream has been enjoyed by diners at Chef Heston Blumenthal's UK-based restaurant The Fat Duck, and taken further with candied bacon bits in a recipe on David Lebovitz's blog (search online for "candied bacon ice cream recipe").

Finally, here are a few tips related to seasonality to keep in mind:

- Use fresh herbs whenever possible, because most dried herbs don't have anywhere near the strength of flavor. The volatile oils that are responsible for so much of the aromas in herbs oxidize and break down, meaning that the dry herbs are a pale substitute. Dried herbs have their place, though; it makes sense to use dried herbs in the dead of winter when annual plants like basil aren't in season. Store dry herbs in a cold, dark place (not above the stove!) to limit the amount of heat and light, which contribute to the breakdown of organic compounds in spices.

- Grind your own spices as much as possible. Fresh-grated nutmeg will be much stronger than preground nutmeg, for the same reasons that many fresh herbs are better than their dried counterparts: the aromatic compounds in a preground spice will have had time to either hydrate or oxidize and disperse, resulting in flavor loss. Most spices also benefit from being bloomed—cooked in oil under moderate but not scorching heat—as a way of releasing their volatile chemicals without breaking them down.

- If you're looking for convenience, commercially frozen vegetables and fruits are actually pretty good. Freezing produce right when it is harvested has a few advantages: nutritional breakdown is halted, and the frozen item is from the peak of the season while the fresh version in your store may have been harvested early or late. Using frozen produce is especially useful if you're cooking for just yourself, because you can pull out a single portion at a time. If you're growing your own food and intend to freeze it, look up online how to use dry ice to pack and quick-freeze the produce; freezing in your home freezer takes too long and leads to mushy veggies.

- When selecting produce at the store, think about when you'll want to use it. For example, if you're buying bananas to eat throughout the week, instead of buying one cluster of mostly green bananas, buy two smaller clusters, one mostly yellow (for sooner) and one mostly green (for later). Picking in-season produce and selecting it so that it will be ripe when you're ready to use it are good ways to guarantee quality.

Organic Versus Conventional

Organic foods are those grown or raised to USDA National Organic Program (NOP) regulations on farms or ranches certified as following those regulations. Organic produce has restrictions on which pesticides can be used; animals butchered for organic meats are required to be given access to the outdoors and are prohibited from being given antibiotics or growth hormones. Because of these restrictions, the cost of producing organic food is typically higher.

Conventional foods are those not certified for sale under the label *organic*, regardless of whether they are grown to the same standards and regardless of their place of origin. They must still be grown to acceptable USDA/FDA standards, though.

When it comes to produce, just because it's organic doesn't mean it's automatically safer, just as software labeled as *open source* isn't necessarily of higher quality than proprietary software. Of course, there are other reasons to buy organic, but if your concern is food safety and pesticides, the benefit of organic isn't necessarily clear-cut: whether exposure to traditional pesticides is always worse for you than exposure to their organic replacements is not yet known. The detectable levels of pesticides in our bodies are well below anything approaching toxic, and as chemists have told me, "it's the dosage that matters." To put some numbers to it, consider what Dr. Belitz et al. wrote in *Food Chemistry* (Springer): "[T]he natural chemicals [in a cup of coffee] that are known carcinogens are about equal to a year's worth of synthetic pesticide residues that are carcinogens."

Given the option, farmers would rather not have to spray any type of pesticide on their crops: it costs money, takes time, and increases their exposure to the chemicals. Just keep in mind that if there were an easy answer—say, if organic practices were always better and always cheaper—everyone would be doing it that way.

If you do feel going organic is for you but are on a tight budget, here are some general rules of thumb. For fruits, if you're going to eat the skin, buy organic. If you're going to peel them, buying organic appears to offer comparatively little advantage when it comes to exposure to pesticides. For veggies, organic bell pepper, celery, kale, and lettuce test as having lower levels of pesticides than their conventional counterparts. Go organic for dairy, eggs, and meats; for seafood, see the previously mentioned Monterey Bay Aquarium's "Seafood WATCH" program at *http://www.seafoodwatch.com*.

If you're interested in seeing for yourself if organic food tastes different from conventional food, try this side-by-side experiment. Make two versions of a simple pasta dish with sautéed chicken and red and yellow bell peppers, using organic ingredients in one and conventional in the other. How does organic chicken compare to conventional chicken? Can you taste the difference in the bell peppers? For that authentic scientific experience, serve the side-by-side meal to a bunch of friends without revealing which bowl contains the organic version to run a true "taste test." You might find the variance in flavor to be greater—or less—than you expect.

Tim Wiechmann's Beet Salad

Chef Wiechmann is the chef and owner of T.W. Food in Cambridge, MA. He creates his menus using local organic produce with a classic French approach.

How you go from planning a dish to putting it on the table?

I start with the ingredients—they all have to be in season. I came up with a dish that was made with leftover cheese from the Pyrénées. Black cherries and beets are in season, so how can I dress up a beet salad? In the Pyrénées, they have cherries with sheep's milk cheese. Most of my stuff comes from cultural things, from traveling all over and having a sound grasp on food in Europe. I study what people make from all over—they do this here, they do that there. And these things are done for thousands of years. I try to have a knowledge of these things and then I just look at my own ingredients here, and I draw them together.

What is your approach in the kitchen and what thoughts do you have on technology in the kitchen?

My menu is actually really difficult. My employees start out with this picture that we just dig out a carrot and boil it. We don't do that. Everything goes through a rigorous, precise set of cooking parameters. With certain preparations, time and temperature are everything. There are things like the water circulator you can use to cook all the meats and fishes perfectly every time. Even for cutting things, we use rulers and metal caramel cutters.

Observation is critical, as is getting experience in knowing what looks good. If you're cooking an onion, it changes color over time. There are certain stages where you want to pull it because the bitterness increases as the caramelization increases. Onions in a tall pot will sweat differently than onions in wide pot. In a tall pot, they release their own water and cook evenly because the water doesn't disappear. We have specific pots that are good for certain things—sweat the onions in this pot; don't use that pot—but a new cook will just grab any pot.

How do you know if something is going to work?

You just try. When you start to play the piano, you don't know where the notes are. You have to have the technique, then you can think about putting the notes together. If I hit this note, then I'll get this sound; if I want onions to be sweet, I'll caramelize them. The technique follows the knowledge. I keep a log of my own recipes and times for each thing. How long to put cherries or apples in a bag and cook in the water circulator, that comes out of experience.

My big thing I always say: "Get into it and go for it." Just buy it and try it. Every time you cook something—even if you burn it and it goes in the trash—it's not a failure, it's just: next time I'm not going to burn it.

Out of all these criteria that make for a good evening, clearly food is an important one, but what do you think people underestimate?

Little things. Maybe they don't know why they don't like something. You know what I mean? "Well, I'm not sure. I just didn't like it." I think very few people know what they like and can identify what they like. That's why I'm pretty good at what I do—I really know what I like. Do you know what you like?

I'll have to give it some thought.

I don't know what I like in the visual art world. I just haven't spent enough time on it.

I can answer that one on the visual art world. Anything that prompts an emotional response. It might not be a positive emotion, but it should stir an emotion or create an experience. Have you seen *Ratatouille*?

Yes.

The scene where the camera zooms into the critic's eye and goes back to his childhood as he's eating ratatouille. He has an experience. For me, food needs to touch on emotions.

Everybody is geared with that, but I think a lot of people don't know how to identify that. They'll say, "I don't like cauliflower." One really great French chef taught me that you don't have to like it, you just have to make it good. He said, "taste this," to which I said, "I don't like it; I don't want to taste it." He yelled at me. "You're going to be a chef and you can't taste it? You have to taste it." I'll never forget him screaming at me.

I think this would apply if you're cooking for friends: keep in mind what your friends are going to enjoy.

That's right. My job is to make something that people will enjoy.

Roasted Red and Candystripe Beet Salad with Almond Flan, Black Cherry Compote, and Ossau-Iraty

Serves 8; Prep time: 2 hours.

Prepare the cherry compote. In a container, measure out and soak overnight:

4 cups (600g) pitted black cherries

1⅔ cups (340g) sugar

1 tablespoon (10g) apple pectin

2 vanilla beans, sliced open lengthwise

After soaking overnight, transfer to a pan, add the zest and juice of a lemon, and cook over medium heat for an hour, until the mixture reaches a jam-like consistency. Transfer to a plastic container or jar and cool.

Prepare the flan. In a blender, combine:

1 cup (150g) almonds, toasted

1 teaspoon (5g) almond extract

6 medium (330g) eggs

2 cups (480g) heavy cream

Nutmeg, salt, pepper to taste

Pour onto a quarter sheet pan (9" × 13" / 23 cm × 33 cm) lined with a Silpat or parchment paper and bake at 300°F / 150°C until the custard sets, about 45 minutes. Cool on the sheet in refrigerator.

Prepare the beets. Preheat oven to 450°F / 230°C. Create a foil pouch containing:

6 medium (500g) red beets

6 medium (500g) candystripe beets (also known as chioggia beets)

Salt, olive oil, and pepper to taste

Roast until tender, about 45 minutes, depending on the size of the beets. Remove from pouch and peel with a knife. Cut the beets into attractive circles or cubes.

To serve. Make a quick salad dressing with oil and vinegar, salt and pepper. Toss the beets and 1 cup / 90g of toasted slivered almonds in the dressing.

Arrange the beets and almonds on large plates. Place a nice slice of flan somewhere among them and drop a few scoops of the cherry compote in various places.

Using a vegetable peeler, shave into long strands (about 4" / 10 cm):

½ pound (225g) Ossau-Iraty (a medium-soft cheese from the French Pyrénées, creamy and complex)

Decorate the salad with the shaved Ossau-Iraty.

Genetically Modified Foods

Regardless of your feelings about or definition of GMO (genetically modified organism) foods, the topic is an intensely charged political and social minefield. Fear of the unknown has a long record of helping to guarantee the survival of our species, so avoiding things until they've established a history of being safe does certainly seem prudent. But this view doesn't consider the potential harms that a GMO-based food might be able to avert.

What if a strain of rice could be produced that was more resilient in the face of floods and droughts? Such a strain of rice would increase crop yields for families in impoverished countries, and the need is only going to increase. The United Nations' food agency expects that worldwide food production will need to increase by 70% between 2010 and 2050. Or how about strains of rice or corn that need fewer pesticides to remain viable crops? Worldwide, some 300,000 deaths a year are attributed to pesticide poisoning.

Then there's "Golden Rice," a golden-yellow rice that has been genetically modified to produce increased amounts of beta-carotene as a way of addressing Vitamin A deficiencies that impact the extremely poor in some nations. Everyone agrees that Vitamin A deficiencies are a serious problem: an estimated 1 to 2 *million* children die every year due to Vitamin A deficiency, according to a 1992 World Health Organization report. Still, Golden Rice has not yet been approved for human consumption; organizations like Greenpeace have opposed it, saying that it's an unproven solution and that other, better solutions exist.

More personally, would you accept genetically engineered cows guaranteed to be free of prions, which cause Bovine spongiform encephalopathy (a.k.a. "mad cow disease")? Or how about a GMO banana that was able to withstand the fungus *Fusarium oxysporum* that threatens to wipe out the banana as we know it? Related to GMO foods, would you accept irradiated chicken if it was guaranteed to be free of salmonella?

This isn't to suggest that you should *seek out* GMO-based foods; but at the very least you should recognize that there are very real trade-offs. Hundreds of Americans die annually from salmonellosis, and while those deaths can be avoided with proper cooking, perhaps we as a society shouldn't blindly fear technologies that could prevent those deaths just because they're unfamiliar.

Sure, it might be reasonable to fear *corporate overlording*—the idea that our food chain might become reliant upon a corporation with a patent on the very food we need to survive—but this is a separate issue from GMO food itself. Another argument against GMO foods claims that the money spent on GMO research would be better spent on other areas of agricultural technology; but again, this is a separate issue from whether genetically modifying food itself should be done.

I personally do not enjoy burgers served by fast food chains, but I recognize that they are able to feed literally millions of American families every day. Around the world, advances in technology have increased crop yields and improved the quality of life for many, although there are still many in starving conditions. What happens to those families who are just barely making ends meet when the prices of food exceed what they can afford?

Non-GMO foods are not inherently more expensive, but the economics to date have tended to make the price of GMO foods cheaper. The quick-serve industry is not saying "we want GMO foods"; they're simply buying what's most economical, because in a price-sensitive market, the chains need to keep prices down to remain in business.

For a glimpse into the interconnectedness of our food system, search online for Louise Fresco's touching TED talk, "On Feeding the Whole World" (*http://www.ted.com/talks/louise_fresco_on_feeding_the_whole_world.html*).

P.S. That nice sandwich you had at lunch with the organic bread made with organic wheat? Probably from a strain of wheat genetically modified around World War II via mutation breeding that relied on thermal neutron radiation or sodium azide. Seems perfectly safe at this point...

Analytical Method

There have been a number of attempts over the years to devise a scientific model for predicting which flavors will work well together. While not particularly well suited for day-to-day cooking, these types of approaches do have a place in helping create new combinations of flavors and they are used by the food industry and some high-end chefs.

A disclaimer: picking pleasing flavors—or at least ones that invoke an emotional response or trigger a memory—is somewhere between an art and a science, so no scientific equation can capture the entire picture. Still, understanding how such a "flavor compatibility algorithm" would work can provide you with a way of organizing your thoughts on food, and for geeks, it's fun to see how far one can take these sorts of things. If you really want to geek out and need a food project to work on, an open source version of this concept would be fun.

To start, we need a model of how to describe individual flavors, before considering how to combine them. Odors can be categorized in a few ways, most commonly either chemically or descriptively.

Chemical taxonomies classify compounds by their odors. Such a taxonomy is essentially a database of chemicals that each map to distinct flavor sensations. For example, Flavornet (*http://www.flavornet.org*), created by two researchers at Cornell (Acree and Arn), describes some 700+ chemical odorants detectable by the human nose. Listing compounds such as *citronellyl valerate* (smells like honey or rose; used in drinks, candies, and ice cream), the database is useful for generating certain flavors artificially, but not so useful outside of laboratory kitchens.

Descriptive taxonomies apply labels to odors as a way of classifying and grouping foods. For example, both lemon and orange are generally classified as "fruity/citrus." Lacking the precision of a chemical taxonomy (the compound is either present or it isn't), descriptive taxonomies suffer from the subjectivity of human judgment. Most of us would agree that a lemon smells "fruity/citrus," but how much does a food like chocolate smell of the odors in celery? Not much, but certainly more than chocolate smells of fish.

The simplest descriptive taxonomy, from the 1950s by J. E. Amoore, proposes just seven primary odors: camphoric (like mothballs), ethereal (like cleaning fluid), floral (like roses), musky (like aftershave), pepperminty, pungent (like acetic acid in vinegar), and putrid (like rotten eggs).

One modern descriptive taxonomy can be found in the American Society for Testing and Materials' *Atlas of Odor Character Profiles – DS61*, by Andrew Dravnieks. While you might not necessarily think of all of the terms included as pleasant, it's certainly a diverse set, which is useful in thinking about smells. With 146 terms, Dravnieks's list also provides enough granularity to begin to form a meaningful model for food flavors.

Common	Sweet, fragrant, perfumy, floral, cologne, aromatic, musky, incense, bitter, stale, sweaty, light, heavy, cool/cooling, warm
Foul	Fermented/rotten fruit, sickening, rancid, putrid/foul/decayed, dead animal, mouse-like
General foods	Buttery (fresh), caramel, chocolate, molasses, honey, peanut butter, soupy, beer, cheesy, eggs (fresh), raisins, popcorn, fried chicken, bakery/fresh bread, coffee
Meats	Meat seasoning, animal, fish, kippery/smoked fish, blood/raw meat, meat/cooked good, oily/fatty
Fruits	Cherry/berry, strawberry, peach, pear, pineapple, grapefruit, grape juice, apple, cantaloupe, orange, lemon, banana, coconut, fruity/citrus, fruity/other
Vegetable	Fresh vegetables, garlic/onion, mushroom, raw cucumber, raw potato, bean, green pepper, sauerkraut, celery, cooked vegetables
Spices	Almond, cinnamon, vanilla, anise/licorice, clove, maple syrup, dill, caraway, minty/peppermint, nut/walnut, eucalyptus, malt, yeast, black pepper, tea leaves, spicy
Body	Dirty linen, sour milk, sewer, fecal/manure, urine, cat urine, seminal/like sperm
Materials	Dry/powdery, chalky, cork, cardboard, wet paper, wet wool/wet dog, rubbery/new, tar, leather, rope, metallic, burnt/smoky, burnt paper, burnt candle, burnt rubber, burnt milk, creosote, sooty, fresh tobacco smoke, stale tobacco smoke
Chemicals	Sharp/pungent/acid, sour/acid/vinegar, ammonia, camphor, gasoline/solvent, alcohol, kerosene, household gas, chemical, turpentine/pine oil, varnish, paint, sulphidic, soapy, medicinal, disinfectant/carbolic, ether/anaesthetic, cleaning fluid/carbona, mothballs, nail polish remover
Outdoors	Hay, grainy, herbal/cut grass, crushed weed, crushed grass, woody/resinous, bark/birch, musty/earthy, moldy, cedarwood, oakwood/cognac, rose, geranium leaves, violets, lavender, laurel leaves

Dravnieks's 146 odor terms, broken down into main categories, provide a good basis for thinking about odors. If you're heading out on a date and want to impress, this list is a pretty good starting point for describing wines!

Another adjective classification system, Allured's *Perfumer's Compendium*, is used by the perfume industry, the fine folks responsible for the smells of products from laundry detergent to toothpaste. Think that new car smell is accidental? Trained employees smell the materials that go into the interior of a new car to make sure that it smells just right. (To quote *The Matrix*: "You think that's air you're breathing now?") Allured's taxonomy uses more descriptive and narrow scents—familiar items such as banana, peach, and pear—but also specific items like hyacinth, patchouli, and muguet (lily of the valley), making it less useful to the layperson.

Let's start by defining a flavor profile as the weighted scores of a collection of terms in a classification system, such as Dravnieks's 146 odor terms. For every term, imagine taking an item of food—say, a pear—and scoring it on a scale from 1 to 5, where a score of 1 indicates "doesn't smell like it at all" and 5 is "the very definition of the word!" Given a pear, how much does it smell like a "heavy" odor? 1. Fruity? Maybe a 3? Fragrant? Say it's a ripe pear, so 4.

The scoring is not asking if it is a compatible smell, just if the odor label describes the smell. Are the odors you sense in a pear (are the chemoreceptors that fire off in your nose) the same as when you smell other things that are considered fragrant? Given the weights of all of these odor descriptions for a pear, you can plot a graph (almost like a histogram) that you can then compare to similar graphs for other foods.

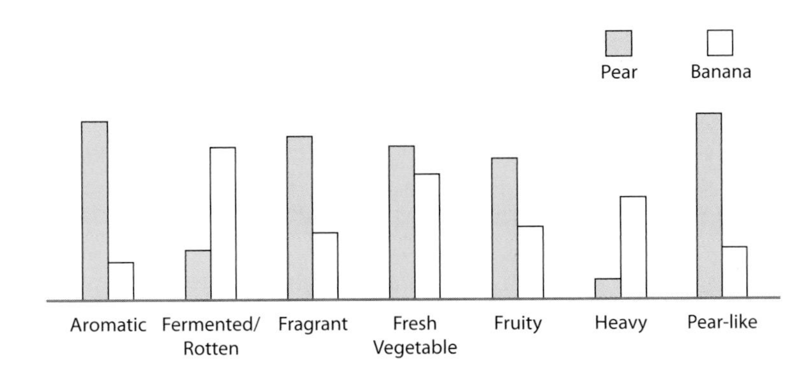

Some of Dravnieks's odor terms associated with banana and pear, as scored by a few thousand Internet voters (taller bars indicate a larger degree of agreement between food and odor). These voters were not trained or verified to be familiar with the definitions of those odors; for these reasons, this graph should be treated as a conceptual demonstration only.

Given such a graph for each individual ingredient in a recipe, you can imagine a combined graph that describes the overall profile of a dish, showing all the "frequencies" present in the smells of each ingredient. Think of it like the various instruments that contribute to a piece of music: each has its own set of frequencies, and the combination of all the instruments makes up the overall song's frequency distribution. When in tune, the frequencies line up and balance one another; when out of tune, the combination of sounds can be jarringly dissonant, even if each sounds fine individually. Of course, this music analogy isn't a perfect fit for thinking about flavors: chemical changes brought about by cooking or by reactions between foods change the histogram, and the music analogy doesn't cover other variables in foods, such as texture, weight, or mouth-feel.

Many chefs—often pros, but also non-pros who've been cooking for years—can imagine flavor combinations in their heads, doing something similar to this process mentally. Just as a composer imagines each voice and track in a piece of music, an experienced cook imagines the profile of the entire dish. Good cooks think about which notes are missing or are too soft and figure out what ingredients can be added to bring up those values.

What about achieving entirely new pairings, combinations that have no precedence in tradition? This same concept of matching up foods by their flavors can be done via the chemical taxonomy method, given enough time. The high end of the luxury restaurant industry spends an inordinate amount of time working on new flavor combinations, often with upward of two years spent working on a concept before it's presented. Chef Heston Blumenthal of The Fat Duck (UK) maintains three distinct kitchens, one of which is devoted to laboratory work and is staffed by individuals holding both masters-level degrees in hard sciences like physics or chemistry *and* degrees from first-tier culinary institutions such as *Le Cordon Bleu*. Here is a partial list of pairings Chef Blumenthal has used: strawberry and coriander, snails and beetroot, chocolate and pink peppercorn, carrot and violet, pineapple and certain types of blue cheese, banana and parsley, harissa and dried apricot. Give them a try!

In addition to conducting their own private research, high-end chefs interested in creating new flavor combinations sometimes work with researchers at universities. Both Flavornet (*http://www.flavornet.org*) and FoodPairing (*http://www.foodpairing.be*) include such research in their sites. If you're interested in exploring some of the chemical commonalities between ingredients, look at FoodPairing, which uses a chemical flavor database in order to suggest what ingredients to try together. (FoodPairing claims to be used by Chefs Heston Blumenthal and Ferran Adrià.)

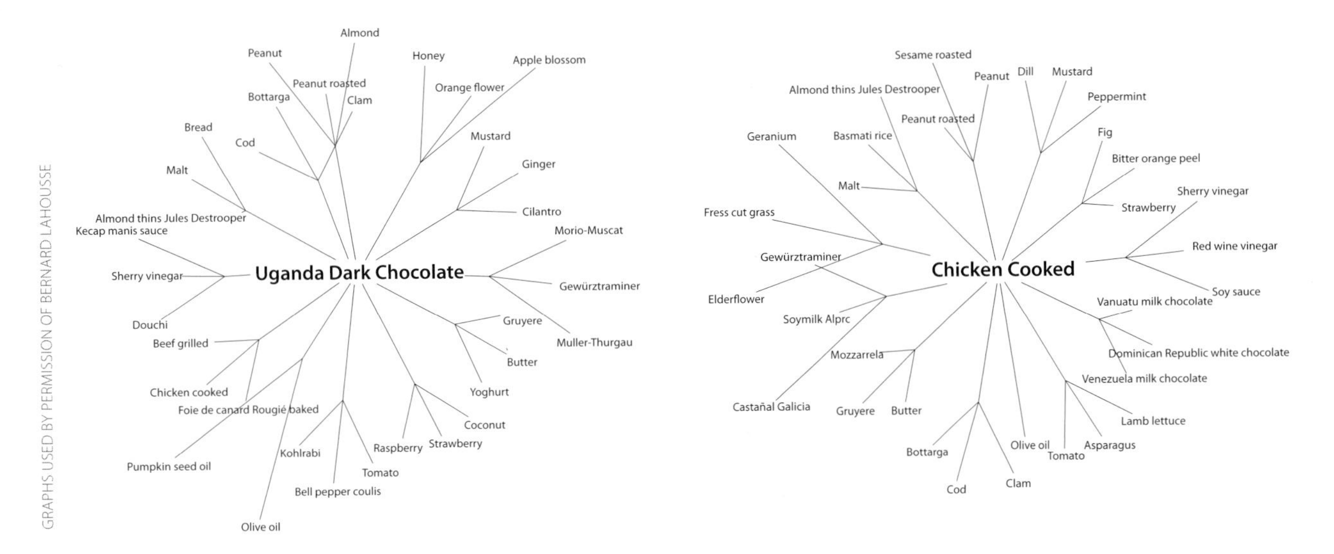

Food Pairing diagrams for chocolate and chicken. Their database is based on chemical analysis, and it gives suggestions based on both chemical similarity and chemicals known to be complementary.

The analytical approach tends to be very abstract. There's little here that helps one select what ingredients to toss into a bowl together to make dinner. For this reason, these sorts of tools have yet to become particularly successful. This technique doesn't generate recipes. While a set of odors might go together from an aromatic perspective, there are other variables in cooking that prevent mixing and matching various ingredients indiscriminately. For example, one ingredient might require cooking, while another might break down in high heat.

You can work around these constraints by separating the two ingredients into different components that are prepared separately and combined on the plate—say, a meat with a sauce. Or try using cooking methods that are, in essence, about conveying the perfume in food. Soups, ice creams, even soufflés: all are methods of transporting the flavors and aromas of ingredients without carrying the texture or volume of the original ingredient. A number of more recent, novel flavor pairings have used this solution. At the very least, you might find these types of tools a fun source of inspiration to try new things. Go experiment!

Harold McGee on Solving Food Mysteries

PHOTO OF HAROLD MCGEE USED BY PERMISSION OF KARL PETZKE

Harold McGee writes about the science of food and cooking. He is the author of On Food and Cooking *(Scribner), described by Alton Brown as "the Rosetta stone of the culinary world." He also writes a column, The Curious Cook, for the* New York Times. *His website is at* http://www.curiouscook.com.

How do you go about answering a food mystery?

It depends on the nature of the mystery. It can start with and mainly involve experiments in the kitchen, doing a particular process several different ways, changing one thing at a time, and seeing what the effect is. Or it can mean going to the food science or technical literature and hunting for information that might be relevant.

A recent example of the latter would be this column I wrote for the *New York Times* about keeping berries and fruits longer than normal. I had been going to the farmers' market and getting way too much fruit. It looked and tasted so good, but I couldn't eat it all, and after a day, it would begin to mold, sometimes even in the refrigerator. I thought there might be a way to deal with this. So I drove up to UC Davis and used their online databases to search the literature for methods of controlling mold growth on produce.

I discovered that back in the 1970s some guys at one of the ARS [USDA Agricultural Research Service] stations here in California came up with a mild heat treatment that didn't damage the fruit but did slow down substantially the growth of mold on the outside. I came back and gave it a try, and it worked. I didn't have the knowledge or the tools to deal with it without doing some library research. I put it to the test because it's one thing to read about something in the literature and another thing to make sure that it actually plays out that way in somebody's kitchen.

Why not do this kind of literature search online? Is there something that UC Davis or an institution like that is able to provide researchers that they can't get directly online back home in front of their computers?

There are wonderful resources that are available at both university and public libraries that an individual just can't afford to subscribe to. In institutions with a food science department, there are resources on the shelf that you would never know about without going and looking, and I enjoy doing that, not necessarily to answer the question "What do people know today about X?" but more "How have people dealt with X over the centuries?"

Centuries? Can you give me an example of something from that kind of historical research?

Tomato leaves are not toxic the way people thought they were. In fact, they're probably beneficial to eat because they bind to cholesterol and prevent us from absorbing it. The question arose: "How did we get this idea that they're toxic if they're not?"

I delved back as far as I could in some pretty obscure literature to try to figure that out, and that included going up to UC Davis and taking a look at a couple of books from the 17th and 18th centuries on Dutch ethnography of the Pacific. I tracked down a reference to people eating tomato leaves on an island in the Indonesian Archipelago in the 17th century. This would have been shortly after tomatoes had been introduced there because they are not native to that part of the world. That fleshes out the story of how this plant found its way around the world, how it developed a reputation, and the kinds of aesthetic judgments that people made about it.

In Europe, people didn't eat the leaves because they thought they stank. In Central and South America, where tomatoes came from, the leaves weren't much eaten, which I still don't understand. Just pulling all of these bits together to me is part of the pleasure of understanding and appreciating the food that I sit down and eat at my table today. There is this tremendous depth of history and complexity that, if you delve into it, can make it even more pleasurable to eat these things.

One of the things I like best about the job I have is not so much the writing; it's the exploring, it's tracking down these books and reading this paragraph about people on this island centuries ago doing this with the leaves, then coming home and trying to get some sense of what that tasted like using leaves from my own backyard and the equivalent of the preserved fish that they were probably using back then to season them.

I imagine that our understanding about food is getting more refined, and we're correcting a lot of previous misconceptions. What do you hope future research will spend time working on?

If I could name one area that I wish people with the equipment, expertise, and resources would pay more attention to and work harder on, it is flavor and the influence of different cooking methods on the ultimate experience of particular preparations. There are so many interesting questions about different ways of doing the same thing where, at the moment, basically you have your own personal experience and the experience of other people but no good, objective yardstick.

What are the real differences? Are we experiencing the same set of compounds differently because we have different sensory systems, or do, in fact, different techniques produce different sets of compounds where you happen to prefer this and I happen to prefer that? An example would be making stocks. There are some people who are real partisans of doing stocks in pressure cookers and others who think that the long, slow, barely-at-a-simmer method gives you a superior result. I've done both, and I like both, but they are different. I'm not sure I can really explain how they are different, so I would love to know what's going on there.

What does the home cook need to understand about what they're doing in the kitchen?

A scale and a good thermometer are absolutely essential if you're going to try to understand things and do experiments carefully enough to draw real conclusions. You need to be able to measure, and temperature and weight are the main variables.

Is there something that really surprised you in the kitchen?

I suppose the one moment in my life that really confounded my expectation was the copper bowl versus glass bowl for beating egg whites. I was reading Julia Child while I was writing the book [*On Food and Cooking*] the first time in the late 1970s. She said that you should whip egg whites in a copper bowl because it acidifies the whites and gives you a better foam for meringue and soufflés, but the chemistry was wrong. Copper doesn't change the pH of solutions, so I thought that since the explanation was wrong, there probably was nothing to the claim either.

Then a couple of years later, when it came time to get ready for publication, I was looking at old graphic sources for illustrations for the book. I looked at a French encyclopedia from the 17th century that had a lot of professions illustrated. One of them was a pastry kitchen. In the engraving, there was a boy beating egg whites, and it said that the boy was beating egg whites in a copper bowl to make biscuits. It specified a copper bowl, and it looked exactly like today's copper bowls: it was hemispherical and had a ring for hanging. I thought if a French book from 200 years ago is saying the same thing that Julia Child said, then maybe I should give it a try.

I tried a glass bowl and a copper bowl side by side, so I could look at them and taste them, and the difference was huge. It took twice as long to make a foam in the copper bowl; the color was different, the texture was different, the stability was different. That was a very important moment for me. You may know that somebody else doesn't know the chemistry, but they probably know a lot more about cooking than you do. That certainly got me to realize that I really did have to check everything I could.

A French chef told me a story. He'd made a million meringues in his life, and one day he was in the middle of whipping the egg whites in a machine. The phone rang—there was some kind of emergency and he had to go away for 15 or 20 minutes—so he just left the machine running. He came back to the best whipped egg whites he'd ever seen in his life. His conclusion from that was, in French, "*Je sais, je sais que je sais jamais.*" It sounds a lot better in French than it does in English, but the English is, "I know, I know that I never know."

Thanks to that experience with the copper bowl, that's been my motto as well. No matter how crazy an idea sounds or how much I distrust my own senses when I do something, and it somehow seems inexplicably different from what it should be, I know that I'm never going to understand everything completely, and there's probably a lot more to learn about whatever it is that's going on.

4

Time and Temperature: Cooking's Primary Variables

EVER SINCE CAVEMEN FIRST SET UP CAMPFIRES AND STARTED ROASTING THEIR KILL, MANKIND HAS ENJOYED A WHOLE NEW SET OF FLAVORS IN FOOD. Cooking is the application of heat to ingredients to transform them via chemical and physical reactions that improve flavor, reduce chances of foodborne illness, and increase nutritional value.

From a culinary perspective, the more interesting and enjoyable changes are brought about when compounds in food undergo the following chemical reactions:

Protein denaturation

The *native* form of a protein is the three-dimensional shape (conformation) assumed by the protein that is required for normal functioning. If this structure is disrupted (typically by heat or acid), the protein is said to be *denatured*. Changes in the shapes of proteins also alter their taste and texture.

Different proteins denature at different temperatures; most proteins in food denature in the range of 120–160°F / 49–71°C. Egg whites, for example, begin to denature at 141°F / 61°C and turn white because the shape of the denatured protein is no longer transparent to visible light. In meat, the protein *myosin* begins to denature around 122°F / 50°C; another protein, *actin*, begins to denature around 150°F / 65.5°C. Most people prefer meat cooked such that myosin is denatured while keeping the actin native.

Maillard reaction

A Maillard reaction is a browning reaction that gives foods an aromatic and mouth-watering aroma. Usually triggered by heat, this occurs when an amino acid and certain types of sugars break down and then recombine into hundreds of different types of compounds. The exact byproducts and resulting smells depend upon the amino acids present in the food being cooked, but as an example, imagine the rich smell of the crispy skin on a roasted chicken.

For culinary purposes, the reaction generally becomes noticeable around 310°F / 154°C, although the reaction rate depends on pH, chemical reagents in the food, and amount of time at any given temperature. Many meats are roasted at or above 325°F / 160°C—at temperatures lower than this, the Maillard reaction hardly occurs.

Caramelization

Caramelization is the result of the breakdown of sugars, which, like the Maillard reaction, generates hundreds of compounds that smell delicious. Pure sucrose (the type of sugar in granulated sugar) caramelizes at between 320–400°F / 160–204°C, with only the middle range of 356–370°F / 180–188°C generating rich flavors.

In baking, those goods that are baked at 375°F / 190° C generally have a noticeably browned exterior, while those baked at or below 350°F / 175°C remain lighter-colored.

"Great," you're probably thinking, "but how does knowing any of this actually help me cook?"

You can tell when something is done cooking by understanding what reactions you want to trigger and then detecting when those reactions have occurred. Cooking a steak? Check the internal temperature with a thermometer; once it's reached 140°F / 60°C, the myosin proteins will have begun to denature. Baking crispy chocolate chip cookies at 375°F / 190°C? Open your eyes and keep your nose online; the cookies will be just about done when they begin to turn brown and you're able to smell the caramelization occurring. Really, it's that simple. Foods are "done" when they achieve a certain state, once they have undergone the desired chemical reactions. As soon as the reactions have occurred, pop the food out; it's done cooking.

A small but critical detail: as we'll discuss elsewhere, proteins don't simultaneously denature at a given temperature. Denaturation is a function of duration of exposure at a given temperature. And there are many different types of proteins in different foods, each with its own temperature/time response rate.

Smell, touch, sight, sound, taste: learn to use all of your senses in cooking. Meat that has been cooked until it is medium rare—a point at which myosin has denatured and actin has yet to denature—will feel firmer and also visibly shrink. The bubbling sound of a sauce that's being boiled and reduced will sound different once the water is mostly evaporated, as bubbles pushing up through the thicker liquid will have a different sound. Bread crust that has reached the temperatures at which Maillard reactions and caramelization occur will smell wonderful, and you'll see the color shift toward golden brown. By extension, this also means that the crust of the bread must reach a temperature of 310°F / 155°C before it begins to turn brown, which you can verify using an IR thermometer. (Bread flour has both proteins and sugars, so both caramelization and Maillard reactions occur during baking.)

This chapter shows you when and how these changes occur so that you can become comfortable saying, "It's done!" We'll start by looking at the differences between the common sources of heat in cooking and how differences in the type of heat and temperatures impact cooking. Since one of the main reasons for cooking is reducing the chances of foodborne illness, we'll also discuss the key issues in food safety, including a look at how to manage bacterial contamination and parasites, along with some example recipes to demonstrate the principles behind food safety. The remainder of the chapter will then examine a number of key temperature points, starting with the coldest and ending with the hottest, discussing the importance of each temperature point and giving example recipes to illustrate the reactions that occur at each of these temperatures.

As with most recipes in this book, the recipes here are *components*, not necessarily entire dishes or meals unto themselves. Create your own combinations as you like! It's usually easier to take each of the components in a dish and cook them separately: veggies in one pan, meats or proteins in another, and starches in a third. This allows you to isolate the variables for each component, then combine them at the end. Eggplant Parmesan might be your favorite dish, but if you're new to cooking, it's probably not the best place to start to learn about the reactions taking place.

Finally, cooking and baking share an axiom with coding and product development: it's done when it's done—not when the timer goes off. One of the best tips I can offer for improving your skills in the kitchen is to "calibrate" yourself: take a guess if something is done and then check, taking note of what your senses, especially smell and sight, notice in the process.

Timers are great guides for reminding you to check on a dish and a good safety net in case you're like me and absentmindedly wander off occasionally. But timers are only a proxy for monitoring the underlying reactions. Given a fillet of fish that is done when its core temperature reaches 140°F / 60°C—which might take about 10 minutes—the primary variable is temperature, not time. If the fish is slow to heat up, regardless of the timer going off at the 10-minute mark, it won't be done yet. Not to knock timers entirely: they're a great tool, especially in baking, where the variables are much more controlled and thus the time needed to cook can be more accurately prescribed. But don't be a slave to the timer.

Cooked = Time * Temperature

Since the primary chemical reactions in cooking are triggered by heat, let's take a look at a chart of the temperatures at which the reactions we've just described begin to occur, along with the temperatures that we commonly use for applying heat to food:

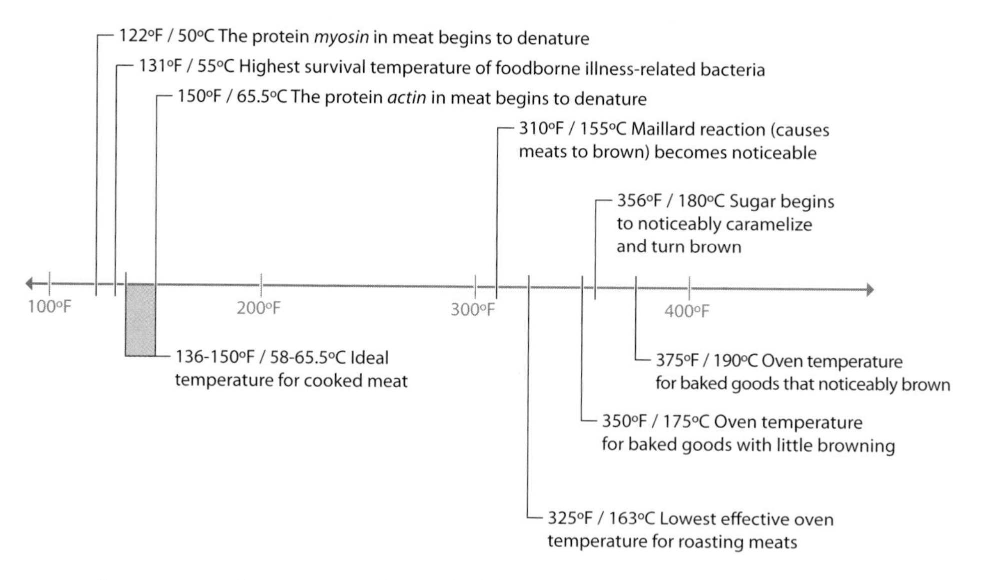

Temperatures of common reactions in food (top portion) and heat sources (bottom portion).

There are a few "big picture" things to notice about these common temperatures in cooking. For one, notice that browning reactions (Maillard reactions and caramelization) occur well above the boiling point of water. If you're cooking something by boiling it in a pot of water or stewing it in liquid, it's impossible for high-heat reactions to occur, because the temperature can't go much above 216°F / 102°C, the boiling point of moderately salted water. If you're cooking a stew, such as the simple beef stew recipe in Chapter 2 (page 67), sear the meats and caramelize the onions separately before adding them to the stew. This way, you'll get the rich, complex flavors generated by these browning reactions into the dish. If you were to stew just the uncooked items, you'd never get these high-heat reactions.

Another neat thing to notice in the temperature graph is the fact that proteins denature in relatively narrow temperature ranges. When we cook, we're adding heat to the food specifically to trigger these chemical and physical reactions. It's not so much about the temperature of the oven, grill, or whatever environment you're cooking in, but the temperature of the item of food itself.

Which brings us to our first major *aha!* moment: the most important variable in cooking is the temperature of the food itself, not the temperature of the environment in which it's being cooked. When grilling a steak, the temperature of the grill will determine how long it takes the steak to come up to temperature, but at the end of the day, what you really want to control is the final temperature of the steak, to trigger the needed chemical reactions. For that steak to be cooked to at least medium rare, you need to heat the meat such that the meat itself is at a temperature of around 135°F / 57°C.

Denaturing Proteins

What's all this talk about "denaturing" proteins? It's all about structure.

Denaturing refers to a change in the shape of a molecule (*molecular conformation*). Proteins are built of a large number of amino acids linked together and "pushed" into a certain shape upon creation. Since the function of a protein is related to its shape, changing the shape changes the protein's ability to function, usually rendering it useless to the organism.

Think of a protein as a bit like the power cable between a laptop and an outlet: while it has a particular primary structure (the cord and wires inside it), the cord itself invariably gets all tangled up and twisted into some secondary structure. (If it's anything like mine, it spontaneously "retangles" itself regardless of attempts to straighten it out, but proteins don't actually do this.)

On the molecular level, the cable is the protein structure, and the tangles in the cable are secondary bonds between various atoms in the structure. Atoms can be relocated to different bonding spots, changing the overall shape of the molecule, but not actually changing the chemical composition. With its new shape, however, the molecule isn't always able to perform its original function. It might no longer fit into places that it used to be able to go, or given the new conformation, other molecules might be able to form new bonds with the molecule and prevent it from functioning as it used to.

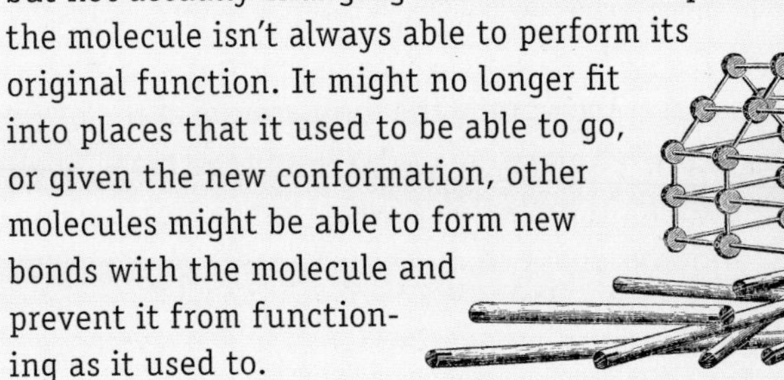

Heat Transfer and Doneness

The idea that you can just cook a steak any old way until it reaches 135°F / 57°C sounds too easy, so surely there must be a catch. There are a few.

For one, how you get the heat into a piece of food matters. A lot. Clearly the center of the steak will hit 135°F / 57°C faster when placed on a 650°F / 343°C grill than in a 375°F / 190°C oven. The hotter the environment, the faster the mass will heat up, thus the rule of thumb: "cooking = time * temperature." Consider the internal temperatures of steak cooked two ways, grilled and oven-roasted:

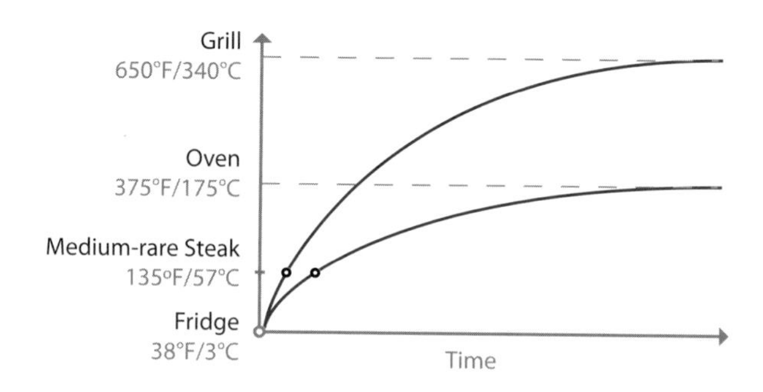

Schematic diagram of temperature curves for two imaginary steaks, one placed in an oven and a second placed on a grill.

Cooking a steak on a grill takes less time than in an oven, because energy is transferred faster in the hotter environment of the grill. Note that the error tolerance of when to pull the meat off the grill is smaller than pulling the meat from the oven, because the slope of the curve is steeper. That is, if t^1 is the ideal time at which to pull the steak, leaving it for t^1+2 minutes will allow the temperature of the grilled steak to overshoot much more than one cooked in the oven.

This is an oversimplification, of course: the graph shows only the temperature at the center of the mass, leaving out the "slight" detail of the temperature of the rest of the meat. (It also doesn't consider things like rate of heat transfer inside the food, water in the meat boiling off, or points where proteins in the meat undergo phase changes and absorb energy without a change in temperature.)

Another thing to realize about heat transfer is that it's not linear. Cooking at a higher temperature is *not* like stepping on the pedal to get to the office faster, where going twice as fast will get you there in half the time. Sure, a hotter cooking environment like a grill will heat up the outer portions of the steak faster than a relatively cooler environment like an oven. But the hotter environment will continue to heat the outer portions of the steak before the center is done, resulting in an overcooked outer portion compared to the same size steak cooked in an oven to the same level of internal doneness.

What's the appeal of cooking on a hot grill, then? For the right cut of meat, you can keep a larger portion of the center below the point at which proteins become tough and dry (around 170°F / 77°C) while getting the outer portion up above 310°F / 154°C, allowing for large amounts of Maillard reactions to occur. That is, the grill helps give the outside of the steak a nice brown color and all the wonderful smells that are the hallmark of grilling—aromas that are the result of Maillard reactions. The outside portion of grilled meat will also have more byproducts from the Maillard reactions, resulting in a richer flavor.

Juggling time and temperature is a balancing act between achieving some reactions in some portions of the meat and other reactions in other parts of the meat. If you're like me, your ideal piece of red meat is cooked so that the outer crust is over 310°F / 155°C and the rest of the meat is just over 135°F / 57°C, with as little of the meat between the crust and the center as possible being above 135°F / 57°C. The modern technique of sous vide cooking can be used to achieve this effect; we'll cover this in Chapter 7.

"Cooking = time * temperature"

This has to be one of the hand-waviest formulas ever. I hereby apologize. To make up for it, here's an actual mathematical model for temperature change as a function of heat being applied. Remember to cook until medium-rare...

$$\begin{cases} t_{i(J+1)} = (q_i \cdot \tau dh + 1_m \cdot d\tau \cdot (t_{i(j-1)} + t_{i(j+1)}) + m_c \cdot c_m \cdot h_{i,j})/2 \cdot 1_m \cdot Fd\tau + m_c \cdot c_m \cdot dh \\ K_{1i} = 0.00836 - 0.001402\ pH + 5.5 \cdot 10^{-7} \cdot t^2 \\ K_{2i} = -0.278 + 7.325 \cdot 10^{-2} pH - 3.482 \cdot 10^{-5} \cdot t^2 \\ K_{3i} = 2.537 \cdot 10^{-3} - 1.493 \cdot 10^{-4} \cdot t_i + 2.198 \cdot 10^{-5} \cdot t^2 \\ K_{4i} = 2.537 \cdot 10^{-2} - 9.172 \cdot 10^{-3} pH + 3.157 \cdot 10^{-5} \cdot t_i^2 \\ m_{1t.i} = m_0^b - (m_o^b - m_t^b) \cdot e^{-K1i \cdot t} \\ m_{2t.i} = m_0^b - (m_o^b - m_t^b) \cdot e^{-K2i \cdot t} \\ m_{3t.i} = m_0^b - (m_o^b - m_t^b) \cdot e^{-K3i \cdot t} \\ m_{4t.i} = m_0^b - (m_o^b - m_t^b) \cdot e^{-K4i \cdot t} \end{cases}$$

SOURCE: M. A. BELYAEVA (2003), "CHANGE OF MEAT PROTEINS DURING THERMAL TREATMENT," *CHEMISTRY OF NATURAL COMPOUNDS* 39 (4)

Temperature gradients

This balancing act—getting the center cooked while not overcooking the outside—has to do with the rate at which heat energy is transferred to the core of a food. Since cooking applies heat to foods from the outside in, the outer portions will warm up faster, and because we want to make sure the entire food is at least above a minimum temperature, the outside will technically be overcooked by the time the center gets there. This difference in temperature from the center to outer edges of the food is referred to as a *temperature gradient*.

Choose the method of cooking to match the properties of the food you are cooking. Smaller items—skirt steak, fish fillets, hamburgers—work well at high heats. Larger items—roasts, whole birds, meatloaf—do better at moderate temperatures.

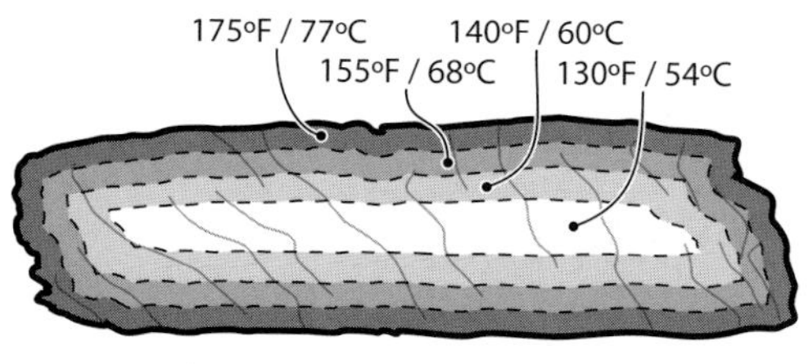

Temperature gradient of steak cooked on a grill (650°F/343°C)

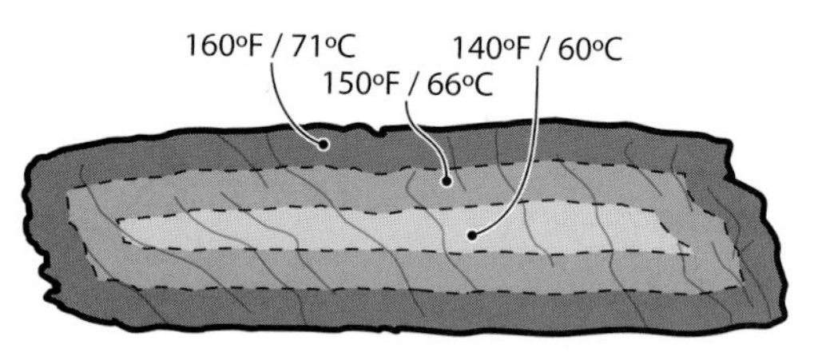

Temperature gradient of steak cooked in an oven (375°F/190°C)

Lower heat sources bring up the temperature of the meat more uniformly than hotter heat sources.

All parts of our example steak are not going to to reach temperature simultaneously. Because grill environments are hotter than ovens, the temperature delta between the environment and the food is larger, so foods cooked on the grill will heat up more quickly and have a steeper temperature gradient.

Carryover

Carryover in cooking refers to the phenomenon of continued cooking once the food is removed from the source of heat. While this seems to violate a whole bunch of laws of thermodynamics, it's actually straightforward: the outer portion of the just-cooked food is hotter than the center portion, so the outer portion will transfer some of its heat into the center. You can think of it like pouring hot fudge sauce on top of ice cream: even though there's no external heat being added to the system, the ice cream melts because the hot fudge raises its temperature.

The amount of carryover depends upon the mass of the food and the heat gradient, but as a general rule, I find carryover for small grilled items is often about 5°F / 3°C. When grilling a steak or other "whole muscle" meat, pull it when it registers a few degrees lower at its core than your target temperature and let it rest for a few minutes for the heat to equalize.

To see how this works, try using a kitchen probe thermometer to record the temperature of a steak after removing it from the grill once it reaches 140°F / 60°C, recording data at 30-second intervals. You should see the core temperature peak at around 145°F / 63°C three minutes into the rest period for a small steak.

Simple Seared Steak

Get a cast iron pan good and hot over medium-high heat. Take a steak that's about 1" / 2.5 cm thick, rub lightly with olive oil, and sprinkle with salt and pepper. Drop the steak onto the cast iron pan and let it cook for two minutes. (Don't poke it! Just let it sit and sear.) After two minutes, flip and let cook for another two minutes. Flip again, reduce heat to medium and cook for five to seven minutes, until the center is about 135°F / 57°C. Let rest on cutting board for five minutes before serving.

Methods of Heat Transfer

There are three methods of transferring heat into foods: conduction, convection, and radiation. While the heating method doesn't change the temperature at which chemical reactions occur, the rate of heat transfer is different among them, meaning that the length of time needed to cook identical steaks via each method will be different. The table below shows the common cooking techniques broken out by their primary means of heat transfer.

Conduction

Conduction is the simplest type of heat transfer to understand because it's the most common: it's what you experience whenever you touch a cold countertop or grasp a warm cup of coffee. In cooking, those methods that transfer heat by direct contact between food and a hot material, such as the hot metal of a skillet, are conduction methods. Dropping a steak onto a hot cast iron pan, for example, causes thermal energy from the skillet to be transferred to the colder steak as the neighboring molecules distribute kinetic energy in an effort to equalize the difference in temperature. For more on thermal conductivity, see the "Metals, Pans, and Hot Spots" sidebar on 59 in Chapter 2.

	Conduction	Convection	Radiation
Description	Heat passes by direct contact between two materials	Heat passes via movement of a heated material against a colder material	Heat is transferred via electromagnetic radiation
Example	Steak touching pan; pan touching electric burner	Hot water, hot air, or oil moving along outside of food	Infrared radiation from charcoal
Uses	Sautéing Searing	*Dry heat methods*: - Baking/roasting - Deep-fat frying *Wet heat methods*: - Boiling - Braising/water bath - Pressure cooking - Simmering/poaching - Steaming	Microwaving Broiling Grilling

Cooking methods listed by type of heat transfer.
(Frying is a dry-heat method because it does not involve moisture.)

Convection

Convection methods of heat transfer—baking, roasting, boiling, steaming—all work by circulating a hot material against a cold one, causing the two materials to undergo conduction to transfer heat. In baking and roasting, the hot air of the oven imparts the heat; in boiling and steaming, it's the water that does this.

Those heat methods that involve water are called wet heat methods; all others fall into the dry heat category. One major difference between these two categories is that wet methods don't reach the temperatures necessary for Maillard reactions or caramelization (with the exception of pressure cooking, which does get up to temperature while remaining moist). The flavorful compounds produced by Maillard reactions in grilled or oven-roasted items won't be present in braised or stewed foods: steamed carrots, for example, won't undergo any caramelization, leaving the food with a subtler flavor. Brussels sprouts are commonly boiled and widely hated. Next time you cook them, quarter them, coat with olive oil and sprinkle with salt, and cook them under a broiler set to medium.

Water is an essential material in cooking, and not just for its heat transfer properties. Rice cookers work by noticing when the temperature rises above 212°F / 100°C. At that point, there's no water left, so they shut off.

Another key difference between most of the dry versus wet methods is the higher speed of heat transfer typical in wet methods. Water conducts heat roughly 23 times faster than air (air's coefficient of thermal conductivity is 0.026, olive oil's is 0.17, and water's is 0.61), which is why hard-cooked eggs finish faster in a wet environment even at a lower temperature.

Try it! Cook one egg for 30 minutes in a 325°F / 165°C oven and another for 10 minutes in a 212°F / 100°C water bath. You need to leave the egg in the oven for 20 minutes longer to get the same results.

One exception to this wet-is-faster-than-dry rule is deep-fat frying. Oil is technically dry (there's no moisture present), but for culinary purposes it acts a lot like water: it has a high rate of heat transfer with the added benefit of being hot enough to trigger a large number of caramelization and Maillard reactions. (Mmm, donuts!)

Wet methods have their drawbacks (including, depending on the desired result, the lack of the aforementioned chemical reactions). While the subtler flavors achieved without browning reactions can be desirable, as in a gently cooked piece of fish, it's also much easier to overcook foods with wet methods. When cooking meat, the hot liquid interacting with it can quickly raise its temperature above 160°F / 71°C, the point at which a significant percentage of the actin proteins in meats are denatured, giving the meat a tough, dry texture. For pieces of meat with large amounts of fat and collagen (such as ribs, shanks, or poultry legs), this isn't as much of an issue, because the fats and collagen (which converts to gelatin) will mask the toughness brought about by the denatured actin. But for leaner cuts of meat, especially fish and poultry, take care that the meat doesn't get too hot! The trick for these low-collagen types of meats is to keep your liquids at a gentle simmer, around 160°F / 71°C, and minimize the time that the meat spends in the liquid.

Even water in its gaseous form—steam—can pack a real thermal punch. While it doesn't conduct heat anywhere nearly as quickly as water in its liquid form, steam gives off a large amount of heat due to the phase transition from gas to liquid, something that air at the same temperature doesn't do. As the steam comes into contact with colder food, it condenses, giving off 540 calories (not to be confused with "food calories," which are technically kilocalories) of energy per gram of water, causing the food to heat up that much more quickly (1 calorie raises the temperature of 1 gram of water by 1°C).

Steamed vegetables, for example, cook quickly not just because they're in a 212°F / 100°C environment, but also because the water vapor condensing on the food's surface imparts a lot of energy. Cheetos, like most "extruded brittle foams" we eat, gain their puff by being expelled under pressure and heat, which causes them to "steam puff." (Think of it as the industrial version of popping popcorn.) There's a lot of energy in steam. For this reason, when pouring boiling water through a colander over a sink, you should be sure to pour away from yourself so that the steam cloud (and any splashed liquid) doesn't condense on your face!

Radiation

Radiant methods of heat transfer impart energy in the form of electromagnetic energy, typically microwaves or infrared radiation. The warmth you feel when sunlight hits your skin is radiant heat.

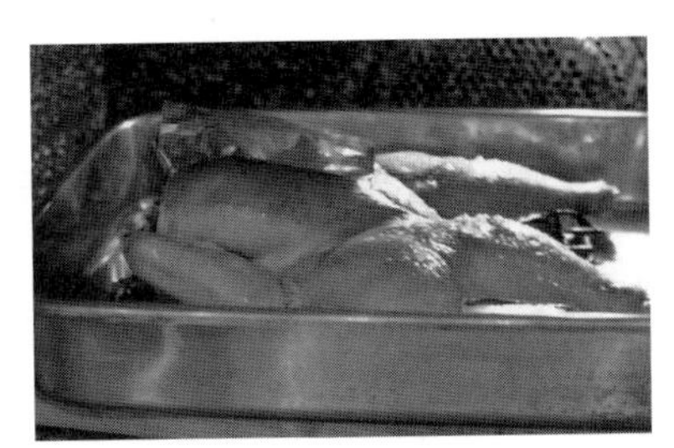

You can create a "heat shield" out of aluminum foil if part of a dish begins to burn while broiling. The aluminum foil will reflect the thermal radiation.

In cooking, radiant heat methods are the only ones in which the energy being applied to the food can be either reflected or absorbed by the food. You can use this reflective property to redirect energy away from parts of something you're cooking. One technique for baking pie shells, for example, includes putting foil around the edge, to prevent overcooking the outer ring of crust. Likewise, if you're broiling something, such as a chicken, and part of the meat is starting to burn, you can put a small piece of aluminum foil directly on top of that part of meat. It might be a hack, but in a pinch it's a decent way to avoid burning part of a dish, and nobody but you, me, and everyone else who reads this book will ever know.

Combinations of heat

The various techniques for applying heat to food differ in other ways than just the mechanisms of heat transfer. In roasting and baking we apply heat from all directions, while in searing and sautéing heat is applied from only one side. This is why we flip pancakes (stovetop, heat from below) but not cakes (oven, heat from all directions). The same food can turn out vastly different under different heat conditions. Batter for pancakes (conduction via stovetop) is similar to that for muffins (convection via baking) and waffles (conduction), but the end result differs widely.

To further complicate things, most cooking methods are actually combinations of different types of heat transfer. Broiling, for example, primarily heats the food via thermal radiation, but the surrounding air in the oven also heats up as it comes in contact with the oven walls, then comes in contact with the food and supplies additional heat via convection. Likewise, baking is primarily convection (via hot air) but also some amount of radiation (from the hot oven walls). "Convection ovens" are nothing more than normal ovens with a blower inside to help move the air around more quickly. All ovens are, by definition, convection ovens, in the sense that heat is transferred by the movement of hot air. Adding a fan just moves the air more quickly, leading to a higher temperature difference at the surface of the (cold) food you're cooking.

To a kitchen newbie, working with combinations of heat might be frustrating, but as you get experience with different heat sources and come to understand how they differ, you'll be able to switch methods in the middle of cooking to adjust how a food item is heating up. For example, if you like your lasagna like I do—toasty warm in the middle and with a delicious browned top—the middle needs to get hot enough to melt the cheese and allow the flavors to meld, while the top needs to be hot enough to brown. Baking alone won't generate much of a toasted top, and broiling won't produce a warm center. However, baking until it's almost done and then switching to the broiler achieves both results.

> The convenience food industry cooks with combinations of heat, too, cooking some foods in a hot oven while simultaneously hitting them with microwaves and infrared radiation to cook them quickly.

When cooking, if something isn't coming out as you expect—too hot in one part, too cold in another—check to see whether switching to a different cooking technique can get you the results you want.

If you're an experienced cook, try changing heat sources as a way of creating a challenge for yourself: adapt a recipe to use a different source of heat. In some cases, the adaptation is already common—pancake batter, when deep-fat fried, is a lot like funnel cakes. But try pushing things further. Eggs cooked on top of rice in a rice cooker? Chocolate cookies cooked in a waffle iron? Fish cooked in a dishwasher? (See 338.) Why not?

It might be unconventional, but heat is heat. Sure, different sources of heat transfer energy at different rates, and some are better suited to transitioning the starting thermal gradient (edge to center) of the food to the target thermal gradient. But there are invariably similar enough heat sources worth trying. And you can push it pretty far: fry an egg on your CPU, or cook your beans and sausage on an engine block like some long-haul truck drivers do! As a way of getting unstuck—or just playing around—it's fun to try.

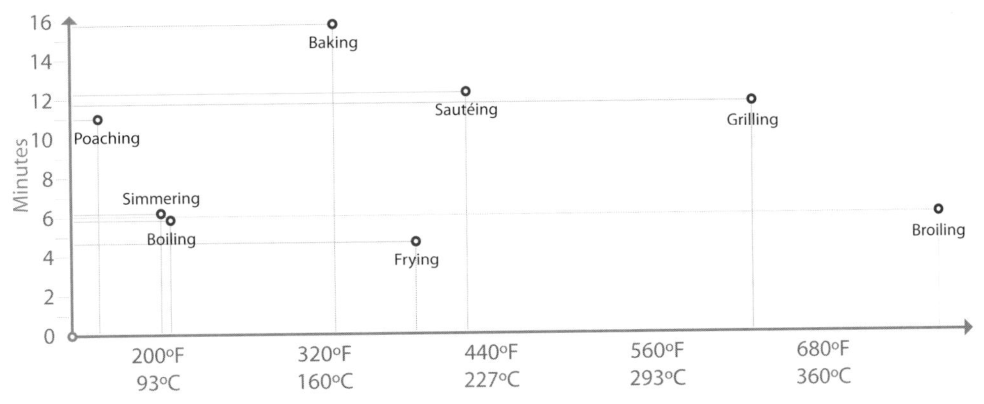

Cooking methods plotted by rate of heat transfer. This plot shows the amount of time it took to heat the center of uniformly sized pieces of tofu from 40°F / 4°C to 140°F / 60°C for each cooking method. Pan material (cast iron, stainless steel, aluminum) and baking pan material (glass, ceramic) had only minor impact on total time for this experiment and are not individually listed.

Foodborne Illness and Staying Safe*

**Well, safer—there's no such thing as 100% safe.*

The American food supply is one of the most interconnected and interdependent ones in the world. As I write this, I'm eating my morning bowl of cereal, yogurt, bananas, and almonds. The muesli cereal is from Switzerland, the yogurt local to New England, the bananas from Costa Rica, and the almonds from California. The only direction from which food hasn't come 3,000 miles is north, and that's probably only because not much grows at the North Pole!

As our food system has become more interconnected, the number of people that can be affected by a mistake in handling food has also increased. Today, a single bad batch of water sprayed onto a field of spinach can sicken hundreds of American consumers because that crop can be transported thousands of miles and make its way into so many dishes before the contamination is noticed.

Handling food carefully—taking note of what has been washed in the case of produce and cooked in the case of meats, and being careful to avoid cross-contamination—is among the easier ways of keeping yourself healthy.

Bacteria related to common foodborne illnesses begin to multiply above 40°F / 4.4°C. The standard food safety rule provided by the FDA for mitigating foodborne illnesses from bacteria states that *food should not be held between the temperatures of 40°F / 4.4°C and 140°F / 60°C for more than two hours*. Below 40°F / 4.4°C, the bacteria remain viable but won't have a chance to multiply to a sufficient quantity to bother us. Above 140°F / 60°C, the bacteria won't be able to survive long. (Bacterial spores, however, can.)

This is called the "danger zone rule," and as you'd probably imagine, a vast simplification of what's really going on in the bacterial world. Still, as an easy safety rule there's really no reason to violate it, because there are few dishes that I can think of that actually need to violate it to be made.

For those recipes that say to marinate meat at room temperature: don't! Let it marinate in the fridge.

Keep in mind that it's cumulative time here that matters. Say you buy a chicken at the store, and that it was kept chilled the entire time before you picked it up. Between the time you put it in the cart and when you stick it in your fridge, it'll be in a warmer environment, and any time it spends above the temperature at which bacteria begin to multiply will increase the bacterial count in the meat.

While cooking food kills off most of the bacteria, a minor (yet safe) number can survive even post-cooking. Given the right temperature range, they can reproduce back up into unsafe quantities. When cooking, stick any leftovers in the fridge right away, as opposed to letting them sit around until post-meal cleanup. The bacterial level is all about exposure—the amount of time and rate of multiplication at a given temperature.

This is why you should defrost large pieces of meat in the fridge overnight. Letting it thaw, even under cold running water, can take too long to be safe—unless your cold water happens to be below 40°F / 4.4°C!

One detail this rule glosses over is that some bacteria can reproduce at lower temperatures. Luckily, most bacteria related to foodborne illness don't multiply very quickly at near-freezing temperatures, but other types of bacteria do. Spoilage-related bacteria, for example, are happy breeding down to freezing temperatures. These are the ones that cause milk to go bad even below 40°F / 4.4°C and break down the flesh in things like raw chicken, causing uncooked meats to go bad after a few days. The danger zone rule addresses only the common pathological bacteria, which don't reproduce very quickly at the temperature of your fridge.

Another area that the danger zone rule glosses over is the different rates of reproduction at different temperatures. Salmonella, for example, is happiest breeding around 100°F / 37.8°C. It's not like the bacteria go from zero multiplication at 40°F / 4.4°C to full-on party mode at 41°F / 5°C; it's a gradual ramp up to an ideal breeding temperature. The two-hour window is given for the worst-case scenario: that the food is being held at the ideal breeding temperature for the most aggressive of the common bacteria, *Bacillus cereus*.

Since food safety codes are currently adopted on a state-by-state level, some states still use a danger zone rule of "40 to 140 for four hours," on the basis that *B. cereus* accounts for only a minor amount of foodborne illness and that a four-hour exposure isn't likely to produce much risk. If you're getting the impression that food safety is a probability game, you're right. The rules reduce the odds to an acceptable level. Still, that lunch you took to work and forgot to toss in the fridge is probably safe, given that the total amount of bacterial multiplication is likely to be well below the level necessary to trigger any sort of foodborne illness.

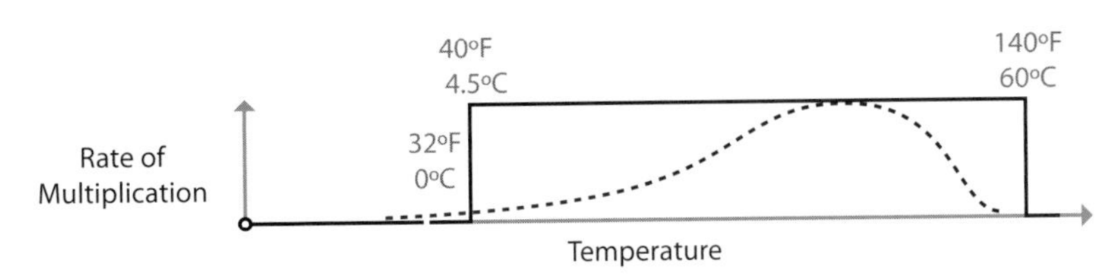

The danger zone rule suggests that bacteria multiply right up until 140°F / 60°C, whereas the real multiplication rates of foodborne illness-related bacteria follow a curve with an ideal breeding range in the middle. Of the foodborne pathogens listed in the FDA's "Bad Bug Book," Bacillus cereus *has the highest survival temperature, at 131°F / 55°C.*

"But wait," you might be saying. "What about all that food in the pantry? Why doesn't it go 'bad'?" There are other factors that bacteria need in order to multiply. The acronym FAT TOM is commonly used to describe the six factors necessary for multiplication:

F = Food

Bacteria need proteins and carbohydrates to multiply. No food, no multiplication.

A = Acidity

Bacteria can only survive in certain pH ranges. Too acidic or basic, and the proteins in the bacteria denature.

T = Temperature

Too cold, and the bacteria effectively sleep. Too hot, and they die.

T = Time

Bacteria have to have enough time to multiply to a sufficient quantity to overwhelm our bodies.

O = Oxygen

As with pH levels, bacteria will reproduce only if sufficient oxygen is present, or, for anaerobic bacteria (e.g., *Clostridium botulinum*), if no oxygen is present. Keep in mind that vacuum-packed bags are not necessarily devoid of oxygen.

M = Moisture

Bacteria need water to reproduce. Food scientists use a scale called *water activity*, which is a measure of the freely available water in a material (from 0 to 1). Bacteria need a water activity value of 0.85 or greater to multiply.

The reason so many pantry goods are "shelf stable" is either low moisture content (crackers, dry goods like beans and grains, oils, even jams and jellies in which the sugar is hygroscopic and "holds on to" the water) or acidity (pickled items, vinegars). Given these six variables, you can see why some foods don't need refrigeration. When in doubt, though, stick it in the fridge, which you should keep on the chilly side (34–36°F / 1–2°C).

Keep basic physics in mind. Placing a large pot of hot soup into the fridge will warm up all the contents of the fridge until the evaporator has a chance to transfer the heat back out. Refrigerators are made to keep things cool, not to chill things, so when you're storing a large quantity of hot food, place it in an ice bath first to chill it and then transfer it to the fridge once it's cooled down.

There's one variable that deserves special mention, because of the potentially fatal consequences, and that's the oxygen level needed for bacterial growth. Specifically, anaerobic environments—ones *without* oxygen—are necessary for some types of bacteria to multiply. Oil creates an anaerobic environment, but by itself doesn't provide any moisture for bacteria to grow. But with the addition of something like a raw clove of garlic, an anaerobic environment is created, and the garlic provides the food and moisture necessary for anaerobic bacteria to thrive when given sufficient time at the right temperatures.

Oil, even though a liquid, is technically dry because there is no water present.

A garlic clove stored in oil can become the perfect breeding ground for botulism, the illness caused by *Clostridium botulinum*. And since *C. botulinum* doesn't produce any noticeable odor, there's little to tip you off that the food is teeming with bacteria and their toxins. The toxins generated by *C. botulinum* are much more heat stable than the bacteria themselves and might remain active through the cooking process. And now to scare you senseless, the botulinum toxin is the most acutely toxic substance known: a dose as small as ~250 nanograms—1/120,000th the weight of a grain of rice—will do you in.

Botox is made from the botulinum toxin.

If you're going to make something like duck confit or homemade jam the traditional way—cooking it and then sealing it for storage at room temperature—make darn sure you get the food hot enough to sterilize it and that you avoid recontaminating it after it's cooked and before it's sealed. To be safe, stick it in the fridge and treat it as perishable. Note that sterilization means completely eradicating any bacteria, as opposed to pasteurization, which reduces bacteria to "safe" levels for near-term consumption. Sterilized milk can be left at room temperature indefinitely; pasteurized milk cannot.

There's one last aspect I should mention in this brief primer on food safety, and that's the risk/consequences equation. While following food safety rules reduces the risk of illness, it doesn't completely eliminate risk. For most of us, the consequence of contracting most foodborne illnesses is *gastrointestinal distress*—diarrhea, vomiting, muscle spasms, and the like. However, for those who are in an *at-risk group*—anyone for whom having foodborne illness can lead to further complications—the consequences of a bout of foodborne illness can be much, much greater, so the acceptable risks are accordingly much lower. If you are cooking for someone who is elderly, extremely young, pregnant, or immunocompromised, be extra vigilant with regard to food safety issues and skip dishes that have higher risks (e.g., raw egg in Caesar salads, unpasteurized cheeses, suspect meats that might be past their expiration dates).

For additional information on food safety and bacteria, check out the FDA's "Bad Bug Book," available online at *http://www.fda.gov/Food/FoodSafety/FoodborneIllness/*. Also, Texas A&M University's Department of Horticultural Sciences maintains a good overview page at *http://aggie-horticulture.tamu.edu/extension/poison.html*.

While bacteria are the most common and easiest-to-manage cause of foodborne illnesses, they're not alone: viruses, molds, toxins, and contaminants are also of concern. Toxins and contaminants are primarily issues for producers of foods, so as a consumer you're (mostly) off the hook for those. (If you grow your own veggies, test your soil for contaminants.)

Besides proper cooking, the best way to combat bacteria, mold, and viruses is good washing and avoiding cross-contamination:

- Wash your hands! And don't double-dip with the same spoon when tasting a dish.
- Nuke your sponges (rinse them and give them two minutes on high) or run them through the dishwasher weekly. Better yet, use towels and allow them to dry completely between uses.
- When working with raw meats, be careful not to use the same towel for wiping your hands that you then use to wipe with post-cooking.
- Wash can lids before opening them. Wash the can opener, too. The blade picks up food while cutting through the lid.

How to Prevent Foodborne Illness Caused by Bacteria

Salmonella: it's the poster child of foodborne illness. And for good reason. Salmonella accounts for a full 30% of the roughly 1,800 deaths due to foodborne illnesses per year in the U.S.—more than any other cause in this category. Salmonella's ideal breeding temperature? Around 100°F / 38°C—close to body temperature. Clearly it likes us. And according to some reports, the most likely food that'll harbor the bacteria in our modern food supply isn't chicken or meats, but produce. Wash your veggies!

The odds of dying from foodborne illness are actually surprisingly low, especially considering the media attention given to it. But the media attention isn't unmerited: 1 in 8 of us will contract illnesses from foods in any given year, and about 1% of those cases will require hospitalization, according to the U.S. Centers for Disease Control and Prevention (CDC).

Contracting a foodborne illness is a game of probabilities: a single bacterium of salmonella isn't likely to cause a problem, but given a few dozen cells, the odds change. *E. coli* is similar: only a few bacteria are necessary for the possibility of infection. A few strains are decidedly nasty, O157:H7 being the most talked about.

It's not always "just a handful" of cells, though. Contracting listeriosis requires ingesting somewhere around 1,000 organisms of *Listeria monocytogenes*, which tends to be present in animal products and multiplies at temperatures as low as 34°F / 1°C. Luckily, listeriosis isn't an issue for many of us, but it can cause complications for at-risk groups—especially pregnant women, where the baby is at risk. This is why pregnant women are told to avoid foods such as soft and surface-ripened cheeses, deli salads, raw milk, hot dogs, and shrimp; to ensure that chicken is thoroughly cooked; and to be careful with previously cooked ready-to-eat foods.

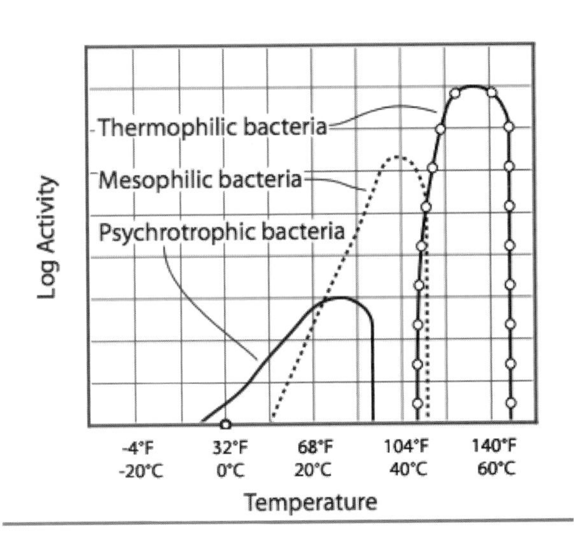

Bacteria can be grouped into three broad categories, based on the temperature at which they are most active. There are bacteria that remain active in food above 122°F / 50°C, but these are only beneficial (e.g., Bacillus coagulans*) or spoilage bacteria, and not related to foodborne illness. From a taste perspective, we're extremely lucky that no thermophilic bacteria cause foodborne illness; otherwise, we'd have to cook foods to higher temperatures to kill them.*

GRAPH BASED ON E. ANDERSEN, M. JUL, AND H. RIEMANN (1965), "INDUSTRIEL LEVNEDSMIDDEL-KONSERVERING," COL. 2, KULDEKONSERVERING, COPENHAGEN: TEKNISK FORLAG)

Salmonella gets most of the limelight in the media for a couple of reasons, though: it's hardy—that is, able to survive in the environment for longer periods of time and at temperatures above what most other common-to-food bacteria can tolerate—and it's surprisingly prevalent, affecting 1.4 million Americans a year on average.

Caliciviruses—a family of viruses, norovirus being the best known—are also getting more attention these days, and deservedly so; these are typically spread by a sick individual preparing food for others. If you've spent a night "praying to the porcelain god"—diarrhea, vomiting, chills, headache—you can probably thank salmonella or norovirus for the experience.

If you're sick, don't cook for others. If you're around someone sick, wash your hands. Often.

Now, pay attention, because this is important. *Salmonella is killed at 136°F / 58°C* ***only when held for a sufficient length of time***. Seeing your thermometer register an even hotter temperature—say, 140°F / 60°C—does not guarantee that the food will be free of salmonella. Think of it like being in a hot desert: you can survive 136°F / 58°C heat for a while, but if you're exposed to it for too long, eventually you will die. The same is true for bacteria like salmonella: given a short amount of time at a particular temperature, the bacteria might survive, but given a longer exposure, they will eventually die.

Salmonella actually lives in a temperature range of 35–117°F / 2–47°C according to the FDA's "Bad Bug Book." The 136°F / 58°C temperature is based on what the FDA Food Code gives as the lower bound for pasteurization.

Back to the desert analogy. Let's say an average human can survive for four hours in 136°F / 58°C heat. Given 100 people in a desert, though, this doesn't mean all 100 people will be alive at 3 hours, 59 minutes and all suddenly drop dead one minute later. The same is true for bacteria that might be hitching a ride on that chicken you're about to cook: the proteins in the bacteria don't all spontaneously denature at a specific temperature. It's a probability thing: as the temperature goes up, the probability of the molecular structure of each kind of protein denaturing increases. There's not an exact temperature at which this occurs, like there is when a solid melts into a liquid.

When talking about reducing the number of bacteria in food, scientists use the term *log_{10} reductions*. A single log_{10} reduction is simply the reduction of the number of bacteria present by a factor of 10; a 7 log_{10} reduction is a 10,000,000-fold reduction. The USDA's Food Safety and Inspection Service (FSIS) division is responsible for providing guidelines relating to the number of log reductions necessary to achieve an acceptable quantity of bacteria. Given that different kinds of meats have different properties—different amounts of fats, water, etc.—the number of log reductions necessary to reduce the bacterial count from a potential starting amount to an acceptable number differs. Hold time for sufficient pasteurization is also affected by variables such as how smooth the surface of the food is and its chemical composition (e.g., nitrite levels).

One important caveat about pasteurization: sometimes it's not the bacteria themselves that are the issue, but the toxins they produce. While appropriate cooking might safely reduce the bacterial count, the toxins themselves, such as those produced by *B. cereus*, can be heat-stable. Refrigeration of meats is therefore critical to prevent the multiplication of bacteria in the meat tissue. Remember the simple food safety rule mentioned earlier: avoid holding foods at temperatures between 40°F / 4°C and 140°F / 60°C for more than two hours. This includes the amount of time it takes to bring the food from fridge temperature to a safe hot temperature! While it is true that the 40-to-140°F / 4-to-60°C rule for two hours is a vast simplification of the real multiplication rates of bacteria, it's a simple rule accepted by the food industry, and rarely is there any need to skirt it.

Who said scientists don't have a sense of humor? Try saying *B. cereus* out loud.

At 140°F / 60°C, a hold time of 35 minutes is necessary for chicken with 12% fat to achieve a 7-$\log_{10}$ reduction. The time drops in leaner chickens; chicken meat with 1% fat requires 25.2 minutes at 140°F / 60°C. Longer hold times are okay; these times are ***minimum*** times. Chicken meats can be infected with salmonella throughout the tissue. While sick birds are supposed to be culled, it's still possible for them to go unnoticed. (Data from *http://www.fsis.usda.gov/OPPDE/rdad/FSISNotices/RTE_Poultry_Tables.pdf.*)

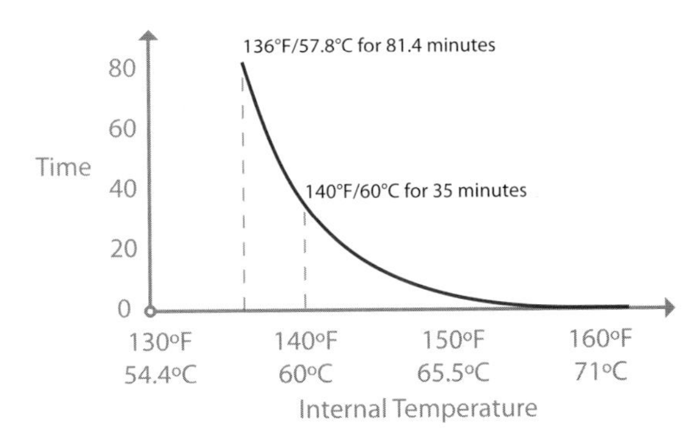

Minimum amount of time in minutes required to cook chicken safely (assuming 7-$\log_{10}$ reduction in chicken with 12% fat).

"Why then," I bet you're thinking, "do 'they' say to cook chicken to a temperature of 165°F / 74°C?" "They" happen to be the fine folks at the CDC, and what they say specifically is:

> *All poultry should be cooked to reach a minimum internal temperature of 165°F [74°C].*
>
> *—http://www.cdc.gov/nczved/divisions/dfbmd/diseases/campylobacter/*

Why 165°F / 74°C? One reason is that this is the temperature at which salmonella dies a quick death. From a "keeping it simple" perspective, seeing 165°F / 74°C on the thermometer is an easy guideline. Even if your thermometer is miscalibrated or you misprobe the meat and it's only reached a temperature of 155°F / 68°C, the pasteurization time for chicken at this temperature is less than a minute, which you're likely to exceed. The 165°F / 74°C guideline effectively removes the variable of time, making it an easier to follow (harder to screw up) rule.

Since none of the bacteria related to foodborne illness can survive, let alone reproduce, at moderate temperatures, holding food above 140°F / 60°C indefinitely is safe. This is why the soup at your local lunch counter can be kept hot all day long in a heat-controlled container and why hot buffets use steam baths to keep the foods warm. While you might be perplexed by the idea of storing foods hot, from a bacterial control perspective, it's actually safer than storing them in the fridge: bacteria are unable to survive in the hot environment, while storing them in the fridge generally only slows their reproduction.

The serving spoons, by the way, are supposed to stay in the food, so that they too stay above 140°F / 60°C. Otherwise, that mashed potato clinging to the serving spoon at room temperature will be a potential hangout spot for bacteria.

In the U.S., the FSIS and the FDA run testing programs to monitor the food supply. Both agencies have the ability to hold foods at processing plants, to request voluntary recalls, and to outright seize product through court order if it comes to that. Still, there's a lot of food going through the system, and lapses in protocol happen (probably more than we want to know about). A lot of work is done in identifying hazard points in the food system (HACCP—Hazard Analysis & Critical Control Points), but still, errors happen. What's a nervous food geek to do?

The most common vector for foodborne illness is surface contamination, either from contaminated water sprayed on vegetables during farming or from fecal contamination in meats during slaughter and processing. How does this affect you when cooking? Since it's the surface of most products that becomes contaminated, it's the surface that needs to be pasteurized. Pan searing a steak heats the outer portion well beyond any temperature that bacteria can survive. Likewise, steaming vegetables thoroughly heats their surface.

When cooking vegetables in the microwave, use a container with the lid mostly closed and with a small amount of water inside: the microwave will boil the water, and the container will keep the steam in contact with the vegetables.

What about hamburgers? Well, they're all outside, in the sense that surface contamination will have been ground throughout the meat. Industry calls things like steak *whole-muscle intact* meat, as opposed to ground meat. When looking at consumer cooking guidelines, the temperatures given are lower for whole-muscle intact than ground meats, presumably because the outside of a whole muscle cut will be well beyond pasteurized by the time the middle comes to temperature.

Simple Cheeseburger

In a clean bowl, work together using your fingers:

1 pound (500g) ground beef or hamburger

1 teaspoon (6g) Worcestershire sauce (optional)

1 teaspoon (5g) salt

½ teaspoon (1g) ground pepper (fresh, not preground)

Form into three or four patties. Using either a grill (radiant heat from below) or broiler (radiant heat from above), cook on each side for about 5 minutes, until the internal temperature registers at 160°F / 71°C.

If grilling, add cheese (try mild cheddar or Provolone) after flipping the first time. If broiling, add the cheese after reaching temperature and return to broiler for half a minute or so, until the cheese has melted.

Notes

- *Yes, you can haz cheezburger. Just cook it properly. Use a digital thermometer and make sure the internal temperature reaches 160°F / 71°C. You can pull it off the grill when it is a few degrees lower, because carryover will take it up to temperature.*
- *Fun fact: "hamburger" can have beef fat added to it; "ground beef" cannot.*

For a little light reading, pull up the FDA's 2009 Food Code (*http://www.fda.gov/Food/FoodSafety/RetailFoodProtection/FoodCode/*) and look at section 3-401.11: Raw Animal Foods.

When cooking a hamburger, the USDA says to heat the meat to 160°F / 71.1°C—high enough to kill any common bacteria but also high enough that both actin and myosin proteins will denature, leading to a drier burger. Since fats help mask dryness in meat, using ground beef that has more fat in it will lead to a juicier burger. Alternatively, if you have a way of cooking your burger to a lower temperature and then holding it at temperature long enough to pasteurize it, you could avoid denaturing the actin proteins while still pasteurizing the meat. Take a look at the section on sous vide cooking on page 333 of Chapter 7 for more on this.

Note that change in color is not an accurate indicator of doneness. Myoglobin, oxymyoglobin, and metmyoglobin can begin to turn grey starting around 140°F / 60°C, and they can also remain pink at 160°F / 71°C if the pH is at or about 6.0. Use a thermometer when cooking ground meats and poultry!

How to Prevent Foodborne Illness Caused by Parasites

Not long ago, I overheard the fishmonger at one of my local grocery stores (which shall remain nameless to protect the guilty) tell a customer that it was okay to use the salmon he was selling for making sushi. Given that the fish wasn't labeled as "previously frozen" and that it was in direct contact with other fish in the case, there wasn't any real guarantee that it was free from harmful parasites or bacteria, two of the biggest concerns that consumers need to manage for food safety. What's a shopper to do in response to the disappearance of the true fishmonger?

For one, start by understanding where the risks actually are. Not all fish and meats share the same set of risks for foodborne pathogens. Salmonella, for example, tends to show up in land animals and improperly handled vegetables, while bacteria such as *Vibrio vulnificus* show up in fish that are exposed to the brackish waters of tidal estuaries, such as salmon. Deep-water fish, such as tuna, are of less concern. Because of these differences, you should consider the source of your ingredients when thinking about food safety, focusing on the issues that are present in the particular food at hand.

With uncooked and undercooked fish, one concern is parasites. Parasites are to fish as bugs are to veggies: really common (if you've eaten fish, you've eaten worms). But on the plus side, most parasites in seafood don't infect humans. However, there are those that do, *Anisakis simplex* and tapeworms (*cestodes*) being the two parasites of general concern. *A. simplex* will give you abdominal pains, will possibly cause you to vomit and generally feel like crap, and will possibly take your doc a while to figure out. It's not appendicitis, Crohn's disease, nor a gastric ulcer, and with only around 10 cases diagnosed per year in the United States, chances are your doc won't have encountered it before. On the plus side, humans are a dead-end host for *A. simplex*. The parasite will die after about 10 days, at which point you'll go back to feeling normal. (Unless you have an extreme infection, in which case, it's off to surgery to remove 'em.) That leaves tapeworms as the major parasitic concern in fish.

For cooked dishes—internal temperature of 140°F / 60°C—there is little risk from these parasites directly. Cooking the fish also cooks the parasite, and while the thought of eating a worm might be unappetizing, if it's dead there's little to worry about other than the mental factor. (Just think of it as extra protein.)

Of course, raw and undercooked seafood is another matter entirely. Cod, halibut, salmon? Fish cooked rare or medium rare? Ceviche, sashimi, cold-smoked fish? All potential hosts for roundworm, tapeworms, and flukes. Fortunately, like most animals, few parasites can survive freezing.

Some parasites do survive freezing. *Trichomonas*—parasitic microorganisms that infect vertebrates—can survive temperatures as cold as liquid nitrogen. Yikes!

For the FDA to consider raw or undercooked fish safe to eat, it must be frozen for a period of time to kill any parasites that might be present:

> FDA 2005 Food Code, Section 3-402.11: *"[B]efore service or sale in ready-to-eat form, raw, raw-marinated, partially cooked, or marinated-partially cooked fish shall be: (1) Frozen and stored at a temperature of –20°C (–4°F) or below for a minimum of 168 hours (7 days) in a freezer; [or] (2) Frozen at –35°C (–31°F) or below until solid and stored at –35°C (–31°F) or below for a minimum of 15 hours…"*

The second concern with undercooked fish is bacteria. While freezing kills parasites, it does not kill bacteria; it just puts them "on ice." Researchers store bacterial samples at –94°F / –70°C to preserve them for future study, so even super-chilling food does not destroy bacteria. Luckily, most bacteria in fish can be traced to surface contamination due to improper handling—that is, cross-contamination from surfaces previously exposed to contaminated items.

Don't put cooked fish or meat on the same plate as the raw food! In addition to being potentially dangerous, that's just gross.

If your grocery store sells both raw and "sashimi-grade" fish, the difference between the two will be in the handling and care related to the chances of surface contamination, and in most cases the sashimi-grade fish should have been previously frozen. The FDA doesn't actually define what "sashimi grade" or "sushi grade" means, but it does explicitly state that fish not intended to be completely cooked before serving must be frozen before being served.

If you don't have access to a good fish market or find the frozen fish available at your local grocery store unappealing, and you plan on serving undercooked fish, you can kill any parasites present in the fish by freezing: check that your freezer is at least as cold as –4°F / –20°C, and follow the FDA rule of keeping the fish frozen for a week. If you happen to have a supply of liquid nitrogen around—you know, just by chance—you can also flash-freeze the fish, which should result in better texture and cut the hold time down to less than a day.

Luckily for oyster lovers, the FDA excludes molluscan shellfish, as well as some types of tuna and some farm-raised fish (those that are fed only food pellets that wouldn't contain live parasites) from the freezing requirement.

Doug Powell on Food Safety

PHOTOS USED BY PERMISSION OF DOUG POWELL

Doug Powell is an associate professor at Kansas State University's Department of Diagnostic Medicine and Pathobiology. His blog, "barfblog: musings about food safety and things that make you barf," is at http://www.barfblog.com.

Is there a tension between safety and quality in cooking, and if so, are there methods to achieve both?

Safety and quality are two very different things. Quality is something that people love talking about, whether it's wine, or organic food, or how it was

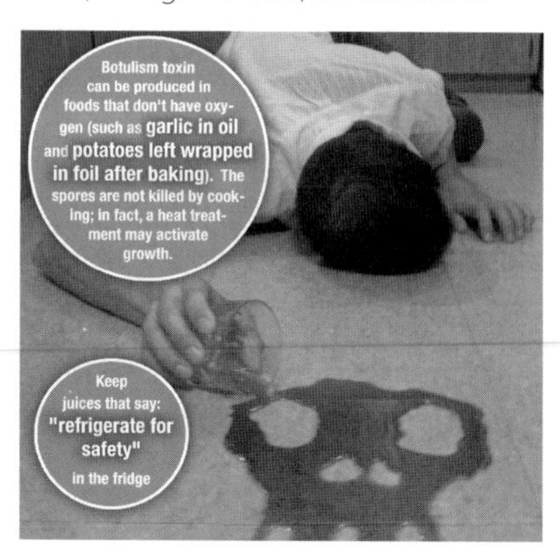

grown, and people can talk themselves to death about all that. My job is to make sure they don't barf.

For somebody cooking at home, it's easy for them to see a difference in quality. It's very hard for them to see a difference in safety until they get sick, I imagine?

There are tremendous nutritional benefits to having a year-round supply of fresh fruits and vegetables. At the same time, the diet rich in fruits and vegetables is the leading cause of foodborne illness in North America because they're fresh, and anything that touches them has the potential to contaminate. So how do you balance the potential for risk against the potential benefits? Be aware of the risks and put in place safety programs, beginning on the farm.

If you look at cancer trends in the 1920s, the most predominant cancers were stomach cancers. All everyone ate during the winter were pickles and vinegar and salt. Now that's almost completely eradicated because of fresh food. But now you have to prevent contamination from the farm to the kitchen, because more food is eaten fresh. There are tradeoffs in all of these things. In preparing hamburger and chicken, there is an issue with cooking it thoroughly and validating that with a thermometer, but most of the risk is actually associated with cross-contamination. Potatoes are grown in dirt, and birds crap all over them, and bird crap is loaded with salmonella and campylobacter. When you bring a potato into a kitchen or a food service operation, it's just loaded with bacteria that get all over the place.

What's the normal time between ingestion and symptoms?

It's around one to two days for salmonella and *E. coli*. For things like listeria, it can be up to two months. Hepatitis A is a month. You probably can't remember what you had yesterday or the day before, so how can you remember what you ate a month ago? The fact that any outbreak actually gets tracked to the source I find miraculous. In the past, if a hundred people went to a wedding or a funeral, they all had the same meal. They all showed up at emergency two days later, and they would have a common menu that investigators would look at to piece it together. Nowadays, through DNA fingerprinting, it's easier. If a person in Tennessee and a person in Michigan and a person in New York have gotten sick from something, they take samples and check against DNA fingerprints. There are computers working 24/7 along with humans looking to make these matches. And they can say these people from all across the country, they actually have the same bug, so they ate the same food.

Think of spinach contamination in 2006. There were 200 people sick, but it was all across the country. How did they put those together? Because they had the same DNA fingerprint and they were able to find the same DNA fingerprint in *E. coli* in a bag of spinach from someone's kitchen. Then they were able to find the same DNA fingerprint from a cow next to the spinach farm. It was one of the best cases with the most conclusive evidence. Normally, you don't have that much evidence.

What to do about it isn't very clear-cut, but when you look at most outbreaks, they're usually not acts of God. They're usually such gross violations of sanitation that you wonder why people didn't get sicker earlier. With a lot of fresh produce outbreaks, the irrigation water has either human or animal waste in it, and they're using that water to grow crops. These bugs exist naturally. We can take some regulatory precautions, but what are we going to do, kill all the birds? But we can minimize the impact.

When farmers harvest crops, they can wash them in a chlorinated water system that will reduce the bacterial loads. We know that cows and pigs and other animals carry these bacteria and they're going to get contaminated during slaughter. So we take other steps to reduce the risk as much as possible, because by the time you get it home and go to make those hamburgers, we know you're going to make mistakes. I've got a PhD, and I'm going to make mistakes. I want the number of bacteria as low as possible so that I don't make my one-year-old sick.

Is there a particular count of bacteria that is required to overwhelm the system?

It depends on the microorganism. With something like salmonella or campylobacter, we don't know the proper dose response curves. We work backward when there is an outbreak. If it's something like a frozen food, where they might have a good sample because it's in someone's freezer, we can find out more. With something like salmonella or campylobacter, it looks like you need a million cells to trigger an infection. With something like *E. coli* O157, you need about five.

You have to take into account the lethality of the bug. For 10% of the victims, *E. coli* O157 is going to blow out their kidneys and some are going to die. With listeria, 30% are going to die. Salmonella and campylobacter tend not to kill, but it's not fun. So all of these things factor into it. A pregnant woman is 20 times more susceptible to listeria. That's why they are warned to not eat deli meat, smoked salmon, and refrigerated, ready-to-eat foods. Listeria grows in the refrigerator and they're 20 times more susceptible and it can kill their babies. Most people don't know that either.

Are there any particular major messages that you would want to get to consumers about food safety?

It's no different than anything else, like drunk driving or whatever other campaign: be careful. The main message about food in our culture today is dominated by food pornography. Turn on the TV and there are endless cooking shows, and all these people going on about all these foods. None of that has anything to do with safety. You go to the supermarket today, you can buy 40 different kinds of milk and 100 different kinds of vegetables grown in different ways, but none of it says it's *E. coli*–free. Retailers are very reluctant to market on food safety, because then people will think, "Oh my god, all food is dangerous!" All they have to do is read a newspaper, and they'll know that food is dangerous.

A lot of the guidelines I see talk about the danger zone of 40–140°F / 4–60°C.

A lot of those guidelines are just complete nonsense. The danger zone is nice and it's important not to leave food in the danger zone, but at the same time it doesn't really get into any details. People learn by telling stories. Just telling people "Don't do this with your food" doesn't work; they say, "Yeah, okay, why?" I can tell you lots of stories of why or why not. The guidelines aren't changing what people do and that's why we do research on human behavior, how to actually get people to do what they're supposed to do. As Jon Stewart said in 2002, if you think those signs in the bathrooms ("Employees must wash hands") are keeping the piss out of your food, you're wrong! What we want to do is come up with signs that work.

I'm wondering what your signs look like?

We have some good ones! Our favorite picture is the skull in the bed of lettuce! The dead person from carrot juice is pretty good, too.

A Final Note on Food Safety

The safest way of preventing bacterial and parasitic infections from seafood and meats is with proper cooking. The USDA recommends cooking fish to a minimum internal temperature of 145°F / 63°C, ground beef to a minimum internal temperature of 160°F / 71°C, and poultry to 165°F / 74°C.

If you enjoy your fish cooked only to a rare point or even raw in the middle and you're concerned about parasites, give frozen fish a chance. I've found distinct differences in the quality of frozen fish. Some stores sell frozen product that's downright bad—mushy, bland, uninspiring—but this isn't *because* the fish was frozen. Some of the best sushi chefs in Japan are finding that quick-frozen tuna is exceptionally good. Frozen at sea right after it's caught (in a slurry of liquid nitrogen and dry ice), the tuna doesn't have much time to break down and so maintains its quality during transportation.

One last comment on keeping yourself safe in the kitchen: the biggest issue isn't contaminated food from the store, but cross-contamination while preparing it at home. Avoid cross-contamination by washing your hands often, especially both before and after working with raw meat. Use hot water and soap, and wash for a good 20 seconds.

Key Temperatures in Cooking

Most discussions of cooking are structured around the different heat transfer methods listed at the beginning of this chapter. Instead of looking at sources of heat, the rest of this chapter is going to take a different approach and talk about what reactions happen when each of the critical temperatures in the following table is reached, briefly touching on cooking techniques that relate to each temperature as they come up.

Temperature	What happens
104°F / 40°C and 122°F / 50°C	Proteins in fish and meat begin to denature
144°F / 62°C	Eggs begin to set
154°F / 68°C	Collagen denatures (Bovine Type I)
158°F / 70°C	Vegetable starches gelatinize
310°F / 154°C	Maillard reactions become noticeable
356°F / 180°C	Sugar (sucrose) begins to caramelize visibly

104°F / 40°C and 122°F / 50°C: Proteins in Fish and Meat Begin to Denature

Chances are, you haven't given much thought to the chemical reactions that happen to a piece of meat when the animal supplying it is slaughtered. The primary change is, to put it bluntly, that the animal is dead, meaning the circulatory system is no longer supplying the muscle tissue with glycogen from the liver or oxygen-carrying blood. Without oxygen, the cells in the muscle die, and preexisting glycogen in the muscle tissue dissipates, causing the thick and thin myofilaments in the muscle to fire off and bind together (resulting in the state called *rigor mortis*).

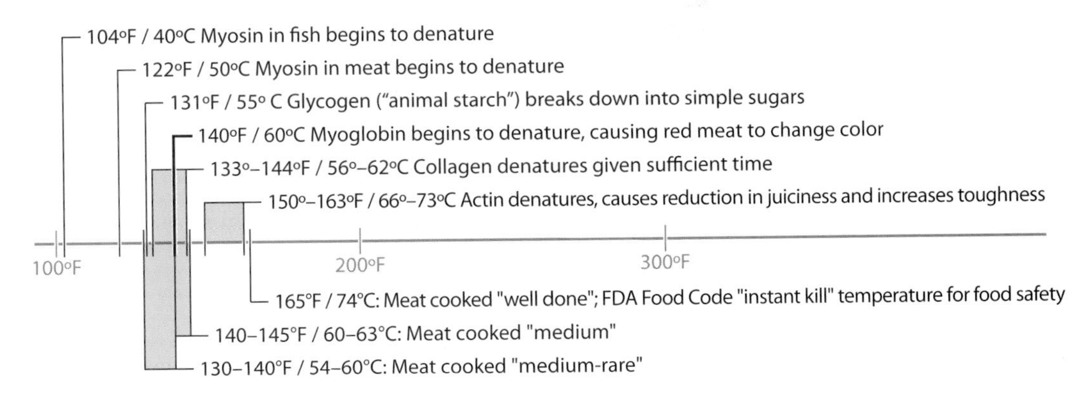

Denaturation temperatures of various types of proteins (top portion) and standard doneness levels (bottom portion).

Somewhere around 8 to 24 hours later, the glycogen supply is exhausted and enzymes naturally present in the meat begin to break down the bonds created during rigor mortis (*post-mortem proteolysis*). Butchering before this process has run its course will affect the texture of the meat. Sensory panels have found that chicken breasts cut off the carcass before rigor mortis was over have a tougher texture than meat left on the bone longer. And since time is money, much mass-produced meat is slaughtered and then butchered straightaway. (I *knew* there was a reason why roasted whole birds taste better!)

Proteins in meat can be divided into three general categories: myofibrillar proteins (found in muscle tissue, these enable muscles to contract), stromal proteins (connective tissue, including tendons, that provide structure), and sarcoplasmic proteins (e.g., blood). We'll talk about myofibrillar proteins here and save the stromal proteins for the section on collagen later in the chapter. (We're going to ignore sarcoplasmic proteins altogether, because understanding them doesn't help in cooking many dishes, blood-thickened soups aside.)

Muscle tissue is primarily composed of only a few types of proteins, with myosin and actin being the two most important types in cooking. About two-thirds of the proteins in mammals are myofibrillar proteins. The amount of actin and myosin differs by animal type and region. Fish, for example, are made up of roughly twice as much of these proteins as mammals.

Lean meat is mostly water (65–80%), protein (16–22%), and fat (1.5–13%), with sugars such as glycogen (0.5–1.3%) and minerals (1%) contributing only a minor amount of the mass. When it comes to cooking a piece of fish or meat, the key to success is to understand how to manipulate the proteins and fats. Although fats can be a significant portion of the mass, they are relatively easy to manage, because they don't provide toughness. This leaves proteins as the key variable in cooking meats.

Of the proteins present in meat, myosin and actin are the most important from a culinary texture perspective. If you take only one thing away from this section, let it be this: denatured myosin = yummy; denatured actin = yucky. Dry, overcooked meats aren't tough because of lack of water inside the meat; they're tough because on a microscopic level, the actin proteins have denatured and squeezed out liquid in the muscle fibers. Myosin in fish begins to noticeably denature at temperatures as low as 104°F / 40°C; actin denatures at around 140°F / 60°C. In land animals, which have to survive warmer environments and heat waves, myosin denatures in the range of 122–140°F / 50–60°C (depending on exposure time, pH, etc.) while actin denatures at around 150–163°F / 66–73°C.

Food scientists have determined through empirical research ("total chewing work" and "total texture preference" being my favorite terms) that the optimal texture of cooked meats occurs when they are cooked to 140–153°F / 60–67°C, the range in which myosin and collagen will have denatured but actin will remain in its native form. In this temperature range, red meat has a pinkish color and the juices run dark red.

The texture of some cuts of meat can be improved by tenderizing. Marinades and brines chemically tenderize the flesh, either enzymatically (examples include bromelain, an enzyme found in pineapple, and zingibain, found in fresh ginger) or as a solvent (some proteins are soluble in salt solutions). Dry aging steaks works by giving enzymes naturally present in the meat time to break down the structure of collagen and muscle fibers. Dry aging will affect texture for at least the first seven days. Dry aging also changes the flavor of the meat: less aged beef tastes more metallic, more aged tastes gamier. Which is "better" is a matter of personal taste preference. (Perhaps some of us are physiologically more sensitive to metallic tastes.) Retail cuts are typically 5 to 7 days old, but some restaurants use meat aged 14 to 21 days.

Soy Ginger Marinade

The salt in the soy sauce and zingibain in the ginger give this marinade both chemical and enzymatic tenderizers. Mix this up, transfer it to a resealable bag, and toss in some meat, such as flank steaks. Allow to marinate for an hour or two in the fridge, and then pan sear the meat.

1 cup (290g) soy sauce

2 tablespoons (15g) grated fresh ginger or ginger paste

1 teaspoon (2g) ground black pepper

Then there are the mechanical methods for "tenderizing," which aren't actually so much tenderizing as they are masking toughness: for example, slicing muscle fibers against the grain thinly, as is done with beef carpaccio and London broil, or literally grinding the meat, as is done for hamburger meat. (Some industrial meat processors "tenderize" meat by microscopically slicing it using very thin needles, a method called jacquarding.) Applying heat to meats "tenderizes" them by physically altering the proteins on the microscopic scale: as the proteins denature, they loosen up and uncurl. In addition to denaturing, upon uncurling, newly exposed regions of one protein can come into contact with regions of another protein and form a bond, allowing them to link to each other. This process is called *coagulation*, and while it typically occurs in cooking that involves protein denaturation, it is a separate phenomenon.

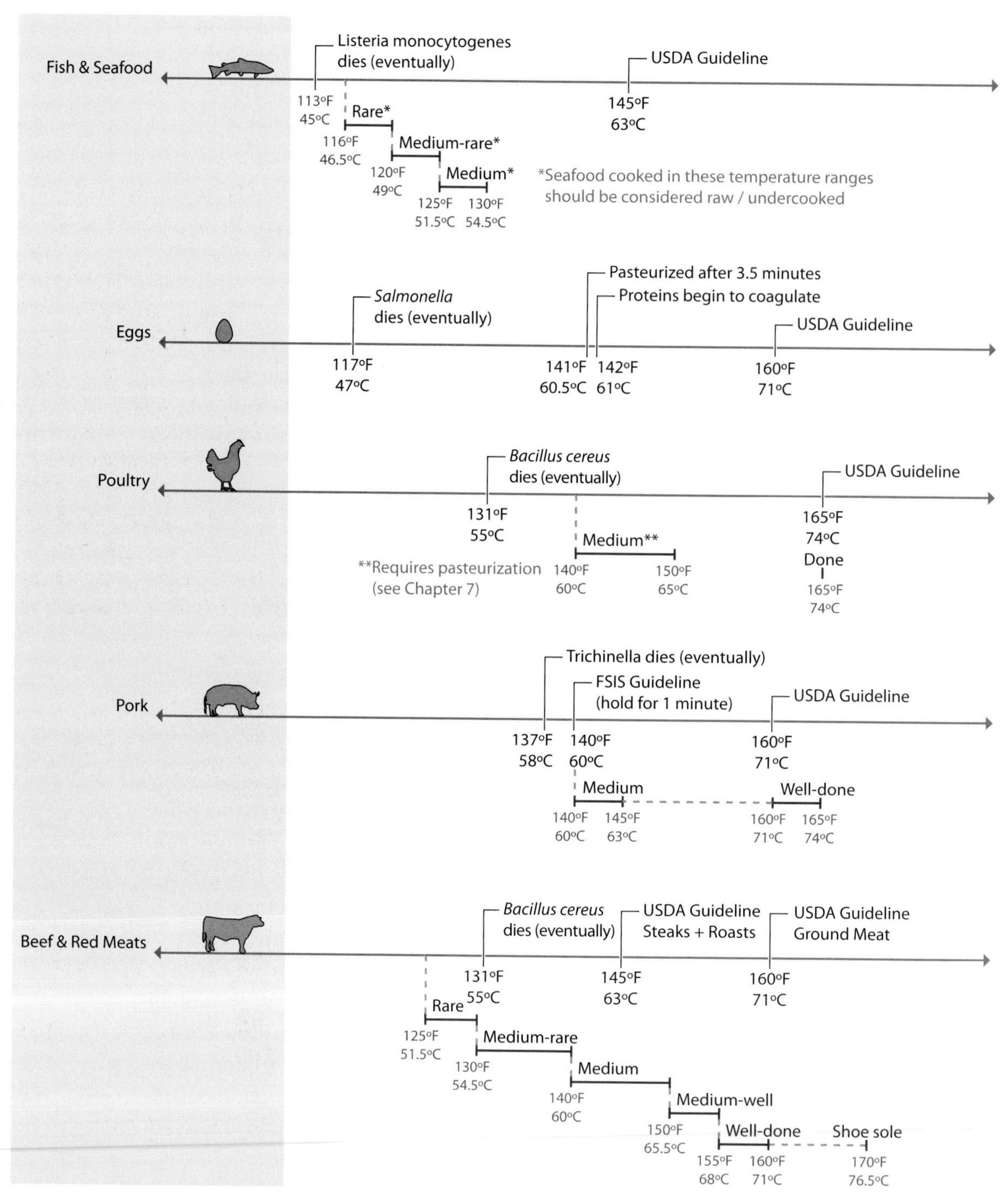

Temperatures required for various levels of doneness. Note that seafood cooked very rare or medium rare and chicken cooked medium must be held for a sufficiently long period of time at the stated temperature in order to be properly pasteurized. See the section on Sous Vide Cooking in Chapter 7 for time and temperature charts.

Salmon Poached in Olive Oil

Fish, such as salmon and Atlantic char, becomes dry and loses its delicate flavor when cooked too hot. The trick with poaching fish is to not overcook it. Poaching fish is an easy way to control the rate of heat being applied, and it is amazingly easy and tasty.

Place a fillet of fish, skin side down, in an oven-safe bowl just large enough for the fish to fit. Sprinkle a small amount of salt on top of fish. Cover with olive oil until the fillet is submerged. Using a bowl that "just fits" the fish will cut down on the amount of olive oil needed.

Place into a preheated oven, set to medium heat (325–375°F / 160–190°C).

Use a probe thermometer set to beep at 115°F / 46°C and remove the fish when the thermometer goes off, letting carryover bring the temperature up a few more degrees.

Consider this fish as raw/undercooked. See Chapter 7's section on sous vide cooking for a discussion on pasteurization and time-at-temperature rules.

Notes

- *Try serving on top of a portion of brown or wild rice and spooning sautéed leeks, onions, and mushrooms on top. (A squirt of orange juice in the leeks is really good.) Or serve with string beans sautéed with red pepper flakes and white rice, with a splash of soy sauce drizzled on top.*
- *Salmon contains a protein, albumin, that generates a white congealed mess on the outside of the flesh, as shown on the bottom piece in the following photo. This is the same protein that leaches out of hamburgers and other meats, typically forming slightly gray "blobs" on the surface. You can avoid this by brining the fish in a 5–10% salt solution (by weight) for 20 minutes, which will set the proteins. The top piece in the first photo below was brined; you can see the difference.*

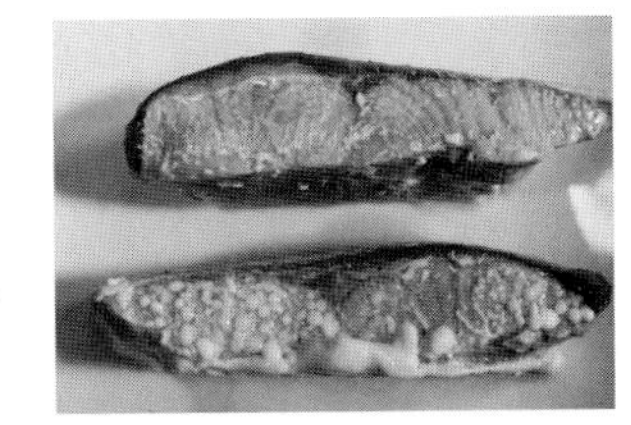

Salmon contains a protein, albumin, that is expressed out of the flesh and leads to an unsightly, curd-like layer forming on the surface of the fish when poached, as shown in the bottom piece in this photo.

If your fish doesn't fit in your pan, you can fold the tail bit over in a pinch, or cut it and poach it face down. This won't win you any Foodie Points, but as long as you don't take a photo and publish it in a book, who's going to know?

Time & Temperature: Cooking's Primary Variables

Seared Tuna with Cumin and Salt

Pan searing is one of those truly simple cooking methods that produces a fantastic flavor and also happens to take care of bacterial surface contamination in the process. The key to getting a rich brown crust is to use a cast iron pan, which has a higher thermal mass than almost any other kind of pan (see the "Metals, Pans, and Hot Spots" sidebar in Chapter 2). When you drop the tuna onto the pan, the outside will sear and cook quickly while leaving as much of the center as possible in its raw state.

Coat all sides of the tuna in cumin seeds and salt by pressing the tuna down onto a plate that has the spice mixture evenly spread out on it.

You'll need 3–4 oz (75–100 grams) of raw tuna per person. Slice the tuna into roughly equal-sized portions, since you'll be cooking them one or two at a time.

On a flat plate, measure out 1 tablespoon cumin seed and ½ teaspoon (2g) salt (preferably a flaky salt such as Maldon sea salt) *per piece of tuna*. On a second plate, pour a few tablespoons of a high-heat-stable oil, such as refined canola, sunflower, or safflower oil.

Place a cast iron pan on a burner set as hot as possible. Wait for the pan to heat up thoroughly, until it just begins to smoke.

Make sure the pan is really hot. Some smoke coming off the fish as it sears is okay!

For each serving of tuna, dredge all sides in the cumin/salt mix, and then briefly dip all sides in the oil to give the fish a thin coating.

Sear all sides of the fish. Flip to a new side once the current facedown side's cumin seeds begin to brown and toast, about 30 to 45 seconds per side.

Slice into ⅓" (1 cm) slices and serve as part of a salad (place fish on top of mixed greens) or main dish (try serving with rice, risotto, or Japanese udon noodles).

Notes

- *Keep in mind that the temperature of the pan will fall once you drop the tuna in it, so don't use a piece of fish too large for your pan. If you're unsure, cook the fish in batches.*
- *Use coarse sea salt, not rock (kosher) salt or the table salt you'd find in a salt shaker. The coarse sea salt has a large, flaky grain that prevents all of the salt from touching the flesh and dissolving.*

Pan-seared tuna will be well-done on the outside and have a very large "bull's eye" where the center is entirely raw.

144°F / 62°C: Eggs Begin to Set

The lore of eggs is perhaps greater than that of any other food item, and more than one chef has gone on record judging others based on their ability—or inability—to cook an egg. Eggs are the wonder food of the kitchen—they have a light part, a dark part, and bind the culinary world together. Used in both savory and sweet foods, they act as binders holding together meatloaf and stuffing; as rising agents in soufflés, certain cakes, and cookies like meringues; and as emulsifiers in sauces like mayonnaise and hollandaise. Eggs provide structure to custards and body to ice creams. And all of this so far doesn't even touch on their flavor or the simple joys of a perfectly cooked farm egg. Simply put, I cannot think of another ingredient whose absence would bring my cooking to a halt faster than the simple egg.

Egg whites are composed of dozens of different types of proteins, and each type of protein begins to denature at a different temperature. In their natural "native" state, you can think of the proteins as curled-up little balls. They take this shape because portions of the molecular structure are *hydrophobic*—the molecular arrangement of the atoms making up the protein is such that regions of the protein are electromagnetically repulsed by the polar charge of water.

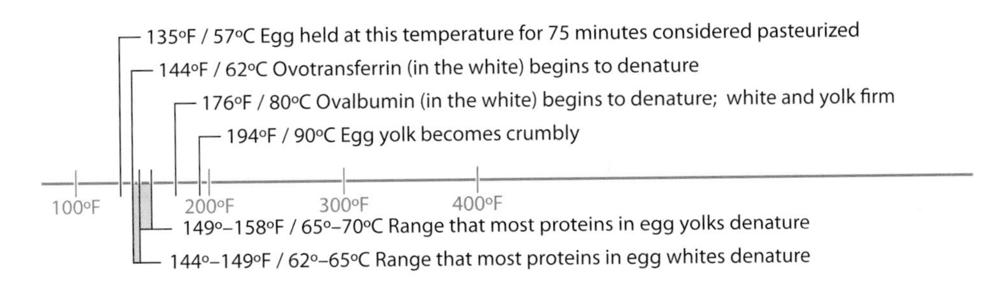

Important temperatures in eggs.

Because of this aversion to water, the protein structure folds up on itself. As kinetic energy is added to the system—in the form of heat or mechanical energy (e.g., whipping egg whites)—the structure starts to unfold as kinetic energy overtakes potential energy. The unfolded proteins then get tangled together, "snagging" around other denatured proteins and coagulating to form a linked structure. This is why a raw egg white is liquid, but once cooked becomes solid. (Well, technically, raw egg white is a gel that coagulates into a solid-like substance when heated. We'll get to gels in Chapter 6.)

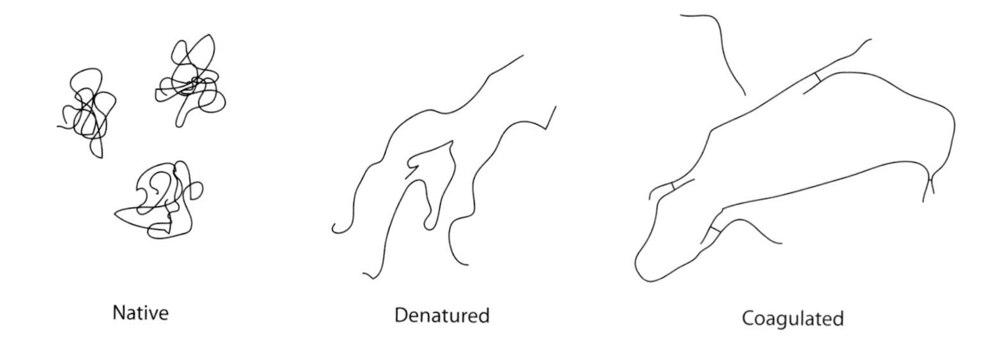

Hydrophobic proteins in their native state (left) remain curled up to avoid interacting with the surrounding liquid. Under heat, they denature (center) and uncurl as the kinetic energy exceeds the weaker level of energy generated by water molecules and regions of the proteins that repel each other. Once denatured and opened up, the hydrophobic parts of the protein that were previously unexposed can interact and bond with other proteins.

The most heat-sensitive protein is ovotransferrin, which begins to denature at around 144°F / 62°C. Another protein, ovalbumin, denatures at around 176°F / 80°C. These two proteins also are the most common in egg whites: ovotransferrin accounts for 12% of the proteins in an egg white and ovalbumin 54%. This explains the difference between soft-boiled and hard-boiled ("hard-cooked") eggs. Get that egg up to about 176°F / 80°C for sufficient time, and voilà, the white is hard cooked; below that temperature, however, the ovalbumin proteins remain curled up, leaving the majority of the egg white in its "liquid" state.

Most of the proteins in egg yolks set at between 149°F / 65°C and 158°F / 70°C, although some set at lower temperatures.

Proteins in foods such as eggs don't denature instantaneously once they reach denaturation temperature. This is an important point. Some cooking newbies have the mental model that cooking an egg or a piece of meat is something like melting an ice cube: all ice below a certain temperature, ice and water at the freezing/melting point, and all water above that temperature. From a practical perspective in the kitchen, it's not an entirely incorrect picture, because heat pours into the foods so quickly that the subtle differences between a few degrees aren't obvious. But as heat is transferred into the food more slowly, the subtleties of these chemical reactions become more noticeable. And unlike melting an ice cube, where increasing the heat transfer by a factor of two causes the ice to melt in half the time, cooking foods do not respond to additional energy in a linear fashion.

You might find it easiest to think of the different proteins in foods as having particular temperatures at which they denature, and try to shoot for a target temperature just above that of the proteins you do want denatured. Just remember: there's more to a piece of meat or egg than one type of protein or connective tissue, and the different proteins have different temperature points at which they're likely to denature.

Here are some examples of cooking eggs that show how to take advantage of the thermal properties of different portions of the egg.

Hard-Cooked Eggs, Shock and Awe Method

There's a silent war of PC-versus-Mac proportions going on over the ideal way to make hard-cooked eggs. Should you start in cold water and bring the water up to a boil with the eggs in them, or should you drop the eggs into already boiling water? The cold-start approach yields eggs that taste better, while the boiling-water approach yields eggs that are easier to peel. But can you have both?

Thinking about the thermal gradient from shell to center of egg, it would make sense that cooking an egg starting in cold water would result in a more uniform doneness. The delta between the center and outer temperatures will be smaller, meaning that the outer portion won't be as overcooked once the center is set compared to the boiling-water method.

The conjecture for ease of peeling in the boiling water approach is that the hot water "shocks" the outer portion of the egg. Into industrial-grade cooking? Steam 'em at 7.5 PSI over atmospheric pressure and quick-release the pressure at the end of cooking to crack the shell. (Hmm, I wonder if one could do this in a pressure cooker…) But what about the rest of us? What if we shock the outside, and then cook in cold water?

Try it. Place your eggs into rapidly boiling water. After 30 seconds, transfer the eggs to a second pot containing cold tap water, bring to a boil, and then simmer. The second-stage cooking time will take about two minutes less than the normal cold-start approach. Cook for 8 to 12 minutes, depending upon how well cooked you like your eggs.

The 30-Minute Scrambled Egg

This method involves ultra-low heat, continuous stirring, and a vigilant eye. I wouldn't suggest this as an everyday recipe, because it takes a while to make, but after however many years of eating eggs, it's nice to have them cooked a new way. Cooking the eggs over very low heat while continuously stirring breaks up the curds and allows for cooking the eggs to a point where they're just cooked, giving them a flavor that can be described as cheese or cream-like. It's really amazing, and while the thought of "cheese or cream-like" eggs might not have you racing off to the kitchen, it's really worth a try!

In a bowl, crack two or three eggs and whisk thoroughly to combine the whites and yolks. Don't add any salt or other seasonings; do this with just eggs. Transfer to a nonstick pan on a burner set to heat as low as possible.

Stir continuously to avoid hot spots so that the eggs are kept at a uniform temperature. If you have an IR thermometer, make sure your pan doesn't exceed 160°F / 71°C.

Stir continuously with a silicone spatula, doing a "random walk" so that your spatula hits all parts of the pan. And low heat means really low heat: there's no need for the pan to exceed 160°F / 71°C, because enough of the proteins in both the yolks and whites denature below that temperature and the proteins will weep some of their water as they get hotter. If your heat source is too hot, pull the pan off the stovetop for a minute to keep it from overheating. If you see any curds (lumps of scrambled eggs) forming, your pan is getting too hot.

Continue stirring until the eggs have set to a custard-like consistency. When I timed myself, this took about 20 minutes, but you might reach this point in as few as 15 minutes or upward of half an hour.

Oven-Poached Eggs

Here's a simple way to cook eggs for a brunch or appetizer. In an individually sized oven-safe bowl (ideally, one that you can serve in), add:

Breakfast version

1 cup (30g) fresh chopped spinach
3 tablespoons (20g) grated mozzarella cheese
3 tablespoons (40g) heavy cream
4 teaspoons (20g) butter

Dinner version

½ cup (100g) crushed tomatoes
¼ cup (50g) black beans (canned are easiest)
½ cup (50g) grated mozzarella cheese

Create a "well" in the center of the ingredients by pushing the food into a ring around the edges of the bowl. Crack two eggs into the well, add a pinch of salt and some fresh ground pepper, cover with aluminum foil, and bake in a preheated oven set to 350°F / 180°C until the egg is set, about 25 minutes. (You can use a probe thermometer set to beep at 140°F / 60°C.) Try adding some crushed red pepper flakes to the breakfast version or sriracha sauce to the dinner version.

Pasteurized Eggs

While salmonella is quite rare in uncooked eggs, with estimates being somewhere around 1 in 10,000 to 20,000 eggs carrying the bacteria, it does occur in the laying hen populations of North America. If you're cracking a few dozen eggs into a bowl for an omelet brunch at your local hacker house every week, let's just say that odds are you'll eventually crack a bad egg. Luckily, this isn't a problem if those eggs are properly cooked and cross-contamination is avoided.

The real risk for salmonella in eggs is in dishes that use undercooked eggs that are then served to at-risk populations (e.g., infants, pregnant women, elderly or immunocompromised people). If you're making a dish that contains raw or undercooked eggs—Caesar salad, homemade eggnog, mayonnaise, raw cookie dough—and want to serve that dish somewhere where there might be at-risk individuals, you can pasteurize the eggs (assuming your local store doesn't happen to carry pasteurized eggs, but most don't). Pasteurized eggs do taste a little different, and the whites take longer to whip into a foam, so don't expect them to be identical to their raw counterparts.

Since salmonella begins to die at a noticeable rate around 136°F / 58°C and the proteins in eggs don't begin to denature until above 141°F / 61°C, you can pasteurize eggs to reduce the quantity of salmonella, should it be present, to an acceptable level by holding the egg at a temperature between these two points. The FDA requires a 10,000-fold reduction (5 $\log_{10}$ in food safety lingo), which can be achieved by holding the egg at 141°F / 61°C for 3.5 minutes (according to Margaret McWilliams's *Foods: Experimental Perspectives,* Fifth Edition, from Pearson Publishing). Most consumers won't have the necessary hardware to do this at home, but if you do have a Sous Vide setup, as described in Chapter 7, you're golden.

The 60-Minute Slow-Cooked Egg

Going back to our earlier discussion of time and temperature, when food is left in an environment long enough, its temperature will come to match that of its environment. Therefore, if we immerse an egg in water held at 145°F / 62.7°C, it follows that the proteins in the white and the yolk that denature at or below that temperature will denature and coagulate, and those that denature above that temperature will remain unaltered.

The added benefit of this method is that the egg *cannot overcook*. "Cooking" is effectively the occurrence of chemical reactions in the food at different temperature points, and holding the egg at 145°F / 62.7°C will not trigger any reactions that don't occur until higher temperatures are reached. This is the fundamental concept of sous vide cooking. We'll cover the details of sous vide in Chapter 7, so you may want to take a peek at that chapter now or make a mental note to come back to this section when you get there. For a sous vide–style cooked egg, immerse an egg in water that is maintained at 145°F / 62.7°C for one hour. As you'll see, sous vide cooking has some incredible properties that greatly simplify the time and temperature rule.

Your average, run-of-the-mill (or is that run-of-the-yard?) chicken laid only 84 eggs per year a century ago. By the turn of the millennium, improvements in breeding and feed had pushed this number up to 292 eggs per year—almost 3.5 times more. And, no, science has not yet figured out which came first.

154°F / 68°C: Collagen (Type I) Denatures

An animal's connective tissues provide structure and support for the muscles and organs in its body. You can think of most connective tissues—loose fascia and ligaments between muscles as well as other structures such as tendons and bones—as a bit like steel reinforcement: they don't actively contract like muscle tissue, but they provide structure against which muscles can pull and contract.

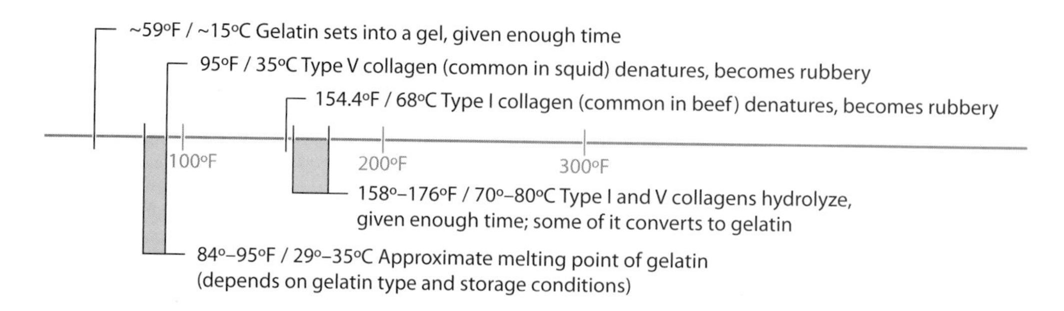

Temperatures related to collagen hydrolysis and the resulting gelatin.

Fun fact: pound-for-pound, collagen is tougher than steel.

The most common type of protein in connective tissue is collagen, and while there are several types of collagen in animals, from a culinary perspective, the main chemical difference between the different types of collagen is the temperature at which they denature. In cooking, collagen shows up in two different ways: either as discrete chunks (e.g., tendons, silverskin) outside of the muscle, or as a network that runs through the muscle. Regardless of its location, collagen is tough (it provides structure, after all) and becomes palatable only given sufficient time at sufficiently high temperatures.

It's easy to deal with collagen that shows up as discrete pieces: get rid of it by cutting it off. For cuts of meats that have a thin layer of connective tissue on them (called *silverskin*, presumably because of its somewhat iridescent appearance), cut off as much as possible and discard it. Beef tenderloin cuts commonly have a side with this layer; trim off as much as possible before cooking.

Chicken breasts also have a small but noticeable tendon connected to the chicken tenderloin. Uncooked, it's a pearlescent white ribbon. After cooking, it turns into that small white rubber-band-like thing that you can chew on endlessly yet never get any satisfaction from. Generally, this type of collagen is easy to spot, and if you miss it, it's easy to notice while eating and can be left on the plate.

However, for the other kind of collagen found in some cuts of meat—collagen that forms a 3D network through the muscle tissue—the only way to remove it is to convert it to gelatin via long, slow cooking methods. Unlike muscle proteins—which in cooking are either in a native (i.e., as they are in the animal), denatured, or hydrolyzed state—collagen, once hydrolyzed, can enter a coagulated (gelled) state. This property opens up an entirely new world of possibilities, because gelatin gives meats a lubricious, tender quality and provides a lip-smacking goodness.

In its native form, collagen is like a rope: it's a linear molecule composed of three different strands that are twisted together. The three strands are held together by weak secondary bonds (but there are a lot of them!) and stabilized by a small number of *crosslinks*, which are stronger covalent bonds.

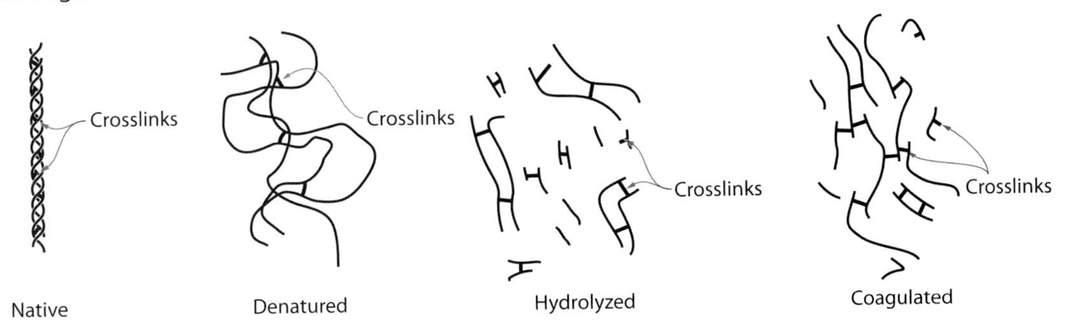

Collagen in its native form is a triple helix, held together in its helical structure by secondary bonds (left) and stabilized by crosslinks. Under heat, the secondary bonds break and the protein becomes denatured, but the crosslinks between the strands continue to hold the structure together (second from left). Given sufficient heat and time, the strands in the triple helix themselves break down via hydrolysis (third from left) and, upon cooling, convert to a loose network of molecules (right) that retains water (a gel).

Covalent bonds are bonds where the electrons from an atom in one location are shared with another atom.

In addition to being crosslinked, the strands also form a helical structure because of secondary bonds between different regions of the same molecules. You can think of it something like a braided rope, where each strand wraps around the other two strands. It has a "curl" to it because the internal structure finds its optimal resting place in that shape.

Under the right conditions—usually, exposure to heat or the right kinds of acids—the native form of collagen denatures, losing its linear structure and untwisting into a random mess. With the addition of sufficient heat, the molecules in the structure will vibrate enough to overcome the electromagnetic energy that caused the structure to twist up in the first place, leading it to lose its helical structure and denature.

Acids can also denature the collagen protein: their chemical properties provide the necessary electromagnetic pull to disrupt the secondary bonds of the helical structure. It's only the twisting that goes away during denaturing in collagen; the crosslinks remain in place and the strands remain intact. In this form, collagen is like rubber—it actually is a rubber from a material science point of view—and for this reason, you'll find its texture, well, rubbery.

Given even more heat or acid, though, the collagen structure undergoes *another* transformation: the strands themselves get chopped up and lose their backbone, and at this point the collagen has no real large-scale structure left. This reaction is called *hydrolysis*: thermal hydrolysis in the case of heat, acid hydrolysis in the case of, you guessed it, acid. (Think ceviche. See the section on Acids and Bases in Chapter 6 for more.)

It's possible to break up the collagen chemically, too: lysosomal enzymes will attack the structure and "break the covalent bonds" in chem-speak, but this isn't so useful to know in the kitchen.

For fun, try marinating a chunk of meat in papaya, which contains an enzyme, papain, that acts as a meat tenderizer by hydrolyzing collagen.

One piece of information that is critical to understand in the kitchen, however, is that hydrolysis takes time. The structure has to literally untwist and break up, and due to the amount of energy needed to break the bonds and the stochastic processes involved, this reaction takes longer than simply denaturing the protein.

Hydrolyzing collagen not only breaks down the rubbery texture of the denatured structure, but also converts a portion of it to gelatin. When the collagen hydrolyzes, it breaks into variously sized pieces, the smaller of which are able to dissolve into the surrounding liquid, creating gelatin. It's this gelatin that gives dishes such as braised ox tail, slow-cooked short ribs, and duck confit their distinctive mouthfeel.

Since these dishes rely on gelatin for providing that wonderful texture, they need to be made with high-collagen cuts of meat. Trying to make a beef stew with lean cuts will result in tough, dry meat. The actin proteins will denature (recall that this occurs at temperatures of 150–163°F / 66–73°C), but the gelatin won't be present in the muscle tissue to mask the dryness and toughness brought about by the denatured actin. Don't try to "upgrade" your beef stew with a more expensive cut of meat; it won't work!

"Great," you might be thinking, "but how does any of this tell me whether I need to slow-cook a piece of meat?" Think about the piece of meat (or fish or poultry) that you're working with and consider what part of the animal it comes from. For a land-based animal, those regions of the animal that bear weight generally have higher levels of collagen. This should make sense: because the weight-bearing portions have a higher load, they need more structure, so they'll have more connective tissue. This isn't a perfect rule of thumb, though, and cuts of meat generally have more than one muscle group in them.

For animals like fish, which don't have to support their weight on land, the collagen levels are much lower. Squid and octopus are notable exceptions to this weight-bearing rule, because their collagen provides the equivalent support that bone structures do for fish.

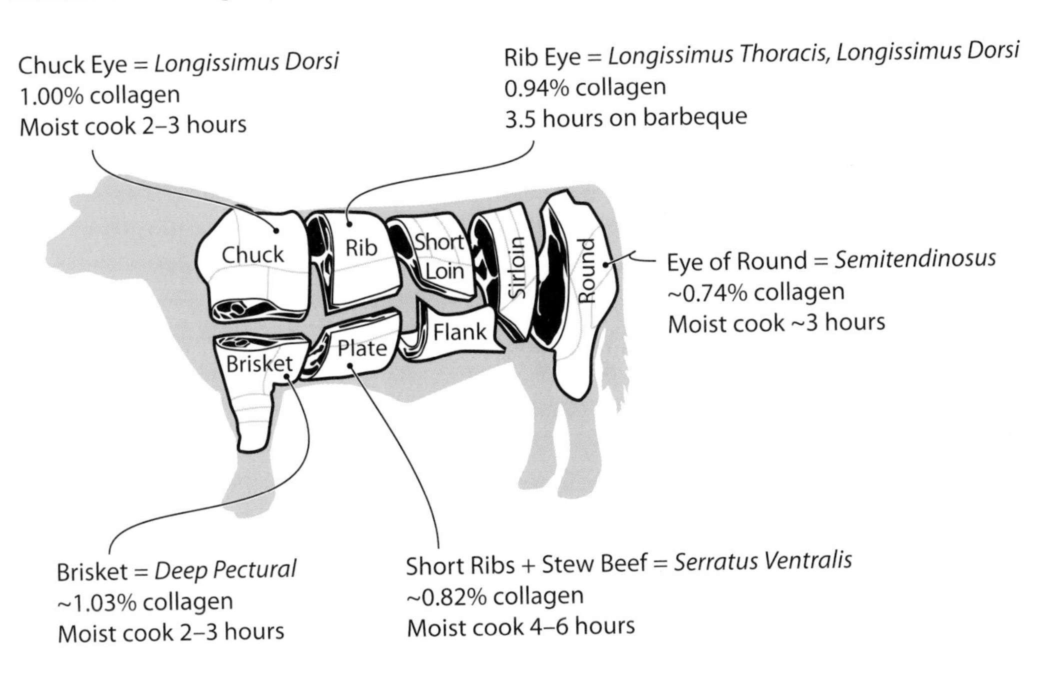

When cooking a piece of meat, if it's from a part that is responsible for supporting the animal's weight (primarily muscles in the chuck, rib, brisket, and round), it'll probably be higher in collagen and thus need a longer cooking time.

Older animals have higher levels of collagen. As animals age, the collagen structure has more time to form additional crosslinks between the strands in the collagen helix, resulting in increased toughness. This is why older chickens, for example, are traditionally cooked in long, slow roasts. (The French go so far as to use different words for old versus young chickens: *poule* instead of *poulet*.) Most commercial meat, however, is young at time of slaughter, so the age of the animal is no longer an important factor.

The other easy rule of thumb for collagen levels is to look at the relative price of the meat: because high-collagen cuts require more work to cook and come out with a generally drier texture, people tend to favor other cuts, so the high-collagen cuts are cheaper.

Squid Bruschetta

Squid was a culinary mystery to me for a long time. You either cook it for a few minutes or an hour; anywhere in between, and it becomes tough, like chewing on rubber bands. (Not that I chew on rubber bands often enough to say what that's like.) Why is this?

The collagen in squid and octopus is enjoyable in either its native state or hydrolyzed state, but not in its denatured state. It takes a few minutes to denature, so with just a quick pan sear it remains in its native state (tossed with some fresh tomatoes and dropped on top of bruschetta, it's delicious). And hydrolysis takes hours to occur, so a slowly simmered braised octopus turns out fine. Braising it in tomatoes further helps by dropping the pH levels, which accelerates the hydrolysis process.

To make a simple squid bruschetta, start by preparing a loaf of French or Italian bread by slicing it into ½″ (1 cm) slices. You can create larger slices by cutting on a bias. (Save the triangular end piece for munching on when no one is looking.) Lightly coat both sides of the bread with olive oil (this is normally done with a pastry brush, but if you don't have one, you can either fold up a paper towel and "brush" with it or pour olive oil onto a plate and briefly dip the bread into the oil). Toast the bread. A broiler works best (the slices of bread should be 4–6″ / 10–15 cm from the heat). Flip as soon as they begin to turn golden brown. If you don't have a broiler, you can use an oven set to 400°F / 200°C. For small batches, a toaster also works.

Once your bread is toasted, place it on a plate and store it in the oven (with the heat off) so that it remains warm.

Prepare the squid:

1 lb (500g) squid (either a mix of bodies and tentacles or just bodies)

Slice the squid with a knife or, better yet, cut it into bite-sized pieces using kitchen shears.

Bring a sauté pan up to medium heat. You want the pan hot enough so that the squid will quickly come to temperature. Add a small amount of olive oil—enough to coat the pan thinly when swirled—and drop the squid into the pan.

Use a wooden spoon or silicone spatula to stir the squid. Take note when it starts to turn white—it should become subtly less translucent—and cook for another 30 seconds or so. Add to the pan and toss to combine:

1 cup (250g) diced tomato (about 2 medium tomatoes, seeds removed)

1 tablespoon (2g) fresh herbs such as oregano or parsley

¼ teaspoon sea salt

Ground pepper to taste

Transfer squid and tomato topping to a bowl and serve with toasted bread.

Try using a pair of kitchen shears to snip the squid into small pieces directly into a hot pan. Add tomatoes and herbs, toss, and serve.

Slow-Cooked Short Ribs

Short ribs and other high-collagen cuts of meat aren't difficult to work with, they just require time at temperature (collagen takes many hours to hydrolyze). The trick is to cook this type of meat "low and slow"—for a long time at low temperature. Too cold, and the collagen won't break down; too hot, and the water in the meat will evaporate, drying it out. Using a slow cooker cooks the meat in the ideal temperature range. After all, this is what they're designed for!

This is an intentionally easy recipe, but don't let this fool you: slow-cooked meats can be *amazingly* good, and if you're cooking for a dinner party, they make for easy work when you go to assemble the dinner. If you have a rice cooker, check to see if it has a "slow-cook" setting. In this mode, the rice cooker will heat foods to a temperature typically between 170–190°F / 77–88°C, which is warm enough to be safe from bacterial contamination and cool enough to not steam-dry the meat.

Pour a bottle of barbeque sauce into the bowl of the rice cooker or slow cooker. Add the short ribs, arranging them in a layer so that the barbeque sauce covers the meat.

Slow-cook for at least four hours (longer is fine). Try starting this in the morning before going to work—the slow cooker will keep the food safe, and the extra time will help ensure that the collagen is fully dissolved.

Notes

- *Ideally, you should pan sear the short ribs (in a cast iron pan) for a minute or two before cooking. As discussed at the beginning of this chapter, this will cause browning reactions, bringing a richness to the final product.*
- *Keep in mind the danger zone rule covered earlier. Don't load up a slow cooker with so much cold meat that the cooker will be unable to raise the temperature above 140°F / 60°C within a two-hour period.*
- *Try adding other ingredients to the sauce, or making your own sauce if you like. I'll often pour a tablespoon or so of wine or port into the empty BBQ sauce jar to "rinse out" the thick sauce, then pour the port-sauce slurry into the slow cooker.*

Duck Confit

Duck confit—duck legs cooked in fat—tastes entirely different from duck cooked almost any other way. It's like bacon and pork—to quote Homer Simpson, they're from "some wonderful, magical animal." Good duck confit is succulent, flavorful, tender, mouth-watering, and perhaps a bit salty. Even if you're not otherwise a fan of duck, give duck confit a chance.

As you can probably tell, I'm a pragmatic cook. Traditional recipes for duck confit prescribe a long, drawn-out affair, which is fine for a leisurely Sunday afternoon spent in the company of friends and a bottle of good wine, but doesn't line up well with my idea of keeping things simple.

Cooking duck "confit-style" is all about converting tough collagen proteins into gelatin. While this isn't a fast chemical reaction, it's a simple one to trigger: hold the meat at a low temperature for long enough, and the collagen proteins denature and eventually hydrolyze.

The secret to duck confit is in the time and temperature, not the actual cooking technique. The upshot? You can make duck confit in a slow cooker or in an oven set at an ultra-low temperature. The fat that the duck is cooked in doesn't matter either; some experiments have shown that duck confit cooked in water and then coated in oil is indistinguishable from traditionally cooked duck confit. Regardless, definitely skip the exotic block of duck fat; duck legs are expensive enough as it is.

Rub salt into the outside of the duck legs, covering both the side with skin and the side with meat exposed. I use roughly 1 tablespoon (18g) of salt per duck leg; you want enough to coat the outside thoroughly.

Place the salted duck legs in a bowl or plastic bag and store them in the fridge for several hours to brine.

Remember: store raw meats in the bottom of the fridge so if they drip the runoff won't contaminate fresh produce or ready-to-eat foods.

Salting the meat adds flavor and draws out a little bit of the moisture, but if you're in a real rush, you can skip this step and just lightly coat the duck legs with a few pinches of salt.

After dry-brining the duck legs, wash off all the salt. At this point, you have a choice of heat sources. Duck confit is about cooking via convection heat with the energy being imparted into the meat by the surrounding fat. Regardless of heat source, the duck legs should be entirely submerged in oil. With careful arrangement and the right size pan, you'll find that it doesn't take much oil to cover them. I generally use olive or canola oil and save the oil after cooking for use in other dishes.

Note that the oil after cooking will be a blend of duck fat and your starting oil. You can also use it for things like sautéing greens and shallow-frying potatoes.

Slow-cooker method

Arrange duck legs in bowl of slow cooker or multipurpose rice cooker. Cover with oil and set to slow-cook mode for at least 6 hours (preferably 10 to 12).

Oven method

Arrange duck legs in an oven-safe pan and cover with oil. Place in oven set at 170°F / 77°C for a minimum of six hours. (200°F / 95°C will work, but avoid anything hotter to prevent steaming the meat.)

The duck legs will become more tender with longer cook times. I've cooked batches of 36 duck legs overnight using a large pot held at temperature in an oven. If you do cook a large batch, remember that the core temperature needs to get to about 140°F / 60°C within two hours. In this case, heat the oil up to ~250°F / 120°C before placing the duck legs in it. This way, the hot oil will impart a solid thermal kick to get the cold legs up to temperature faster.

Duck leg that has been cooked at low heat for a long time falls apart easily, because most of the collagen and connective tissues that normally hold muscles together are gone.

After cooking, the duck skin will still be flabby and, frankly, gross. But the meat should be tender and yield with a bit of poking. You can either remove the skin (pan sear it by itself for duck lardons!) or score the skin with a knife and then pan sear the skin side of the duck to crisp it up.

If you are not going to use the duck legs straightaway, store them in the fridge.

Notes

- *Traditional recipes call for duck fat instead of olive oil. One advantage to the duck fat is that, upon cooling to room temperature, it solidifies, encasing and sealing the duck leg in a sterilized layer of fat, somewhat like how some jams are preserved with a wax seal. If you were living in France a century ago, this would've been a great way to preserve duck legs for a long winter, but with the invention of refrigeration and modern grocery stores, there's no need for the duck fat to store the meat safely for the few days it might last. Use olive oil. It's cheaper and healthier.*
- *If you pour off the oil and liquid into another container, a layer of gelatin will separate out on the bottom once it cools. Use that gelatin! Try tossing it into soups.*

Collagen Experiment

If all this talk about collagen and texture isn't gelling for you, do the following experiment.

Take a few pieces of beef stew meat, and proceed as though you're making beef stew. (See page 67 in Chapter 2.) Once your beef is in the slow cooker, set a timer for 30 minutes.

After 30 minutes, remove a few pieces of the beef. Use a probe thermometer on one to record the internal temperature; it should register somewhere around 160–180°F / 71–82°C, although it'll depend on your slow cooker. Stash the 30-minute sample in a container in your fridge.

After six hours of stewing, repeat the procedure: remove a few pieces, verify that the temperature is about the same, and stash the second batch in a second container in the fridge. (You could heat up the 30-minute batch, but then we'd be changing more than one variable: who's to say that reheating doesn't change something?)

Once both samples are cold, do a taste comparison. Got kids? Do a single-blind experiment to remove the placebo effect: blindfold the kids and don't let them know which is which. Got a spouse and kids? Do a double-blind experiment to control for both placebo effect and observer bias: have your significant other scoop the beef into the containers and label them only "A" and "B," not telling you which is which, and then go ahead and administer the blindfold test to your kids.

158°F / 70°C: Vegetable Starches Break Down

Whereas meat is predominately proteins and fats, plants are composed primarily of carbohydrates such as cellulose, starch, and pectin. Unlike proteins in meat, which are extremely sensitive to heat and can quickly turn into shoe leather if cooked too hot, carbohydrates in plants are generally more forgiving when exposed to higher temperatures. (This is probably why we have meat thermometers but not vegetable thermometers.)

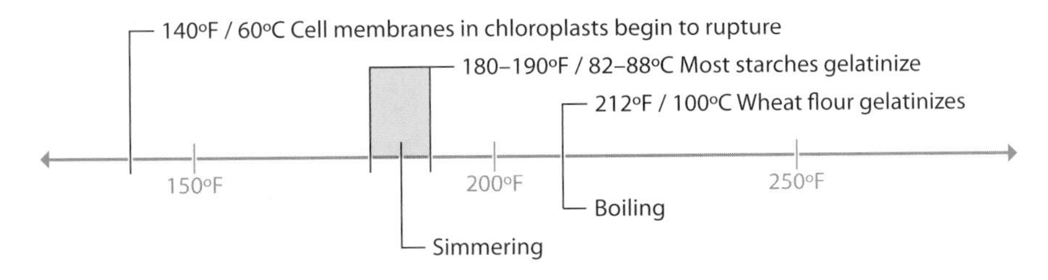

Temperatures related to plants and cooking.

Cooking starchy vegetables such as potatoes causes the starches to gelatinize (i.e., swell up and become thicker). In their raw form, starches exist as semicrystalline structures that your body can only partially digest. Cooking causes them to melt, absorb water, swell, and convert to a form that can be more easily broken down by your digestive system.

As with most other reactions in cooking, the point at which starch granules gelatinize depends on more than just the single variable of temperature. The type of starch, the length of time at temperature, the amount of moisture in the environment, and processing conditions all impact the point at which any particular starch granule swells up and gelatinizes. See the section Making gels: Starches on page 306 in Chapter 6 for more about starches and gelatinization.

Leafy green vegetables also undergo changes when cooked. Most noticeably, they lose their green color as the membranes around the chloroplasts in the cells rupture. This same rupturing and damage to the cell structure is what improves the texture of tougher greens such as Swiss chard and kale.

For starchy plants (think potatoes), cook them so that they reach the temperature at which they gelatinize, typically in the range of 180–190°F / 92–99°C. For green leafy plants, sauté the leaves above 140°F / 60°C to break down the plant cell structure.

Cellulose—a.k.a. fiber—is completely indigestible in its raw form and gelatinizes at such a high temperature, 608–626°F / 320–330°C, that we can ignore it while discussing chemical reactions in cooking.

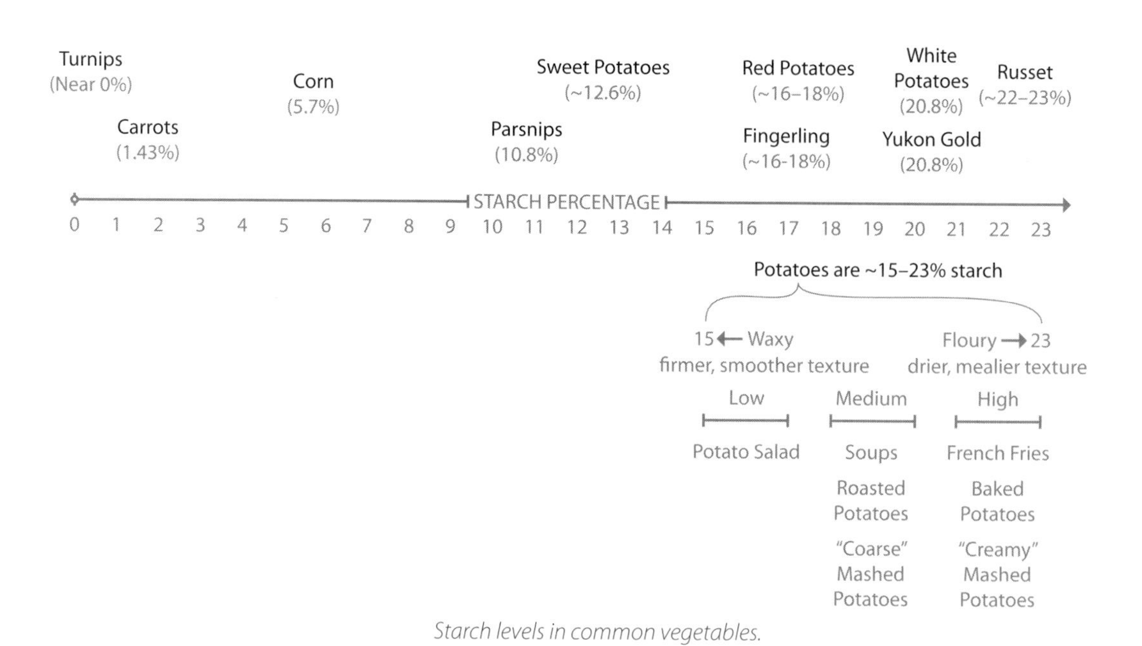

Starch levels in common vegetables.

Quick-Steamed Asparagus

Microwave ovens make quick work of cooking veggies. In a microwave-safe container, place asparagus stalks with the bottoms trimmed or snapped off, and add a thin layer of water to the bottom. Put the lid on, but leave it partially open so that steam has a place to escape. Microwave for two to four minutes, checking for doneness partway through and adding more time as necessary.

Notes

- *This technique cooks the food using two methods: radiant heat (electromagnetic energy in the form of microwaves) and convection heat (from the steam generated by heating the water in the container). The steam circulates around the food, ensuring that any cold spots (areas missed by the microwave radiation) get hot enough to both cook the food and kill any surface bacteria that might be present.*
- *Try adding lemon juice, olive oil, or butter and sautéed, crushed garlic to the asparagus.*

Make Your Own Pectin

Pectin is a polysaccharide found in the cell walls of land plants that provides structure to the plant tissue. It breaks down over time, which is why riper fruits become softer.

Cooking also breaks down pectin, and as a kitchen chemistry experiment, you can capture the pectin from cooked fruits. It's an easy way to see that some food additives aren't so industrial after all, at least not in their sources.

The pectin we use in cooking—primarily in jams and jellies, as a thickener—is divided into two broad types: low- and high-methoxyl. High-methoxyl pectin requires a high concentration of sugar in order to gel; low-methoxyl pectin will gel in the presence of calcium. (The difference between the two types has to do with the number of linkages in the molecular structure.)

If you're making jams or jellies, using a low-methoxyl pectin (such as Pomona's Pectin) removes the variable of sugar concentration. See Tim O'Reilly's tips for making jam in Chapter 5 on page 250.

Making your own pectin is similar to making your own gelatin: start with a couple of pounds of tissue, boil away, and then filter it out. Instead of animal bones, pectin comes from the "bones" of cell walls in plant tissue.

Start with a few pounds of crisp apples. (The firmer the better! They don't need to be ripe.) Chop them into quarters and place the pieces in a stockpot. Cover with water and simmer on low for several hours, stirring occasionally. (This is exactly the way stock is made.) After several hours, you should have a slushy sauce. Filter this through a strainer. (See the section on filtration in Chapter 7 for tips.) The slimy liquid that you filter out is the pectin.

Using homemade pectin will be a bit trickier than Pomona's Pectin, for two reasons. First, it's high-methoxyl pectin, so you'll need to have a proper balance of sugar in whatever you're attempting to gel. And secondly, the concentration of pectin to water will be unknown, so you will have to experiment some. Add a small quantity and test if it gels; if not, add more. If the liquid pectin seems too thin, you can boil it down further to create a more concentrated pectin.

For more ideas and tips on testing homemade pectin, see *http://www.wildflowers-and-weeds.com/The_Forager/pectin.htm*.

Sautéed Greens

In a sauté or nonstick pan preheated over medium heat, add:

1 bunch Swiss chard, collard greens, or other hardy leafy green vegetable; stems and thick veins removed, and sliced into 1" strips

2 tablespoons (26g) olive oil (enough to coat the pan)

Using tongs, quickly toss the greens to coat them with oil. Your pan should be reasonably hot so that the greens quickly heat, but not so hot that the oil burns. Continue tossing while cooking so that the greens wilt evenly. Add salt and pepper to taste.

Notes

- *Depending upon your tastes, extend this by adding one of these combinations:*
  - *5 cloves garlic, minced; juice from half a small lemon (about a teaspoon)*
  - *2 teaspoons balsamic vinegar, and possibly a pinch of sugar*
  - *1 teaspoon sherry vinegar, 1/4 teaspoon crushed red pepper flakes, 1 can cannellini beans, 3 cloves garlic*
  - *1/4 red onion, sliced thin and sautéed to cook; 1/2 apple sliced into bite-sized pieces and cooked; handful of chopped walnuts, toasted*
- *The same technique can be applied to spinach (it's great with sesame seeds). Or try cooking some strips of bacon, removing the bacon, and then sautéing the spinach in the rendered bacon fat, adding a teaspoon or so of balsamic vinegar. Dice the bacon and combine the two, and optionally add some blue (or other) cheese. Quantity of ingredients is really down to personal preference, so experiment!*
- *You can strip the stem and thicker veins from hardy greens such as Swiss chard by grabbing the stalk with one hand and the green leaf portion with the other hand, and pulling the stalk out.*

Poached Pears in Red Wine

Poached pears are easy, tasty, and quick. And, at least compared to most desserts, they're relatively healthy, or at least until the vanilla ice cream and caramel sauce are added. Much of our enjoyment of fruit comes from not just their flavor but also their texture. Consider an apple that's lacking in crispiness or a banana that's been bruised and become mushy: without their customary texture, their appeal is lost. But this isn't always the case. Poaching fruits such as pears causes similar changes in the structure of the fruit's flesh, breaking down cell walls and affecting the bonds between neighboring cells to create a softer texture that's infused with the flavor of the poaching liquid.

In a shallow saucepan or frying pan, place:

2 medium (350g) pears, sliced lengthwise (longitudinally) into eighths or twelfths, and core removed

1 cup (240ml) red wine

¼ teaspoon ground pepper

Set the pan over low to medium heat, bringing the wine to a simmer and then poaching the pears for 5 to 10 minutes, until soft. Flip them halfway through, so that both sides of the slices spend some time facedown in the liquid. Remove the pears and discard the liquid. (You can also reduce the liquid down into a syrup.)

Notes

- *Fun chemistry fact: the boiling point of wine is lower than that of water. The exact temperature depends upon the sugar and alcohol levels, and as the wine simmers, the ratios shift. It'll start somewhere around 194°F / 90°C. It's doubtful that this will actually help you avoid overcooking the pears, though.*
- *Pears are one of those fruits that are underripe until you look away and then go rotten before you can look back. To encourage them to ripen, you can keep underripe pears in a paper bag so the plant tissue will be exposed to the ethylene gas they give off. I find I can get away with poaching pears that are a little more underripe than I might want to eat fresh, but your pears should be at least a little soft.*
- *Try serving this with caramel sauce (see page 212) and vanilla ice cream. Or try poaching other fruits, like fresh figs, and using other liquids. Figs poached in port or a honey/water syrup with a small amount of lemon juice and lemon zest added after poaching are sweet and tasty.*
- *You don't need to actually measure out the ingredients. As long as the pears have enough liquid to poach in, they'll turn out great. Add freshly ground pepper to suit your tastes.*

Don't use preground pepper. Preground pepper quickly loses its complex aromatic flavors—well before it makes it into your hands—leaving it with just a hot spicy kick but none of the subtlety of peppercorns.

Grilled Vegetables

Grilling is as American as apple pie, which is to say that while it's part of our culture, its roots can be traced back to somewhere on the other side of the Atlantic Ocean. Grilling became an American tradition after World War II, when one of the owners of Weber Brothers Metal Works came up with the Weber Grill and ignited a backyard pastime. What self-respecting geek wouldn't have a good time playing with fire?

Whether a propane or charcoal grill is "better" depends on your usage. Propane grills are easier to fire up if you just want to cook a quick burger or roast a few veggies. Charcoal grills, on the other hand, take a bit more work to get going but create a hotter cooking environment that can lead to better flavor development (more Maillard reactions). Regardless of what you go for, grilling is a great way to cook relatively thin items such as skirt steaks, burgers, or sliced vegetables. You can also slow-cook larger items on a grill—I've enjoyed a few summertime afternoons drinking with friends while waiting for an entire hog to cook.

The second major difference between propane and charcoal grills is temperature. While propane itself burns at somewhere around 3100°F / 1700°C, by the time the heat dissipates around the grill, it cools down to about 650°F / 340°C. Using a generous but reasonable amount of wood or charcoal generates a heat source with a much higher amount of thermal radiation. When I've metered wood and charcoal grills, I've gotten temperatures around 850°F / 450°C.

Grilled Summer Vegetables

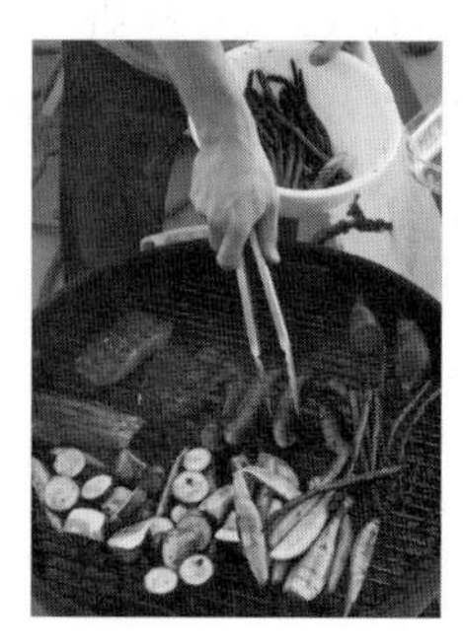

Grilled veggies are a fantastic treat, and easy, too. While someone might yet find a way to make cucumber/lettuce kabobs work, it's easier to stick with the classics: choose sturdy vegetables low in water content (e.g., asparagus, squash, bell peppers, onions).

Slice your vegetables into large pieces and toss them into a bowl with a small quantity of olive oil and a few pinches of salt. You can get fancy with marinades and sauces, but if you're working with great produce, it seems like a shame to mask the flavor.

I generally grill my burgers or whatever meat I'm cooking first, and grill the vegetables while the meat rests. Grill the veggies for a few minutes, flipping halfway through.

Grilled Sweet Potato "Fries"

Slice a sweet potato into wedges. (Can't find sweet potatoes? Look for yams. Americans use the word *yam* when they really mean *Ipomoea batatas*.)

Coat the outside with olive oil and sprinkle with coarse sea salt. Place on the grill for 10 minutes, flip, and grill until tender, about another 10 minutes. Serve while hot.

Instead of the olive oil/salt coating, you can make a sweet coating by brushing the wedges with a mixture of roughly equal parts of butter and honey melted together. Or try sprinkling red pepper flakes on the cooked wedges for a spicy version.

Rosemary Mashed Potatoes

This simple mashed potato recipe uses the microwave for cooking the potatoes. If you're in the anti-microwave category, consider this: cooking a potato—or any other starchy root vegetable—requires gelatinizing the starches in the vegetable. For this to occur, two things need to happen: the starch granules need to get hot enough to literally melt, and they need to be exposed to water so that the granules absorb and swell up, which causes the texture of the tissue to change. Luckily, the temperature at which most starches undergo the gelatinization process is below the boiling point of water, and there's enough water naturally present in potatoes for this to happen without any intervention needed. Try popping a sweet potato in your microwave for a few minutes—fast, easy, and healthy!

Microwave until cooked, about six minutes:

3 to 4 medium (600g) red potatoes

After cooking, cut the potatoes into small pieces that can be mashed with the back of a fork. Add and mash together:

½ cup (120g) sour cream

⅓ cup (85g) milk

4 teaspoons (20g) butter

2 teaspoons (2g) finely chopped fresh rosemary leaves

¼ teaspoon (1g) salt (2 large pinches)

¼ teaspoon (1g) ground pepper

Notes

- *For a tangy version, trying substituting plain yogurt for a portion of the sour cream.*
- *Different types of potatoes have different amounts of starch. Varieties with high starch content (e.g., russets, the brown ones with rough skin) turn out lighter and fluffier when baked and are generally better for baked or mashed potatoes. Lower-starch varieties (red or yellow potatoes, typically smaller and smooth-skinned) hold their shape better and are better suited for applications in which you want the potato to stay intact, such as potato salad. Of course, there's still a lot of room for personal preference. When it comes to mashed potatoes, I prefer a coarse texture to the creamy, perfectly smooth potatoes so often seen in movie scenes associated with Thanksgiving, so I tend to use red potatoes.*

Aki Kamozawa and Alex Talbot's Sweet Corn and Miso Soup

Aki Kamozawa and Alex Talbot write about their experiences with food on their blog (http://www.ideasinfood.com). *A husband-and-wife team, they met while working at one of Boston's premier restaurants, Clio, in 1997, and in recent years have run their own consulting business, educating chefs in new techniques and creative ideas.*

How has having the blog changed your cooking?

It made us more meticulous about paying attention to what we were doing and recording recipes. People would constantly ask us questions, so we needed to have a good answer. Just throwing things in a pan and trying to explain that to someone else really didn't work. Also, we get questions about a lot of different techniques, I think because we work with a lot of chefs. They're more interested in how things work than they are with specific recipes.

Where does the inspiration for some of the more unusual approaches you blog about come from, such as using liquid nitrogen to freeze and shatter beets?

When you have stuff in your kitchen and you're working with it, you just try. You try to figure out what you can do with it and what's possible. With liquid nitrogen, back in science class they did the demonstration where they put the ball in the liquid nitrogen and then smashed it. So when you have it in your own kitchen, you try to break everything with it.

What advice would you give somebody who wants to learn to cook?

Really just to get in there and start cooking and not be afraid to fail. You probably learn more from failure than you do from success because when something goes right you don't really think about how you did it or why it worked. But when something goes wrong and you have to fix it, then you learn a lot more about what's happening.

Why do you think people have a fear of failure in the kitchen?

Because they have a fear of failure in life. Nobody wants to fail. Most people who are cooking at home and trying a new recipe are usually cooking for someone else. And if you screw up in the kitchen, it's expensive—to ruin a whole thing of food and then you don't have anything to eat? You have to have a sense of humor in the kitchen. You have to be able to laugh at yourself. You can always order a pizza.

Sweet Corn and Miso Soup

Cleaned squid

3 pounds (1360g) whole squid

Separate the heads and bodies of the squid. Rinse the bodies under cold water to clean the interior and exterior. Remove the cartilage piece from inside each squid body. When the bodies are clean, pat dry and reserve in the refrigerator. Rinse the squid heads under cold water to remove any gritty material. Lay each head on a cutting board with the tentacles extending to the left. Cut the eyes and interior beak off the righthand side of the head. Discard the beak and eyes. Reserve the tentacles in the refrigerator.

Calamari crackling

~2/3 cup (145g) cleaned squid tentacles
~1/3 cup (80g) cleaned squid bodies
1 teaspoon (5g) squid ink
2 1/2 cups (260g) tapioca flour
6 1/2 cups (1.5 liters) canola oil for frying
Salt

Purée the squid tentacles, bodies, and ink in a food processor until it forms a smooth paste. Add the tapioca flour and pulse the mixture to evenly combine the tapioca into the squid paste. Turn the machine on and purée the mixture to form a sticky dough. Divide the dough between two large vacuum bags and seal on high pressure.

Use a rolling pin to spread the dough to the inside edges of the bag so that a uniform thickness is achieved. The dough should be about 1/10" / 2 mm thick. Place both bags in a steamer large enough to hold them and gently steam the dough for 25 minutes. After 25 minutes, remove the bags and cut them open. Carefully pull the dough out of the bags and lay it on dehydrator trays.

Dehydrate the sheets for several hours until the dough is completely dry and brittle. The dough will take on a shiny matte appearance. When the dough is dry, remove it from the dehydrator and break the sheet into pieces roughly 1 1/2" / 4 cm wide and 3" / 8 cm long. This recipe makes more calamari crackers than are needed for the dish. You can reserve the dried cracker base in its dry form in a zip-top bag for several weeks.

In a medium-sized pot, heat 6 cups (1.5 liters) of canola oil to 350°F / 177°C. Slide the crackers two at a time into the oil. The crackers will sink to the bottom of the oil and then begin to puff and expand. Fry the crackling until it is completely puffed and there are no dark spots of unexpanded cracker dough. Remove the puffed cracklings from the oil and drain on a paper-towel-lined tray. Sprinkle with salt while still hot from the fryer.

Calamari couscous

~3 1/2 cups (725g) cleaned squid bodies
5 cloves (20g) fresh garlic

Use a microplane grater to zest the garlic. Place the zested garlic and the cleaned squid bodies in a food processor. Purée the mixture into a coarse paste. When the mixture takes on a creamy texture and is almost homogenous, stop processing.

Heat a large, nonstick pan on medium heat. When the pan is heated, add the squid mixture. Stir the squid paste in the pan and continue to cook. The mixture will begin to stick to the pan. Use a heat-proof rubber spatula to scrape the bottom of the pan and keep the mixture from sticking.

Continue to cook and stir the mixture. The squid will begin to lose its creamy texture and begin to firm up and form small squid pieces, which will resemble cooked sausage. As the squid continues to cook, it will exude liquid. Continue to cook the squid, allowing the moisture to evaporate until the mixture is dry.

Remove from heat. Place the squid in a shallow pan sitting on an ice bath to cool it quickly. When the cooked squid is cold, place it a food processor. Pulse the food processor to chop the squid nuggets into fine granules resembling couscous. Reserve the calamari couscous in the refrigerator.

Broiled corn planks

8 ears of corn, with husks

Lay the corn out on a sheet tray and broil on high heat for five minutes on each side. The husks will blacken. Keep watch over the corn to make sure that they do not catch on fire. Let the corn cool on top of the stove for 10 minutes.

Peel the husks and the silk from the corn. Cut each piece of corn in half. Stand each piece up vertically on the cut end and, using a sharp knife, slice the kernels away from the cobs. They will come away in large pieces and loose kernels. Set aside the 16 largest chunks.

Cut the large pieces of corn into planks ¾″ / 2 cm wide and 2″ / 5 cm long. The width may be adjusted so that the planks are three kernels wide. Place on a small plate or tray, cover with plastic wrap, and place in the refrigerator. Reserve the trim with the rest of the corn kernels in a covered bowl in the refrigerator. Reserve the cobs for corn stock.

Corn stock

8 corncobs

½ cup (150g) white miso paste

~2 cups (250g) sliced onion

6½ cups (1.5 liters) water

Cut the top tip and bottom end off each corncob and discard. Cut each corncob in half. Combine the corncobs, miso, onion, and water in a 6-quart pressure cooker. Cook at high pressure for 25 minutes. Allow the pressure to dissipate naturally.

Alternatively, combine all ingredients in a heavy-bottomed pot and bring to a simmer over medium heat. Cook for one hour, skimming as needed. Remove from the heat, cover, and let steep for 30 minutes. Strain the finished stock through a fine mesh conical strainer. Chill and reserve until needed.

Prepare the corn soup:

5 cups (1150g) corn stock

~7 cups (975g) broiled corn

⅔ (200g) white miso paste

Combine the cold corn stock and miso in a bowl and whisk gently to blend. Add the broiled corn. Transfer batches of this mixture to the blender. Purée each batch until it is completely smooth. Strain the soup through a fine mesh conical strainer. Refrigerate the soup in a covered container until needed.

Sliced chives

1 cup (50g) chives

Slice the chives into very thin (1 mm) rounds. Reserve.

Assembly

Place the soup in a large pot and gently bring to a simmer, stirring occasionally. Place the corn planks in a sauté pan over low heat and spoon several spoonfuls of warm soup over the planks. Gently cook until the planks are heated through, flipping once to make sure the top and bottom are both hot. In a small pot, heat the calamari couscous. Stir occasionally to prevent the couscous from sticking. When the couscous is hot, fold in the sliced chives.

In each bowl, stack two of the corn planks at the nine o'clock position extending into the center of the bowl. Spoon two spoonfuls of the calamari couscous on the inside edge of the bowl at the five o'clock position. Pour the soup into the bowls, leaving the second half of the top corn plank exposed. Place a calamari crackling on the edge of the bowl and the corn planks so that the corn planks are partially exposed and the crackling rests between the edge of the bowl and the corn planks. Serve immediately.

310°F / 154°C: Maillard Reactions Become Noticeable

The Maillard reaction turns foods brown and generates mostly pleasant volatile aromatic compounds. You can thank Maillard reactions for the nice golden-brown color and rich aromas of a Thanksgiving turkey, Fourth of July hamburger, and Sunday brunch bacon. If you're still not able to conjure up the tastes brought about by Maillard reactions, take two slices of white bread and toast them—one until just before it begins to turn brown, the second until it has a golden-brown color—and taste the difference.

The nutty, toasted, complex flavors generated by the Maillard reaction are created by the hundreds of compounds formed when amino acids and certain types of sugars combine and then break down. Named after the French chemist Louis Camille Maillard, who first described it in the 1910s, the Maillard reaction is specifically a reaction between amino acids (from proteins) and reducing sugars, which are sugars that form aldehydes or ketone-based organic compounds in an alkaline solution (which allows them to react with the amines). Glucose, the primary sugar in muscle tissue, is a reducing sugar; sucrose (common table sugar) is not.

Maillard reactions aren't solely dependent on temperature. Besides temperature, there are a number of other variables that affect the reaction rate. More alkaline foods undergo Maillard reactions more easily. Egg whites, for example, can undergo Maillard reactions at the lower temperatures and higher pressure found in a pressure cooker. The amount of water and the types and availability of reactants in the food also determine the rate at which Maillard reactions will occur. It's even possible for Maillard reactions to happen at room temperature, given sufficient time and reagents: self-tanning products work via the same chemical reaction!

All things considered, though, in culinary applications—cooking at moderately hot temperatures for short periods of time—the 310°F / 154°C temperature given here serves as a good marker of when Maillard reactions begin to occur at a noticeable rate, whether you're looking through your oven door or sautéing on the stovetop.

Butterflied Chicken, Broiled and Roasted

You might be the type who prefers to let the butcher do the butchering, but it's worth learning how to butterfly a chicken (this is also known as spatchcocking*), even if you're squeamish about raw meat. A butterflied chicken is easy to cook, and the crispy brown skin of a well-cooked chicken has a very satisfying flavor from the Maillard reactions. It's economical, too, yielding four to six meals for not much money and a few minutes of surgery.*

A chicken that's been cleaned and gutted is topologically a cylinder. It's basically a big, round piece of skin and fat (outer layer), meat (middle layer), and bone (inner layer). Cooking a whole bird intact is harder than a butterflied bird, because invariably that cylinder is going to get heated from different directions at different rates. That is, unless you have a rotisserie grill, which heats the outside uniformly, cooks it uniformly, and makes it uniformly yummy.

By snipping the spine out of the chicken, you transform that cylinder into a plane of chicken—skin on top, meat in the middle, bone on the bottom. And the topology of such a surface is well suited to heat coming from a single direction (i.e., broiling), meaning it's much *easier to cook to develop a nice, brown, crispy skin.*

1. Prepare your working space. I do this in a roasting pan, because it's going to get dirty anyway. Unwrap the chicken, removing the organ meats (discard or save for something else), and fetch a pair of heavy-duty kitchen scissors. The chicken should be dry; if it's not, pat dry with paper towels.

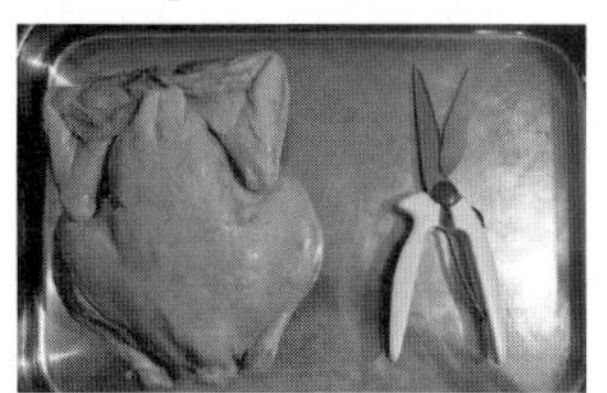

2. Flip the bird around so that the neck flap is facing you. With the scissors, cut down to the right side of the spine (or left side, if you're left-handed). You shouldn't have to apply that much force. Make sure you're not cutting the spine itself, just to the side of it.

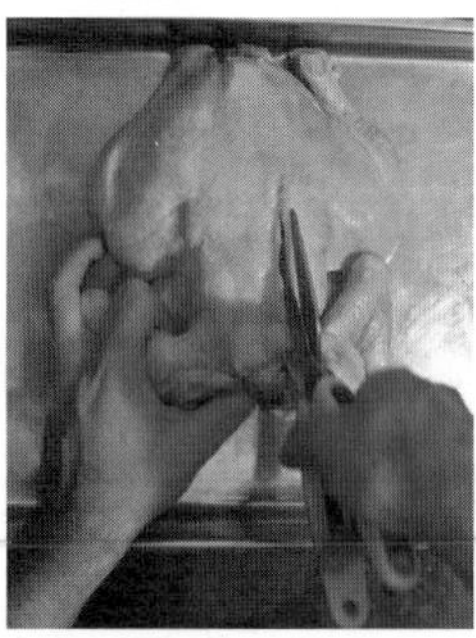

3. Once you've made the first cut, flip the bird around again—it's easier to cut on the outer side of the spine—and cut down the second side.

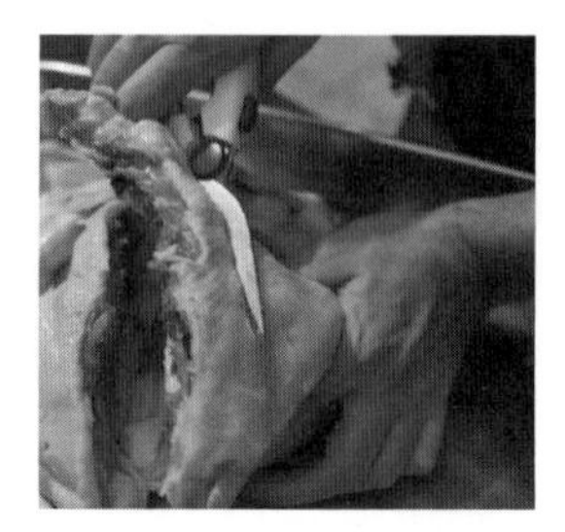

4. Once the spine is removed (trash it, or save it in the freezer for making stock), flip the bird over, skin side up, and using both hands—left hand on left breast, right hand on right breast—press down to break the sternum so that the chicken lies flat. Formally speaking, you should remove the keel bone as well, but it's not necessary. (The keel bone is what connects the two halves of the butterflied chicken together.)

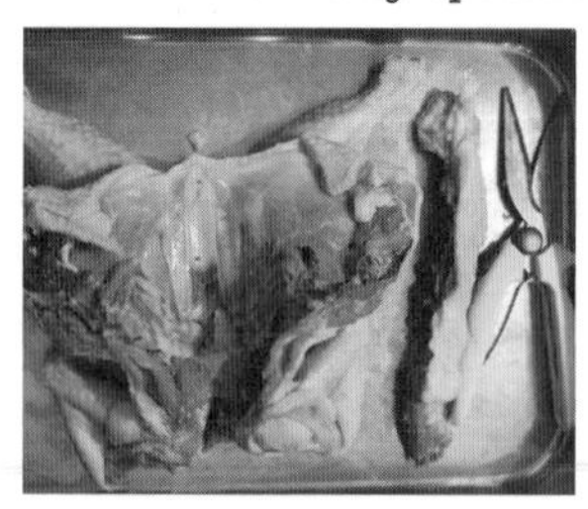

Now that you have a butterflied chicken, cooking it is straightforward. Because the skin is on one side and the bone on the other, you can use two different heat sources to cook the two sides to their correct level of doneness. That is, you can effectively cook the skin side until it's brown from Maillard reactions, and then flip the bird over and finish cooking until a probe thermometer or manual inspection indicates that it is done.

Rub the outside of the butterflied chicken with olive oil and sprinkle it with salt. (The oil will prevent the skin from drying out while cooking.) Place the bird on top of a wire roasting tray in a roasting pan, skin side up. (The wire tray raises the bird up off the pan so that it doesn't stew in the drippings that come out.) Tuck the wings up, over and under the breasts so that they're not exposed to the broiler.

Broil at medium heat for about 10 minutes, or until the skin develops a nice level of brownness. Keep a good 6" / 15 cm between the bird and the heating element of your oven. If your broiler is particularly strong and parts of it begin to burn, you can create a "mini-heat shield" with aluminum foil.

Once the skin side has browned, flip the bird over (I use folded-over paper towels instead of tongs to avoid tearing the skin). Switch the oven to bake mode, at around 350°F / 177°C. Ideally, use a probe thermometer set to beep at 160°F / 71°C (carryover will take it up to 165°F / 74°C). If you don't have a probe thermometer, check for doneness after around 25 minutes by cutting off one leg and checking that the juices run clear and the flesh looks cooked. If it's not done, set the two halves back together and return it to the oven, checking periodically.

Notes

- *Some people like to brine their chickens. At the very least, it adds salt into the meat, changing the flavor. Try brining the chicken in a salt solution for half an hour or so (½ cup / 150g salt, 2 liters ice water—but really, you can just dump salt in water until it's saturated). If you're going to brine it for longer than an hour or so—longer times yield saltier chicken—use cold water (add ice!) and store it in the fridge to keep the chicken below 40°F / 4°C while it brines.*
- *Alton Brown's TV show* Good Eats *has an episode on butterflying a chicken. He creates a garlic/pepper/lemon zest paste to stuff under the skin, and roasts the chicken above a bed of cellar veggies (carrots, beets, potatoes). It's a great recipe, as the paste brings a lot of flavor to the bird and the cellar roots pick up the chicken drippings. For another variation, try putting chopped garlic and aromatic herbs such as rosemary under the skin.*
- *For further inspiration, look at Julia Child et al.'s* Mastering the Art of French Cooking, Volume 2 *(Knopf), which has an excellent description of* Volaille Demi-Désossée*—half-boned chicken—starting on page 269. She removes the breastbone (leaving the spine intact), stuffs the bird (*foie gras, *truffles, chicken livers, and rice), sews it back up, and roasts it. As discussed in Chapter 3, looking at historical recipes—both recent and older—is a great way to understand food better.*

Time & Temperature: Cooking's Primary Variables

Seared Scallops

Scallops are one of those surprisingly easy but often-overlooked items. Sure, fresh scallops can be expensive, but you only need a few for a quick appetizer or part of a meal.

Prepare the scallops for cooking by patting them dry with a paper towel and placing them on a plate or cutting board. If your scallops still have their bases attached, peel them off using your fingers and save them for some other purpose.

Not sure what to do with those little side muscles attached to the main body of the scallop (scallop bases)? Pan fry them after you cook the scallop bodies and nibble on them when no one is looking.

Place a frying pan over medium-high heat. Once the pan is hot, melt about 15g / 1 tablespoon of butter—enough to create a thick coating—in the pan. Using a pair of tongs, place the scallops, flat side down, into the butter. They should sizzle when they hit the pan; if they don't, turn the heat up.

Let them sear until the bottoms begin to turn golden brown, about two minutes. Don't poke or prod the scallops while they're cooking; otherwise, you'll interfere with the heat transfer between the butter and scallop flesh. Once the first side is done cooking (you can use the tongs to pick one up and inspect its cooked side), flip the scallops to cook on the second flat side, again waiting until golden brown, about two minutes. When you flip them, place the scallops on areas of the pan that didn't have scallops on them before. These areas will be hotter and have more butter; you can take advantage of this to cook the scallops more readily.

Once cooked, transfer the scallops to a clean plate for serving.

Notes

- *Try serving these scallops on top of a small simple salad—say, some arugula/rocket tossed with a light balsamic vinegar dressing and some diced shallots and radishes.*
- *If you're not sure if the scallops are done, transfer one to a cutting board and cut it in half. You can hide the fact that you checked for doneness by slicing all of the pieces in half and serving them this way. This lets you check that they're all done as well.*
- *You can dredge the uncooked scallops in breadcrumbs or another light, starchy coating. If you have wasabi peas, use either a mortar and pestle or blender to grind and transfer them to a plate for dredging the scallops.*

Try crushing wasabi-coated peas and dredging the scallops in them before searing.

Sautéed Carrots

Sautéing vegetables will bring a pleasant nutty, toasted flavor to dishes such as braised short ribs.

In a skillet, cook at medium heat until browned, about 5 minutes:

Carrots, sliced into thin rounds or wedges no thicker than around ¼"/ 0.5 cm

Olive oil or butter to coat pan generously

Notes

- *Don't overcrowd the pan. You need the outside of the carrots to get hot enough for the sugars to caramelize. If you put too many in the pan, they'll end up steaming.*
- *The olive oil or butter helps transfer heat. The oil creates a thin layer between the carrot and the pan surface, convecting heat between the two within that very thin layer.*
- *You'll probably want to add a pinch of salt as well. Try grinding on a hefty dose of black pepper. Glazed carrots are made by cooking them with sugar (try adding a tablespoon of brown sugar and a tablespoon of water), or by finishing the carrots with maple syrup. Fresh sage or other aromatic herbs can be julienned and tossed in at the end as well.*

Skillet-Fried Potatoes

Frying potatoes in a heavy cast iron pan develops rich flavors from the starches breaking down and caramelizing. Try serving these potatoes with the butterflied chicken or as part of breakfast accompanied by eggs and bacon.

In a medium-sized pot, bring salted water to a boil and cook for 5 minutes:

3–4 medium (700g) potatoes, diced into "forkable" bite-sized pieces

Drain the potatoes and transfer to a heavy cast iron or enamel pan on a burner set to medium heat. Add:

2–4 tablespoons (25–50g) olive oil or other fat (leftover chicken, duck, or bacon fat tastes great)

1 teaspoon (6g) kosher salt

Stir every few minutes, flipping the potatoes so that the face-down sides have enough time to brown but not burn. Once most of the potatoes are browned on most sides, about 20 minutes, turn the heat down to low, add more oil or fat if necessary, and add:

2 teaspoons (4g) paprika

2 teaspoons (2g) dried oregano

1 teaspoon (2g) turmeric powder

Notes

- *This recipe uses two types of heat: first, boiling, to raise the temperature of the entire potato to quickly cook the starches, and then sautéing, to raise the temperature of the outside.*
- *If you're cooking this as part of a breakfast or brunch, try adding diced red bell peppers, yellow onions, and small chunks of bacon.*

No, par-cooking isn't just about making even more dirty dishes. The par-cooking step speeds up cooking time because the water imparts heat into the potatoes faster. You can skip the par-cooking step and just cook the potatoes in the pan, but they will take an extra 30 minutes or so to cook.

Time & Temperature: Cooking's Primary Variables

356°F / 180°C: Sugar Begins to Caramelize Visibly

Unlike the Maillard reaction, which requires the presence of both amino acids and sugars and has a number of interdependent variables influencing the particular temperature of reaction, *caramelization* (the decomposition via dehydration of sugar molecules such as sucrose) is relatively simple, at least by comparison. Pure sucrose melts at 367°F / 186°C; decomposition begins at lower temperatures (somewhere in the range of 320–340°F / 160–170°C) and continues up until around 390°F / 199°C. (Melting is not the same thing as decomposition—sucrose has a distinct melting point, which can be used as a clever way of calibrating your oven. For more, see The Two Things You Should Do to Your Oven RIGHT NOW on page 42 of Chapter 2.)

Like the Maillard reaction, caramelization results in hundreds of compounds being generated as a sugar decomposes, and these new compounds result in both browning and the generation of enjoyable aromas in foods such as baked goods, coffee, and roasted nuts. For some foods, these aromas, as wonderful as they might be, can overpower or interfere with the flavors brought by the ingredients, such as in a light gingersnap cookie or a brownie. For this reason, some baked goods are cooked at 350°F / 177°C or even 325°F / 163°C so that they don't see much caramelization, while other foods are cooked at 375°F / 191°C or higher to facilitate it.

When cooking, ask yourself if what you are cooking is something that you want to have caramelize, and if so, set your oven to at least 375°F / 191°C. If you're finding that your food isn't coming out browned, it's possible that your oven is running too cold. If items that shouldn't be turning brown are coming out overdone, your oven is probably too hot.

Fructose, a simpler form of sugar found in fruit and honey, caramelizes at a lower temperature than sucrose, starting around 230°F / 110°C. If you have other constraints on baking temperature (say, water content in the dough prevents it from reaching a higher temperature), you can add honey to the recipe. This will result in a browner product, because the largest chemical component in honey is fructose (~40% by weight; glucose comes in second at ~30%).

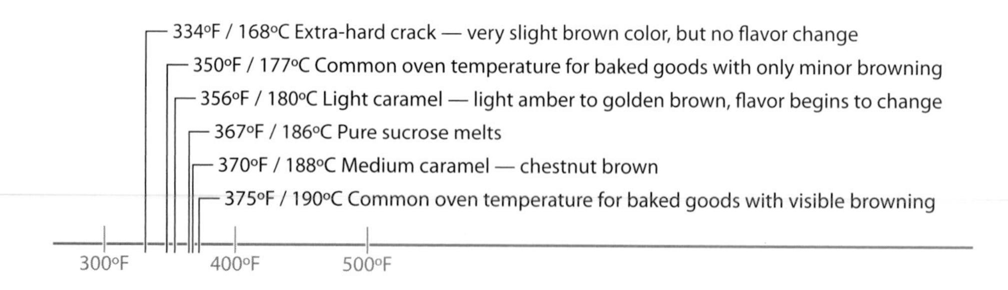

Temperatures related to sucrose caramelization and baking.

Seeing Caramelization with Sugar Cookies

Here's an easy experiment to do with kids (or on your own), and regardless of the results, the data is delicious! Since sugar caramelizes in a relatively narrow temperature range, foods cooked below that temperature won't caramelize. Thus, when making sugar cookies, you can determine whether they will come out a light or dark brown.

Try cooking four batches of sugar cookies at 325°F, 350°F, 375°F, and 400°F (163°C, 177°C, 190°C, and 204°C). Those cooked below the 356–370°F / 180–188°C range will remain light-colored, and those cooked at a temperature above sucrose's caramelization point will turn a darker brown. It's nice when science and reality line up!

This isn't to say hotter cooking temperatures make for *better* results than cooler ones. It's a matter of personal preference. If you're like some of my friends, you may think sugar cookies are "supposed" to be light brown and chewy, maybe because that's the way your mom made them when you were growing up. Or maybe you like them a bit browner on the outside, like a rich pound cake.

Note that the flour used in sugar cookies contains some amount of proteins, and those proteins will undergo Maillard reactions, so cookies baked at 325°F / 163°C and 350°F / 177°C will develop some amount of brownness independent of caramelization.

325°F / 163°C   350°F / 177°C   375°F / 190°C   400°F / 204°C

Cross-section (top piece) and top-down (bottom piece) views of sugar cookies baked at various temperatures. The cookies baked at 350°F / 177°C and lower remain lighter in color because sucrose begins to shift color as it caramelizes at a temperature slightly higher than 350°F / 177°C.

Goods baked at 325–350°F / 163–177°C	Goods baked at 375°F / 191°C and higher
Brownies Chocolate chip cookies (chewy) Sugary breads: banana bread, pumpkin bread, zucchini bread Cakes: carrot cake, chocolate cake	Sugar cookies Peanut butter cookies Chocolate chip cookies Flour and corn breads Muffins

Temperatures of common baked goods, divided into those below and above the temperature at which sucrose begins to visibly brown.

Caramel Sauce

Caramel sauce is one of those components that seems complicated and mysterious until you make it, at which point you're left wondering, "Really, that's it?" Next time you're eating a bowl of ice cream, serving poached pears, or looking for a topping for brownies or cheesecake, try making your own.

Traditional methods for making caramel sauce involve starting with water, sugar, and sometimes corn syrup as a way of preventing sugar crystal formation. This method is necessary if you are making a sugar syrup below the melting point of pure sucrose, but if you are making a medium-brown caramel sauce—above the melting point of sucrose—you can entirely skip the candy thermometer, water, and corn syrup and take a shortcut by just melting the sugar by itself.

In a skillet or large pan over medium-high heat, heat:

1 cup (240g) granulated sugar

Keep an eye on the sugar until it begins to melt, at which point turn your burner down to low heat. Once the outer portions have melted and begin to turn brown, use a wooden spoon to stir the unmelted and melted portions together to distribute the heat more evenly and to avoid burning the hotter portions.

Once all the sugar is melted, slowly add while stirring or whisking to combine:

1 cup (240g) heavy cream

Notes

- *This thing is a calorie bomb: 1,589 calories between the cup of heavy cream and cup of sugar. It's good, though!*
- *Some recipes call for adding corn syrup to the sugar as you heat it. This is because the sucrose molecules, which have a crystalline structure, can form large crystals and chunk up in the process of heating. The corn syrup inhibits this. If you heat the plain sugar with a watchful eye and don't stir it until it gets hot enough, the corn syrup isn't necessary. (It would be necessary, however, if you were only heating the sugar to lower temperatures—temperatures below the melting point—for other kinds of candy making.)*
- *Try adding a pinch of salt or a dash of vanilla extract or lemon juice to the resulting caramel sauce.*
- *Different temperature points in the decomposition range yield different flavor compounds. For a more complex flavor, try making two batches of caramel sauce, one in which the sugar has just barely melted and a second where the caramel sauce is allowed to brown a bit more. The two batches will have distinctly different flavors; mixing them together (once cooled) will result in a fuller, more complex flavor.*
- *Sucrose has a high latent heat—that is, the sugar molecule is able to move and wiggle in many different directions. Because of this, sucrose gives off much more energy when going through the phase transition from liquid to a solid, so it will burn you much, much worse than many other things in the kitchen at the same temperature range. There's a reason pastry chefs call this stuff "liquid Napalm."*

Michael Laiskonis on Pastry Chefs

PHOTO USED BY PERMISSION OF MICHAEL LAISKONIS

Michael Laiskonis is the executive pastry chef at Le Bernardin, *one of only four three-star Michelin restaurants in New York City. A self-proclaimed "accidental pastry chef," he traveled around the United States extensively before working in a bakery, where he had his first big "aha" moment working with bread and discovered a passion for cooking.*

What turned out to matter more than you expected? Just in the process of actually learning to cook.

I guess that when I started cooking it was just something to do and once I developed a passion for it, I realized that—and I don't want to overromanticize it or attach some sort of Anthony Bourdain sort of thing to it, but you kind of enter a culture and it's a completely different culture. I'm sure other professions have it. I'm sure software guys have it. It's just a weird subculture and once you kind of enter that it becomes a lifestyle, not really a job.

That's truly how I feel. With other professional cooks there are obviously colloquialisms and certain physical characteristics that they could have. And then there's also the reality of long hours, bad hours. You're working when everyone is playing; I've come to embrace it and now it's just ingrained in the fabric of my being that it's just—I'm a cook before anything else. It kind of informs everything I do and everything I see. I see through that lens of food. For an outsider that might sound a little creepy, but it's the truth. So when I started cooking, I had no idea that it would take over my life or present so many opportunities to experience other things. I can't imagine giving anything up.

Being from a software background—from one weird subculture to another weird subculture—I hear you. I would be curious how you would describe your weird subculture.

And actually I've spent time thinking about this: what is it about the actual craft of cooking or the act of cooking that does it, and a lot of it is the stress. Granted, it's a self-imposed stress, meaning we're not brain surgeons. We're making people dinner, but dinner is important to a lot of people and especially at the highest ends there is a constant quest for perfection. You're never going to attain perfect, but you can always push further. So I think it's more of the environment of restaurant worlds that kind of informs a lot of that.

I think there is a lot to be said for the power of almost the meditative state that you get, even if you're cooking alone, because you're connecting with nature. You're connecting with things. You're making something with your hands. You're hopefully making something greater than the sum of its parts. It's something that you can't fully describe in words.

It's just what I do. My wife works in a different restaurant. She runs the front of the house, so my work and home life—there's really no separation. We have the same schedule, we come home, and we talk about the business. We wake up and we talk about the business. So it's a lifestyle.

As a pastry chef, are you more of a "by the recipe, exact measurements" type of cook or one who adds an ingredient and tastes, and makes course corrections as you go?

Both. I started in bread and kind of worked in pastry, but I bounced back and forth between each side of the kitchen, between sweet and savory, for a little over five years before I decided to stick with the pastry thing. There is a cliché that pastry chefs are the calm, measured, exacting, precise kind of person and the line cook or the savory chef is the spontaneous one. There is some truth to that. I think the lines are blurring a little bit, but it's really cross-training that gave me a solid foot in both, being spontaneous and being precise. Too much spontaneity, and it's just cook-and-see and you're ultimately lucky if you get the results that you want, but there is that joy in being spontaneous or even taking it further and taking an attitude of well, if it's not broken, let's break it and see what happens. That curiosity and spontaneity are not quite the same thing, but to me, they're of the same spirit.

So if someone is learning how to cook, it's not really a question of them thinking about their own temperament and trying to match it up with baking or cooking; they should really do a bit of both to balance things out?

Yes. It almost sounds like I'm talking out of both sides of my mouth. Because I rely on a recipe, especially in a restaurant situation, consistency is king. Everything has to be the same from batch to batch, day to day. Recipes are useful.

I actually just finished reading *The Craftsman*, by a sociologist, Richard Sennett, who wrote a whole chapter on how-to manuals in the form of recipe writing. I think recipe writing is ultimately flawed. Compare your recipe with "how to set up a computer" or "how to build a shelf" or whatever: you have to tailor the instruction to the experience, to the emotional state, to the personality of the person who is going to be reading and following your recipe.

Recipes are important, but they're also just guidelines or can serve as inspiration. I think it's a natural evolution for a cook—whether it's a professional cook obviously or a home cook—that with confidence, the recipe means less and less, that it can be used as simple inspiration. I still pull books off the shelf all the time, but rarely do I actually write it down. I'll try to wrap my head around what somebody was trying to do, but that really can only come with confidence and experience.

What are your favorite books that you go to?

I would have to say that the Internet is probably my "go to" source right now. It almost feels like I'm being lazy, but I think the Internet has changed everything. It has certainly changed professional cooking—the evolution of it and the speed at which things have progressed. Granted, there is a lot of static you have to sift through to find something of use, increasingly so. But in terms of instant access and comparing and asking different things, I've almost come to favor just an Internet search.

What would you tell somebody who is just learning to bake to keep in mind?

First and foremost: cleanliness and organization are key, and they're always going to save you. Pay attention, especially with baking, which is firmly dictated by the chemical and physical realms that you can't always undo. Also, have that sense of fun and that sense of play and learn from mistakes rather than stressing out about them. It sounds kind of mystical, but I do have this belief that happy people make better-tasting food. I also tell young people: just absorb as much information as you can. It doesn't feel like it's sinking in or you're comprehending it all. Cleanliness, organization, a sense of fun, a sense of play, and always reminding yourself that there is more to learn.

There are certain ways of doing things that ultimately find their way into the dish, whether they're perceivable or not. Sometimes it's about things like cleanliness and organization. When you're eating a dish in a dining room, you're not going to know whether the cook who made it has a dirty apron, but I like to think that that does work its way into it.

Can you give me an example of how you go about thinking about a recipe and putting a dish together?

I have two. They both go toward understanding your ingredients and composition.

We used to make brown butter ice cream, but to give it enough brown butter flavor, we would have to add a ton of fat to the ice cream, which makes the ice cream really texturally challenging. Then I learned about the reaction and the composition of different kinds of dairy products. It's actually just the milk solids in the butter that give us the flavor, but the butter by weight is only 2% solids. We stepped back and looked at heavy cream, which we produce butter from. That heavy cream has three times the milk solids that give us the flavor of butter. So if we take heavy cream and reduce it down to the point where we're left with milk solids and clarified butter, we actually produce more extractable and arguably better-tasting brown butter solids than we could from butter. Then we separate it from the fat and add that to the ice cream.

We also do a lot with caramelized white chocolate. Sometimes I describe it as "roasted chocolate." It sounds kind of counterintuitive, that you'd never want to scorch your chocolate. But if you do it in a controlled way you get an almost *dolce de leche*–like flavor. *Dolce de leche* is usually made by cooking condensed milk; usually, people just boil the can for three or four hours. This gives you more complex flavors, because the proteins and the sugars in the milk and the added sugar are cooking together. If you look at the composition of white chocolate, it's about 40% sugar and 23% milk solids. I researched the composition of condensed milk; the proportion of milk solids and the proportion of sugar are nearly identical. This was a huge connection for me to make personally in terms of substituting ingredients. From there, we've gone on to do all kinds of stuff with caramelized chocolate.

Caramelized White Chocolate

Inspired by Valrhona's L'Ecole du Grand Chocolat

The extent to which the white chocolate is "roasted" will determine the color and flavor of the finished cream. Also, depending on the final application, the amount of gelatin needed will vary. Add more gelatin for a freestanding component, less for a cream that will be put into a shell or glass. Like many similar preparations, the blending phase is vital for achieving the ideal texture.

Caramelize 1 cup (170g) of white chocolate by placing the white chocolate in a sauté pan and heating it over medium-low heat, keeping a watchful eye on it. Stir occasionally, taking care to prevent any bits from turning darker than medium brown. Remove from heat. Add 1.5 teaspoons (10g) of glucose (or corn syrup).

In a separate pan, bring ½ cup (125g) whole milk to a boil. Stir in 2 to 3 sheets of bloomed gelatin (i.e., presoaked in cold water; you can use 2 teaspoons of powdered gelatin, although sheet gelatin is of higher quality). Remove from heat and slowly incorporate into the white chocolate mixture.

Add ¾ cup (175g) of heavy cream (36% fat) to the white chocolate mixture. Emulsify for a few minutes with an immersion blender. Transfer to a container and chill, allowing to crystallize, or dispense into desired forms.

Beurre Noisette Ice Cream

Create a batch of browned milk solids by reducing a quart of heavy cream in a saucepan over low heat, stirring occasionally. After a while—possibly as long as an hour—the heavy cream will separate into clarified butter and the milk solids. Save the clarified butter for some other purpose.

In a clean saucepan, measure out, whisk together to rehydrate the dry milk, and bring to a boil:

1 quart (1000g) skim milk
50g browned milk solids
½ cup (60g) nonfat dry milk
¾ cup (150g) granulated sugar
60g glucose powder
40g trimoline (inverted sugar syrup)

In a separate bowl, measure out and whisk together:

¼ cup (50g) granulated sugar
8g ice cream stabilizer
200g egg yolks (yolks of about 3 large eggs)

Temper the hot milk into the yolk mixture by pouring a quarter of the hot liquid into the yolk mixture and whisking to combine. Add another quarter and whisk to combine. Pour the yolk mixture back into the saucepan, mix thoroughly, and return to low heat and cook, stirring, until slightly thickened (184°F / 84°C).

Remove from heat and whisk in:

⅔ cup (150g) heavy cream

Chill the ice cream base in an ice-water bath, and then transfer to your fridge and allow to mature for at least 12 hours. Transfer the base to an ice cream maker and follow the manufacturer's instructions.

5

Air: Baking's Key Variable

IF TIME AND TEMPERATURE ARE THE KEY VARIABLES IN COOKING, AIR IS THE KEY VARIABLE IN BAKING. While few of us would list air as an ingredient, it's critical to many foods. Most baked goods rely on air for their texture, flavor, and appearance. Baking powder and baking soda generate carbon dioxide, giving rise to cakes and quick breads. Air bubbles trapped in whisked egg whites lift soufflés, lighten meringues, and elevate angel food cakes. And yeast provides texture and adds complex flavors to bread and beer alike.

Unlike cooking, in which the chemical reactions are almost always in balance from the start—a chef rarely needs to tinker with ratios to get a protein to set—baking requires a well-balanced ratio of ingredients from the get-go to trigger the chemical reactions that create and trap air. Achieving this balance is often about precise measurements at the beginning, and unlike most meat and potato dishes, it's virtually impossible to adjust the composition of baked goods as they cook. And as a further challenge, the error tolerances involved in baking are generally much tighter than those in cooking.

If you're the meticulous type—methodical, enjoy precision, prefer a tidy environment—or the type of person who likes to express affection through giving food, you'll probably enjoy baking more than cooking. On the other hand, if you have a wing-it-as-you-go, adapting-on-the-fly style, cooking is more likely to be your thing. But even if baking isn't your thing, the engineering behind it can be fascinating, and plenty of applications in the "winging it" category can benefit from understanding the techniques discussed here.

In this chapter, we'll start with a brief discussion of gluten and then cover the three primary methods of generating air in both savory and sweet applications. We'll also discuss the ingredients associated with each of the three primary methods, giving examples and notes for how to work with them and why they work:

Biological
: Yeast

Chemical
: Baking powder and baking soda

Mechanical
: Egg whites, egg yolks, sugar, whipped cream, and steam

Gluten

Light, fluffy foods need two things: air and something to trap that air. This might seem obvious, but without some way of holding on to air while cooking, baked goods would be flat. This is where gluten comes in.

Gluten is created when two proteins, glutenin and gliadin, come into contact and form what chemists call *crosslinks:* bonds between two molecules that hold them together. In the kitchen, this crosslinking is done by kneading doughs, and instead of talking about crosslinks, bakers speak of developing the gluten: the two proteins bind and then the resulting gluten molecules begin to stick together to form an elastic, stretchy membrane. The same stretchy, elastic property is also responsible for helping trap air bubbles in bread doughs: the gluten forms a 3D mesh that traps air generated by organisms such as yeast and chemicals like baking powder.

Regardless of the rising mechanism, understanding how to control gluten formation will vastly improve your baked goods. Do you want air bubbles to be trapped in the food, or do you want them to escape as the food is cooked? Breads and cakes rely heavily on air for texture, while cookies need less.

The easiest way to control the amount of gluten developed is to use ingredients that have more (or less) of the glutenin and gliadin proteins. Wheat, of course, is the most common source of gluten; rye and barley also have these proteins in small quantities. For practical purposes, though, wheat flour is the primary source of gluten.

While rye has both glutenin and gliadin, it also contains substances that interfere with their ability to form gluten.

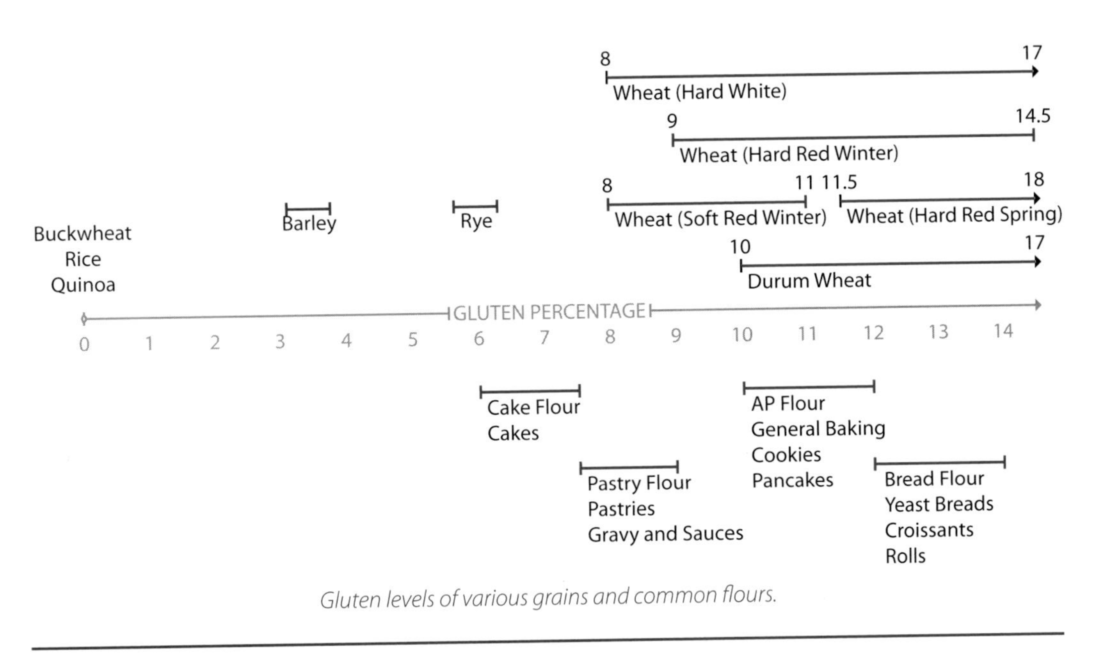

Gluten levels of various grains and common flours.

Gluten levels will vary by both manufacturer and region. Since growing climate impacts gluten levels—colder weather yields higher-gluten wheat—flour in, say, France, won't be identical to that grown in the U.S. Try working with a couple of different brands.

Here are three important things to keep in mind when working with gluten:

Use the appropriate type of flour

Different types of wheat flours have different levels of gluten. Cake flour is low in gluten; bread flour is high in gluten. (All-purpose flour should really be called "general compromise" flour: it just takes the middle ground, which is fine when gluten levels aren't so important.) If you're baking something that would suffer from the elastic texture brought about by gluten—that should have a crumbly texture such as a chocolate cake—use cake or pastry flours, and definitely avoid bread flour.

Fat inhibits gluten formation; water aids it

Fats interfere with the formation of gluten. This is why cookies, which have a lot of flour but also a lot of butter, still manage to crumble. And the opposite is true for water, which helps with gluten formation. The more water there is—up to a point, we're not talking soup here—the more likely it is that glutenin and gliadin will bind.

Mechanical agitation and time develop gluten

Mechanical agitation (a.k.a. kneading)—physically ramming the glutenin and gliadin proteins together—increases the chances for those crosslinks to form and thus increases the amount of gluten in the food. Time, too, develops gluten, by giving the glutenin and gliadin the opportunity to eventually crosslink as the dough subtly moves.

Flaky, crumbly baked goods = low levels of gluten.
Stretchy, elastic baked goods = high levels of gluten.

Flour = Starch + Gluten

Even though gluten is the key variable in wheat flour and baking, it's worth stepping back and looking at what else is hanging out in flour:

- Protein: 8–13%
- Starch: 65–77%
- Fiber: 3–12%
- Water: ~12%
- Fat: ~1%
- Ash: ~1%

The two main compounds in flour are protein (primarily glutenin and gliadin) and starch. Warmer growing climates lead to lower levels of protein and higher levels of starch. Fiber is similar to starch in that both are carbohydrates—saccharides to biochemists—but our bodies don't have a mechanism to digest all forms of saccharides; those that we can't digest get classified as fiber (sometimes called nonstarch polysaccharides). As for ash, this is the broad term given to trace elements and minerals such as calcium, iron, and salt.

Gluten is the most important reason for using flour in baking. Try this simple experiment to separate out and "see" the gluten made by the proteins in flour.

Start with about 1 cup (120g) of bread flour in a bowl and add just enough water so that you can form a ball. Drop the ball of flour into a glass of water for an hour or so, long enough for it to absorb water and allow the gluten to develop.

After the ball has soaked, rinse the starches out by working the ball in your hands, kneading it with your fingers, under slowly running tap water. Keep working the ball until the water runs clear; only about a third of the original mass will be left. At this point, all the starch has washed away. Notice how the part of the flour that remains has a very elastic, stretchy quality to it: this is the gluten. You can drop the ball of gluten into a glass of rubbing alcohol to separate out the glutenin and gliadin proteins—the gliadin will form long, thin, sticky strands, and the glutenin will resemble something like tough rubber.

For comparison, try doing this with cake flour. You'll find it almost impossible to hold on to the ball under the running tap water—there's just not enough gluten present in cake flour to provide any structure to work with while washing away the starch molecules.

P.S. One food additive, transglutaminase, can be used to increase the gluten strength in baked goods by physically increasing the crosslinks within wheat gluten. See page 324 in Chapter 6 for more.

When making breads, gluten impacts the texture not just with its stretchy, elastic quality, but also with its ability to trap and hold on to air. If you're making a loaf of bread using whole wheat flour or grains low in gluten, adding some bread flour (start with 50% by weight) will result in a lighter loaf. You can also add gluten flour, which is wheat flour that has had bran and starch removed (yielding a 70%+ gluten content). Try making a loaf of whole wheat bread with 10% of the flour (by weight) replaced with gluten flour (sometimes called vital gluten flour).

In addition to managing texture, gluten can also be used directly as an ingredient. Consider the following recipe for seitan, a high-protein vegetarian ingredient often used as a substitute for chicken or beef in vegetarian cooking. Seitan is like tofu, in that it is a formed block or roll of proteins, in this case from wheat flour instead of soya beans.

Seitan

Mix together in a large bowl:

¾ cup (175g) water
2 tablespoons (35g) soy sauce
1 teaspoon (5g) tomato paste
½ teaspoon (5g) garlic paste, or 1 clove mashed and finely diced

Add, and use a spoon to mix to a thick, elastic dough:

1 cup (160g) gluten flour (wheat flour that has had bran and starch removed)

Shape the dough into a log and place into a saucepan. Add:

6 cups (1.5 liters) water
½ cup (144g) soy sauce

Bring to a boil and then simmer for an hour. Allow to cool before using.

Notes

- *The gluten flour—also called* vital wheat gluten*—will take a few seconds to absorb the liquid. If you're quick, you can form the dough into a more shapely log and roll it a few times on a cutting board. When cooking the seitan, if it comes out gluey, it wasn't simmered long enough. If you're going to fry the resulting seitan, this is okay, but otherwise you should return it to the simmering liquid and cook longer.*
- *Not sure what to do with seitan? Try thinking of it like tofu: slice off pieces and pan-fry in oil; or shred the seitan, fry, and toss with a quick sweet-and-sour sauce and serve with rice.*

Error Tolerances in Measuring

Measuring out too much (or not enough) butter when making mashed potatoes won't lead to disaster. But with baking, the *error tolerance* in measurement—the amount you can be off by and still have acceptably good results—is much smaller.

How can you learn what measurements are important? Besides trying lots of experiments and keeping detailed notes, you can look at differences between recipes. (Look back at page 22 in Chapter 1 for a discussion of comparing recipes.) By looking at the differences, you can also see what doesn't matter so much.

Consider the ingredients for the following two pie dough recipes.

Joy of Cooking (8" / 20 cm pie)			*Martha Stewart's Pies & Tarts* (10" / 25 cm pie)		
100%	240g	flour	100%	300g	flour
60%	145g	shortening (Crisco)	–	–	(no shortening)
11.25%	27g	butter	76%	227g	butter
25%	59g	water	19.7%	59g	water
0.8%	2g	salt	2%	6g	salt
–	–	(no sugar)	2%	6g	sugar

The numbers in the first column are "baker's percentages," which normalize the quantities to the quantity of flour by weight; the second column gives the gram weights for one pie's worth of dough.

Just comparing these two recipes, you can see that the ratio of flour to fats ranges from 1:0.71 to 1:0.76, and that a higher percentage of water is called for in the *Joy of Cooking* version.

However, butter isn't the same thing as shortening; butter is about 15–17% water, whereas shortening is only fat. With this in mind, look at the recipes again: the Martha Stewart version has 76g of butter (per 100g of flour), for about 64g of fat; the pie dough with shortening has 69g of fat per 100g of flour. The quantity of water is also roughly equal between the two once the water present in the butter is factored in.

You won't always find the ratios of ingredients between different recipes to be so close, but comparing recipes is a great way to learn more about cooking and a good way to determine which recipe to use when trying something new.

There are two broad types of pie doughs: flaky and mealy. Working the fat into the flour until it is pea sized and using a bit more water will result in a flakier dough well suited to prebaked pie shells; working it until it has a cornmeal-like texture will result in a more water-resistant, mealy, crumbly dough, which makes it better suited for uses where it is filled with ingredients when baked.

Simple Pie Dough

Measure and combine all the ingredients for either the *Joy of Cooking* or the Martha Stewart recipe into a mixing bowl or the bowl of a food processor, cutting the butter into small cubes (½" / 1 cm). You should preferably use pastry flour, but AP flour is okay. Chill in the freezer for 15 to 30 minutes. Chilling the ingredients prevents the butter from melting, which would allow the water in the butter to interact with the gluten in the flour, resulting in a less flaky, more bread-like dough.

Pulse the ingredients in a food processor in one- to two-second bursts. Continue pulsing the dough until the ingredients are combined into a coarse sand-like or small pebble-like consistency. If you do not have a food processor, use a pastry blender, a couple of knives, or your fingers to crumble the fats into the flour. Make sure if you use your hands not to let the temperature of the dough rise much above room temperature.

Once the dough is at a coarse sand- or pebble-like consistency, dump the dough out onto a floured cutting board and press it into a round disc. Using a rolling pin, roll the dough out into a sheet, then fold it over on itself and roll it out again, repeating until the dough has been compressed and has enough structure that it can be transferred to a pie tin.

Don't have a rolling pin? A wine bottle will work in a pinch.

Prebaked Pie Shell

Some pies, such as lemon meringue pie (see page 307 in Chapter 6), call for the pie shell to be prebaked. To prebake a pie shell (also called *blind baking*), roll out the dough and transfer it to your pie tin or mold. You'll need to bake the pie with pie weights (no need to be fancy—beans or rice work perfectly); otherwise, the pie dough will slide down the edges and lose its shape. Once it's baked enough to hold its shape, remove the pie weights so that the pie shell has a chance to crisp up and brown.

Set oven to 425°F / 220°C. Bake pie shell with pie weights for 15 minutes (use parchment paper to separate the pie weights from the dough, so that you can pick up the paper and remove the weights). Remove pie weights and bake for another 10 to 15 minutes, until shell is golden brown.

I *hate* the taste of uncooked flour; it burns the back of the mouth. If you're not sure whether your pie dough is done, err on the side of leaving it in longer.

When prebaking—also called "blind baking"—a pie shell, make sure to fill the shell with weights. Otherwise, the sides will collapse. Line the pie shell with a piece of parchment paper or foil and fill it with dried beans or rice.

Martin Lersch on Chemistry in the Kitchen

PHOTO USED BY PERMISSION OF MARTIN LERSCH

Martin Lersch blogs about food and molecular gastronomy at http://blog.khymos.org, *which includes the excellent collection of recipes, "Texture: A hydrocolloid recipe collection," which demonstrates many uses of food additives. (We'll cover food additives and molecular gastronomy in Chapter 6.)*

I see from your online bio that you have a PhD in organometallic chemistry. How did you get interested in chemistry in cooking?

My whole food interest is in no way related to my studies or my work, apart from chemistry. It was when I was a student at the University of Oslo, almost 10 years ago, that I found *On Food and Cooking* by Harold McGee in the faculty library. It was very interesting.

So I started looking for more information, but at that time there wasn't really very much out there. At university, they often have students visiting from high schools, so at one point I was given the opportunity to talk about everyday chemistry; I think the title was something like "Everyday Chemistry in the Kitchen." Then I put up a web page, and when I finished my PhD many years later, the page had grown, so I figured I would continue. I moved everything to *http://khymos.org* and started blogging.

The whole time, it's only been a hobby. I've always liked cooking. Every chemist should actually be a decent cook, because chemists, at least organic chemists, are very used to following recipes. It's what they do every day at the lab. I often tease my colleagues, especially if they claim that they can't bring a cake to the office for a meeting, I say, "Well, as a chemist, you should be able to follow a recipe!" As a chemist, I've always had, in a way, curiosity. I bring that curiosity back home into the kitchen and wonder, "Why does the recipe tell me to do this or that?" That's really the case.

How has your science background impacted the way that you think about cooking?

I think about cooking from a chemical perspective. What you do in cooking is actually a lot of chemical and physical changes. Perhaps the most important thing is temperature, because many changes in the kitchen are due to temperature variations. Searing meat and sous vide are also good places to start. With sous vide, people gradually arrive at the whole concept themselves. If you ask them how they would prepare a good steak, many people would say you should take it out of the refrigerator ahead of time, so you temper the meat. While you temper it, why not just put it in the sink—you could use lukewarm water? Then if you take that further, why not actually temper the meat at the desired core temperature? Most people will say that's a good idea, then I say that's sous vide. It becomes obvious for people that that's actually a good idea.

I'm very fascinated by the hydrocolloids. One of the reasons I spent so much time putting the recipes together was that when I bought hydrocolloids, maybe one or two recipes would be included, but I found them not to be very illustrative. Everyone is familiar with gelatin, less so with pectin, but all the rest are largely unfamiliar. People don't know how they work, how you should disperse them and hydrate them, or their properties. The idea was to collect recipes that illustrate as many of the ways to use them as possible. You can read a couple of the recipes and then can go into the kitchen and do your own stuff. That's what I hope it will enable people to do.

See page 303 in Chapter 6 for an explanation of colloids.

I think it's a fantastic recipe collection, having used it myself for exactly the purpose that you describe. Out of curiosity, is there a favorite hydrocolloid of yours?

No, I haven't even tried them all—I don't have all of them in my kitchen.

Really?

I think the reason is more lack of time. With a full-time job, children, family... there's simply not enough time. It's a lot easier to skip the practical part and concentrate on the theory.

Is there a particular recipe from which you've learned the most or found interesting or unexpected in some way?

It's hard to think of one recipe. When talking about molecular gastronomy, it's easy to focus too much on the fancy applications like using liquid nitrogen or hydrocolloids. It's important to emphasize that this is not what molecular gastronomy is about, although many people think that; many people associate molecular gastronomy with foams and alginate.

I always try to include basic things to get down to earth. One thing that comes to mind is bread. It is really fascinating the great variety that you can achieve by using only water, flour, and salt. With the flour and water, you already have the wild yeast present, so you have everything set up for a sourdough. Then it depends on how you prepare your starter, the ratios involved, how you proof your dough, and how you bake it. Of course, this is not something new; bakers know this. But from a scientific viewpoint, it's very interesting to think about that. The no-knead bread illustrates a lot of chemistry; you're probably familiar with that?

I am, but go on.

Glutamine and gliadin, the two proteins that make gluten, can combine all by themselves once you have a dough that is wet enough. The typical hydration for no-knead bread would be somewhere in the 75% to 77% range. You bake the bread in a preheated pot, where you simulate a steam oven. Moist air is a much better heat conductor than dry air, and the moisture condenses on the surface of the bread. It enhances the crust formation and helps the gelatinization of the starch. It also prevents the crust from drying out and limiting the rise of the bread, so you get a much better oven spring this way. Once you remove the lid, everything is set for the Maillard reaction as the crust dries out. So there is a lot about both the way you make dough and the way you bake the bread that exemplifies basic chemistry and physics.

Bread—No-Knead Method

Weight	Volume	Baker's %	Ingredient
390g	3 to 3¼ cups	100%	All-purpose white flour
300g	1¼ cups	77%	Water
7g	1 teaspoon	1.8%	Salt
~2g	½ teaspoon	–	Fresh yeast (a pea-sized lump); you can substitute 1 teaspoon (5g) instant yeast

Mix everything until the flour is completely moistened. This should take only about 30 seconds. Cover and let rest at room temperature for 20 hours.

Place a medium-sized cast iron pot in your oven and preheat both to 450°F / 230°C. While the oven is heating, transfer the dough onto a floured surface and fold three or four times. Leave for 15 minutes. Shape rapidly into a boule—a round loaf—and place on a generously floured cloth towel. Proof until doubled in size. Dump into the preheated cast iron pot and bake with the lid on for 30 minutes. Take the lid off and bake until the crust has a dark golden color, about 15 minutes.

ADAPTED BY MARTIN LERSCH FROM JIM LAHEY'S *NEW YORK TIMES* RECIPE

Mill Your Own Flour

Milling flour is a lot easier than you might imagine: snag some wheat berries—which are just hulled wheat kernels, with bran, germ, and endosperm still intact—from your local health food store or co-op, run them through a mill, and you've got fresh flour.

Why bother? Well, for one, the taste is fresher; volatile compounds in the wheat won't have had time to break down. Then there are the health aspects. Most commercial whole wheat flours have to heat-process the germ to prevent it from going rancid, but this heat-processing also affects some of the fats in the flour.

On the downside, freshly milled flour won't develop gluten as well as aged flour. For a rustic loaf of bread, this is probably fine, but it's not so good if you're trying to make whole wheat pasta, in which the gluten helps hold the pasta together. Of course, you can always add in some gluten flour to boost the gluten levels back up.

You have a couple of options for mills. KitchenAid makes a mill attachment for its mixers. If you do spring for a KitchenAid attachment, though, be warned that it can put quite a strain on the mixer. Set it to low speed and run your grain through in two passes, doing a first pass to a coarse grind before doing a fine grind. Alternatively, take a look at K-Tec's Kitchen Mill, which is in roughly the same price range but is designed specifically for the task.

You can run other grains, such as rice and barley, through a mill as well. Too-moist grains and higher-fat items such as almonds or cocoa nibs are a no-go, though: they'll gum up the grinder.

Wheat berries.

First pass: coarse grind.

Second pass: fine grind.

P.S. Don't expect to be able to mill things like cake flour. Cake flour is bleached with chlorine gas to mature it. *Maturing*—the process by which flour is aged—would eventually happen naturally due to oxidation, but chlorine treatment speeds it up. It also modifies the starch in the flour so that it can absorb more water during gelatinization (see page 306 in Chapter 6 for more on gelatinization of starches) and weakens the proteins in the flour, reducing the amount of gluten that can be formed. Additionally, chlorination lowers the temperature of gelatinization, so batters that include solids—nuts, fruits, chocolate chips—perform better because there's less time for the solids to sink before the starches are able to gel up around them.

Biological Leaveners

Biologically based leaveners—primarily yeast, but also bacteria for salt-rising breads—are surely the oldest method for generating air in foods. Presumably, a prehistoric baker first discovered that a bowl of flour and water left out will begin to ferment as yeast from the surrounding environment settles in it.

Yeast

Yeast is a single-celled fungus that enzymatically breaks down sugar and other sources of carbon to release carbon dioxide, ethanol, and other compounds, giving drinks their carbonation, spirits their alcohol, and beer and bread their distinctive flavors. Even making chocolate involves yeast—the cocoa beans are fermented, which generates the precursors to the chocolate flavor.

Different strains of yeast create different flavors. Over the years we've "domesticated" certain strains by selective breeding—from common baker's yeast for bread and wine (*Saccharomyces cerevisiae*) to those for beer (usually *S. carlsbergensis*, a.k.a. *S. pastorianus*).

Since there's plenty of yeast literally floating around, you don't have to directly spike your brew or seed your bread with yeast. New strains of yeast usually start out as wild hitchhikers, and sometimes they taste great. Traditionally winemakers relied on ambient yeasts present in their cellars or even on the grapes themselves (this is the origin of the traditional European *le goût de terroir* approach to winemaking).

However, the "Russian roulette yeast method" might not end so well when you're working in your kitchen: there's a decent chance you'll end up with a nasty and foul strain of yeast that'll generate unpleasant-tasting sulfur and phenol compounds. This is why you should add a "starter" strain: providing a large quantity of a particular strain ensures that it will outrace any other yeasts that might be present in the environment.

There's nothing magical about the strains of yeast we use other than someone taking notice of their flavor and thinking, "Hey, this one tastes pretty good, I think I'll hang on to it!"

Like any living critter, yeast prefers to live in a particular temperature zone, with different strains preferring different temperatures. The yeast commonly used in baking breads—aptly named *baker's yeast*—does best at room temperature (55–75°F / 13–24°C). In brewing beer, ales and stouts are made with a yeast that is similar to baker's yeast; it also thrives at room temperature. Lagers and steam beer use a bottom-fermenting yeast that prefers a cooler environment around 32–55° F / 0–13°C. Keep in mind the temperature range that the yeast you're using likes, and remember: too hot, and it'll die.

Yeast in beverages

Wine, beer, and traditional sodas all depend on yeast to ferment sugar into alcohol and generate carbonation. Consider the following equation:

Fermentation = Water + Carbon (usually Sugar) + Yeast + Optional Flavorings

Selecting the appropriate strain of yeast and controlling the breeding environment—providing food, storing at proper temperatures—allows for the creation of our everyday drinks:

Wine = Grape Juice[Water + Sugar] + Yeast

Beer = Water + Barley[Sugar] + Yeast + Hops[Flavoring]

Mead = Water + Honey[Sugar] + Yeast

Soda = Water + Sugar + Yeast + Flavorings

Some of these are easier processes to control than others. Wine, for example, is relatively straightforward, with few variables: vary the sugar level to control the amount of yeast activity and choose the grapes and strain of yeast per your desired type of wine (trace elements in the grapes themselves are usually responsible for the flavor and aromas in wine). Beer has more variables to play with: in addition to sugar levels, proteins and saccharides have to be controlled to correctly balance viscosity and head, and the bitterness of the hops has to be managed.

Hops—the flowers of a herbaceous perennial—are a recent addition to beer making. The earliest beers were flatter and sweeter, and would quickly spoil. Around the eighth century, brewers discovered adding hops extended storage times by acting as a preservative.

Ginger Lemon Soda

Making your own homebrew doesn't have to be a long, drawn-out process. You can make your own soda with just a few minutes of work, and it's rewarding to see the curtain pulled back on an everyday drink, as the following recipe illustrates.

Start with an empty two-liter soda bottle. Add water, sugar, yeast, and flavorings, let it set at room temp for two days to give the yeast a chance to do its thing, and you'll have soda.

Create a ginger syrup by bringing to a boil and simmering for at least 15 minutes:

1 cup (240g) water
¾ cup (150g) sugar
1 cup (90g) ginger, finely chopped

Strain simple syrup to remove ginger pieces and transfer into a two-liter soda bottle. Add:

30 oz (900g) water
4 oz (120g) lemon juice
½ teaspoon (1g) yeast

Screw on lid, shake to combine, let rest at room temperature for two days, and then transfer to fridge and drink.

Dark & Stormy Cocktail

In a highball filled with ice, pour:

6 oz (180ml) Ginger Lemon Soda
2 oz (60ml) Dark Rum

Garnish with a slice of lime.

Notes

- *For yeast, try using Lalvin's #1118 yeast (available online at* http://www.lalvinyeast.com*) or look for a local brewing shop. Baker's yeast can be used, but it'll contribute a slightly odd flavor.*
- *Try adding cayenne pepper or other spices to the simple syrup, or making other flavors, such as mint lime. The method is the same—create a flavored simple syrup (say, ½ cup mint leaves simmered instead of ginger), and use lime juice in the place of lemon juice. Like mojitos? In a tall glass filled with ice, pour 3 oz (90 ml) mint lime soda, 2 oz (60 ml) white rum, and 1 oz (30 ml) lime juice. To be proper, muddle fresh mint leaves with sugar in the glass before adding the ice.*
- *You can skip peeling the ginger, since it will be strained out. However, the ginger will become tender and sugary—this is how to make candied ginger!—so slice off the skin if you do want to save the pieces.*
- *Want to go all out, DIY-Soda-Company style? "Recycle" some beer bottles and cap them yourself with a handy-dandy beer bottle capper, available for about $20 online.*

The Four Stages of Yeast in Cooking

You've just added starter yeast to bread dough or a liquid such as wort (beer liquid before it's beer). What happens next?

Respiration. A cell gains and stores energy. No oxygen? No respiration. During this stage, the yeast builds up energy so it can reproduce.

Reproduction. The yeast cell multiplies via budding or direct division (fission) in the presence of oxygen. Acidic compounds get oxidized during this stage, with the quantity and rate depending upon the strain of yeast, resulting in different pH levels in the food.

Fermentation. Once the yeast has utilized all the available oxygen, it switches to the anaerobic process of fermentation. The cell's mitochondria convert sugar to alcohol and generate CO_2 ("yeast farts"!) and other compounds in the process. You can control the level of carbonation and alcohol in beverages by controlling the amount of sugar.

Sedimentation. Once the yeast is out of options for generating energy—no more oxygen and no more sugar—the cell shuts down, switching to a dormant mode in the hope that more oxygen and food will come along some day. In brewing, it conveniently clumps together (called *flocculation*) and settles to the bottom, where it'll stay if you're careful when pouring out the liquid. Commercial beverages filter out and remove this sedimentation before bottling, but if you make your own brew, don't be surprised at the thick layer of gunk that forms.

While each yeast cell goes through these stages, different cells can be in different stages at the same time. That is, some cells can be reproducing while others are respiring or fermenting.

Yeast in breads

Baker's yeast comes in three varieties: instant, active dry, and fresh. All three types are the same strain: *Saccharomyces cerevisiae*. The instant and active dry versions have been dried so as to form a protective shell of dead yeast cells surrounding some still-living cells. Fresh yeast—also called *cake yeast* because it is sold in a compressed cake form—is essentially a block of the yeast without any protective shell, giving it a much shorter shelf life (well, fridge life): cake yeast is good for about two weeks in the fridge, whereas instant yeast is good for about a year and active dry yeast is good for about two years in the cupboard.

Instant and active dry yeast are essentially identical, with two differences. First, active dry yeast has a thicker protective shell around it. This gives it a longer shelf life, but it also means it must be soaked in water before use to soften up the protective shell. The second difference is that the quantity of active yeast cells in active dry yeast is lower than in instant yeast, because the thicker protective shell takes up more space: when a recipe calls for 1 teaspoon (2.9g) of active dry yeast, you can substitute in ¾ teaspoon (2.3g) of instant yeast.

Instant yeast is the easiest to work with: add it directly into the dry ingredients and mix. Unless you have reason to work with active dry or cake yeast, use instant yeast. Remember to store it in the fridge!

The recipes in this chapter assume that you are using instant yeast. Check in the refrigerated section of your grocery store: SAF Instant and Red Star are two of the more common brands.

If you have active dry yeast instead, you will need to proof it first. Proofing—soaking in lukewarm water—softens the hard shell around the active dry yeast granules. Use lukewarm water (105°F / 40°C). If the water is below 100°F / 38°C, an amino acid called glutathione will leak out from the cell walls and make your dough sticky; if it's above 120°F / 49°C, the yeast will show very little activity.

Don't be worried about too-hot tap water killing your yeast. Yeast actually dies somewhere above 131°F / 55°C, so too-hot water from the tap shouldn't be able to kill the yeast; it just slows down reproduction. You can confirm this by filling a glass with your hottest tap water, dumping in some yeast, waiting a few minutes to give the yeast time to come up to temperature, and then adding some flour and watching the yeast still do its thing.

You can skip all this proofing and temperature stuff by just using instant yeast.

Check Your Yeast!

In baking, *proofing* can refer to a few different things: checking that your yeast is alive, allowing the dough to rise, or allowing the shaped loaf to rest and rise before baking.

Whatever you call it, you should make sure that your yeast is alive before proceeding to work with it. Measure out 2 teaspoons (10g) of the yeast and 1 teaspoon (5g) of sugar into a glass and add ½ cup (120g) of lukewarm water (105°F / 40°C). Stir and let rest for two to three minutes.

You should see small bubbles forming on the surface. If you don't, your yeast is dead—time to head to the store.

You probably don't need to check your yeast every time you use it, especially if you're using instant yeast and storing it in your fridge. If you notice that your doughs aren't rising as expected, though, give the yeast a quick check.

Proofed yeast will bubble up and foam (left); dead yeast will separate out and not foam (right).

Bread—Traditional Method

If you've never made bread before, a simple loaf is easy enough to make, and perfecting it will keep you busy for many years. This is one of those recipes that's worth making several days in a row, making one change at a time to understand how your changes impact the final loaf.

In a large bowl, whisk to thoroughly combine:

1½ cups (180g) bread flour

1½ cups (180g) whole wheat flour

3 tablespoons (30g) gluten flour (optional)

1½ teaspoons salt (2 teaspoons if using kosher or flake salt)

1½ teaspoons instant yeast (*not* active dry yeast)

Add:

1 cup (240g) water

1 teaspoon (7g) honey

Stir just to incorporate—maybe 10 strokes with a spoon—and allow to rest for 20 to 30 minutes, during which the flour will absorb the water (called *autolysing*).

After the dough has undergone autolysis, knead it. You can do this against a cutting board, pressing down on the dough with the palm of your hand, pushing it away from yourself, and then folding it back up on top of itself, rotating the ball every few times. I sometimes just hold the dough in my hands and work it, stretching it and folding it, but this is probably unorthodox. Continue kneading the dough until it passes the "stretch test": tear off a small piece of the dough and stretch it. It shouldn't tear; if it does, continue kneading.

Form the dough into a ball and let it rest in the large bowl, covered with plastic wrap (spray it with nonstick spray to avoid it sticking), until it doubles in size, normally about 4 to 6 hours. Try to store the dough someplace where the temperature is between 72°F / 22°C and 80°F / 26.5°C. If the dough is kept too warm—say, if you're in a hot climate, or it's too close to a heating vent—it will double in size more quickly, so keep an eye on it and use common sense. Warmer—and thus faster—isn't necessarily better, though: longer rest times will allow for better flavor development.

After the dough has risen, give it a quick second kneading—more of a quick massage to work out any large gas bubbles—and form it into a tight ball. Coat it with a light dusting of flour, place it on a pizza peel (or piece of cardboard), cover it with plastic wrap again, and allow it to rest for another hour or two.

Yeast produces both acetic and lactic acid at different rates depending upon temperature. Ideal rising temperature is between 72°F / 22°C and 80°F / 26.5°C.

If kept too cold, dough will be tough and flat due to insufficient gas production, and the final loaf will have uneven crumb, irregular holes, and a too-dark, hard crust.

On the other hand, dough risen in an environment too warm will be dry, lack elasticity, and break when stretched, and the final loaf will have sour-tasting crumbs, large cells with thick walls, and a pale/whitish crust.

While waiting for the dough to proof, place either a pizza stone or a baking stone in your oven and set it to 425°F / 220°C. (No pizza stone? Use a cast iron griddle or cast iron pan, flipped upside down.) Make sure that the oven is fully heated before baking—a full hour of preheating is not unreasonable.

Just before transferring the dough to the oven, pour a cup or two of boiling water into a baking pan or cookie sheet and set it on a shelf below the baking stone. (Use an old cookie sheet; the water may leave a hard-to-clean residue on it.) Alternatively, you can use a spray bottle to squirt the inside of the oven a dozen or so times to increase the humidity. (Be careful not to hit the light bulb inside: it can shatter.) Upping the humidity will help impart heat into the bread faster and will also prevent the outside of the loaf from setting prematurely, giving the bread better *oven spring*—the rise that occurs as the loaf heats up in the oven before the outside of the loaf sets and becomes, essentially, an exoskeleton.

With a serrated knife, lightly slash the top of the loaf with an "X" and then place it into the oven. Bake until the crust is golden brown and the loaf gives a hollow sound when rapped on the bottom with your knuckles, about 30 minutes. You can also check for doneness using a thermometer; the internal temperature should be around 210°F / 98.5°C the temperature at which starches in flour break down (the bread needs to bake a bit longer to be done, though) (see page 306 in Chapter 6 for more about starch gelatinization).

Allow the bread to cool for at least 30 minutes or so before slicing; it needs to cool sufficiently for the starches to gelatinize and set.

Notes

- *If even at the ideal rising and baking temperatures your bread is still coming out too dense, try reducing the amount of whole wheat flour to 1 cup (120g) and increasing the bread flour to 2 cups (240g).*
- *For an even simpler bread, see the interview with Martin Lersch on page 224 earlier in this chapter, or search online for "no-knead bread." Mark Bittman of the* New York Times *describes a technique used by Jim Lahey, a baker in New York, in which the dough is left to sit for a day, during which the gluten forms without kneading.*
- *For a slightly more complicated method, try starting with a sponge: a prefermentation of flour, water, and yeast that allows for better flavor development. Instead of adding all the flour and water together at the beginning, mix half of the flour (180g) with 4/7 (140g) of the water (ideally, at 75°F / 24°C—if it's any warmer, oxidation will impact the flavor) and all of the yeast (7g), and allow that to rise until bubbles start to form on the surface and the sponge starts to fall. Once this stage is reached, mix the sponge up with the rest of the water (100g), add the rest of the flour (180g) and salt (7g), and allow the mixture to rise per the earlier instructions. For more details, see Edward Espe Brown's* The Tassajara Bread Book *(Shambhala).*
- *While the exact science of what causes bread to go stale is still unknown, a couple of different mechanisms are reasonable suspects. One thought is that, upon baking, starches in flour convert to a form that can bind with water, but that they slowly recrystallize after baking and in doing so release the water, which then gets absorbed by the gluten, changing the texture of the crumb. Then there's the crust, which draws away some moisture from the middle of the bread, causing the texture of the crust to change. Regardless of the exact mechanism, storing bread in the fridge speeds up these changes in texture while freezing does not, so keep your bread at room temperature or freeze it. (The only benefit to storing bread in the fridge is that it slows the growth of some types of mold.) Toasting the bread above the temperature at which starches gelatinize reverses some of these changes.*
- *Try adding rosemary, olives, or diced and sautéed onion during the second kneading. Or, use only bread flour and add some large chunks of bittersweet chocolate.*

Yeast Waffles

Baker's yeast contains a number of enzymes, one of which, zymase, converts simple sugars (dextrose and fructose) into carbon dioxide and alcohol. It's this enzyme that gives yeast its rising capabilities. Zymase doesn't break down lactose sugars, though, so doughs and batters made with milk will end up tasting sweeter. This is why some bread recipes call for milk and why foods like yeast waffles come out with a rich, sweet flavor.

At least two hours in advance, but preferably the night before, measure out and whisk together:

1¾ cups (450g) milk (whole, preferably)

½ cup (115g) melted butter

2 teaspoons (10g) sugar or honey

1 teaspoon (6g) salt (table salt—not the kosher or flaky type)

2½ cups (300g) flour (all-purpose)

1 tablespoon (10g) instant yeast (*not* active dry yeast)

2 large (120g) eggs

Cover and store at room temperature. Make sure to use a large bowl or container with enough headspace to allow the batter to rise.

Briefly stir the batter and then bake in your waffle iron per instructions of your waffle iron manufacturer.

Notes

- *In baking, use table salt, not kosher or flake salt, because the finer-grained salt will mix more uniformly into the batter.*
- *Try using honey, maple syrup, or agave nectar instead of sugar, and try substituting whole wheat flour or oat flour for half of the all-purpose flour.*
- *If your waffles come out not as crispy as you like, toss them in an oven preheated to 250°F / 120°C—hot enough to quickly evaporate out water, cold enough to avoid caramelization and Maillard reactions.*

Pizza

If there's one stereotypical geek food, it would have to be pizza: ubiquitous, cheap, and cheesy. But the stuff sold in your local strip mall is far inferior to what you can make at home. It's like the difference between canned fruit and the fresh thing: both can be good, but the fresh version is distinctly more nuanced.

Start by making pizza dough. You can also buy pizza dough at your grocery store, although I find I get better results when I make it from scratch (see 238 for a simple no-knead pizza dough recipe, or 248 for a yeast-free pizza dough recipe).

Set out a large cutting board and sprinkle a handful of flour in the center area. Preheat your oven to at least 450°F / 230°C. Take about 1 lb (450g) of the dough and form it into a ball between your hands, kneading and folding it over. The dough should be just slightly sticky, but not so much that it actually remains stuck to your hand. If it's too sticky, add more flour by dredging it in the flour on the cutting board. Continue to work the dough until it reaches a firm consistency and has good elasticity when stretched. Begin to work the dough into a flat, round disc, and then roll it into a round pizza shape.

Par-bake the pizza dough by baking it on a pizza stone in a hot oven. You can transfer the pizza dough by carefully picking it up and laying it onto the stone; don't burn yourself! If you don't have a pizza stone (although I highly recommend them—see 42 in Chapter 2 for how they can be used to improve your oven), you can use a cast iron pan, upside down, to similar effect. Let the pizza bake for three to five minutes, until the dough has set. If the dough puffs up in one place, use a chef's knife to poke a small hole in the bubble and then use the flat side of the knife blade to push the puffed portion back down. Par-baking the dough isn't traditional, but it'll help avoid soggy, undercooked dough and also makes transferring the topped pizza into the oven a heck of a lot easier. It simplifies the cooking of the pizza, too: cook the dough until it's effectively ready, and then cook the toppings until they melt and fuse, as opposed to trying to get both to occur at the same time.

Once the pizza dough has been par-baked, remove it from the oven and place it on your cutting board. Add sauce and toppings. The sauce can be anything from a thin coating of olive oil to traditional tomato sauce. Or make a white cheese sauce, as described on page 116 in Chapter 3. For toppings such as onions and sausage, sauté them before placing them on the pizza. Cooking the dough and toppings separately removes all the constraints associated with the various ingredients needing varying cooking times, leaving just three goals: melting the cheese to fuse the ingredients together, browning the edge of the crust, and browning the top surface of the toppings. Finish cooking by transferring the dressed pizza into the oven (using a pizza peel or, in a pinch, a piece of cardboard) and baking it until any cheese is melted and the pizza has begun to turn golden brown, about 8 to 12 minutes.

Wet ingredients such as scallops can make for soggy pizzas. You can avoid this by cooking ingredients individually, draining off any excess liquid, and then adding them to the pizza.

Jeff Varasano on Pizza

PHOTO USED BY PERMISSION OF JEFF VARASANO

Jeff Varasano moved from New York to Atlanta, where a lack of New York–style pizza drove him to years of experimenting—to the point where he clipped the lock on his oven so that he could bake pizza in a super-hot oven set to its cleaning cycle. He eventually quit his job as a C++ programmer and opened Varasano's Pizzeria in Atlanta.

How did you go from C++ programming to making pizza?

I moved from New York to Atlanta. Like a lot of people transplanted from the Northeast, I started to seek out the best pizza. A lot of places claim to be like New York, and you go there and you're like, "Hmm, have these guys ever been to New York?" So I started to bake at home. At first I would just call up all of my friends and say, "Look, I'm making pizza tonight. It's going to be pretty terrible, but why don't you come try it?" And it really was pretty bad.

I started experimenting. I did all the flours. I experimented with different methods of heating my oven. I tried to do it on the grill. I tried to wrap my oven in aluminum foil to keep all the heat in. Then I moved to a new house and I had an oven with a cleaning cycle. I didn't really know what a cleaning cycle was. I had never had an oven with a cleaning cycle, but I ran it and I realized that it was basically just incinerating the contents. It was like, "Aha, I've got to get in there!" So that's where the whole idea of clipping the lock came from.

I threw up this website (now at *http://www.varasanos.com/PizzaRecipe.htm*). I really didn't think too much about it. For a year and a half the counter was at about 3,000 and in a day it jumped from 3,000 to 11,000 and crashed my server. I realized that people were pounding that page and pretty much from that day forward I started to get email. That's what started me down the whole tunnel of thinking about giving up the software stuff and going into pizza.

In the process of learning how to do your pizza, what turned out to matter more than you expected, and on the other side, what turned out to matter less?

Well, clearly what mattered less was the flour. Everyone is looking for the piece of equipment or secret ingredient that they can buy which will all of a sudden transform their pizza into something great. It's not that. This is one of the things I realized early on. There is no magic bullet. If you look at the top five pizzerias on my list, you'll see they use five different ovens: gas, wood burning, coal burning, electric, and believe it or not, an oil burning oven. Not only do they use different fuels, they're different shapes, they're different temperatures, some bake their pizza for two minutes, some seven. So what is it then? The answer is that it's an art, it's everything all together at that one moment. That's what I realized, learning the basics and the fundamentals, you come into style and artistry and that's much more difficult to define. It's not going to be a single secret.

A lot of geeks who are learning to cook get hung up on the very small details and miss the big picture of just getting in there and trying something and playing with it.

Yeah. I've always been an experimenter. But I've always had sort of a different way of approaching problems. I don't make very many assumptions about the way things should be done. Most people assume that knowing how things should be done is the best way, so they keep struggling within a very small circle, whereas I have a tendency to just try a much wider variety of things that may work and may not work.

So when you get stuck on one of these problems even though you're working in a wider circle, how do you go about getting unstuck?

That's an interesting question. Let me deviate from that slightly and then I'll come back. Most people are familiar with the scientific method, which is holding everything exactly the same and changing this one thing. This reminds me of people trying to do one side of the Rubik's Cube. Most of the good methods don't involve getting any side. That's the last thing you do. So people get stuck because they don't want to toss in the towel on the progress they think they've made so far. So if you want to make it past one level, you may have to scrap your whole methodology and just start over. And you see that with pizzas.

Art begins where engineering ends. Engineering is about taking what's known and carrying it to its logical conclusion. So what do you do when you have exploited everything you know, but you want to go to the next level? At that point, you have to start opening your mind up to completely random ways of thinking through something. That might involve taking multiple steps at a time. It might be that you don't abandon one thing, but you have to abandon five things.

As an example using pizza, as soon as I switch flour, I can't just keep the same hydration because if I change the flour then I may also have to change the water, or the dough may have a different consistency. Well, guess what, when I increase the hydration then the heat penetration into the dough is going to be slower because more of that water has to boil off. So now all of a sudden I might have to change the oven temperature, too. I'd love to conduct a controlled experiment that would conclude that Flour B is better than Flour A, holding all other variables constant. But in the real world such a test is somewhat meaningless. This is why it's an art.

This makes a lot of sense. I think a lot of geeks out there would say that this would be a multivariate approach to finding one of these optimal points of pizza recipes and techniques.

That's right. And you have to work on the underlying forces and begin to understand them independently, but in the end the results are not going to be a set of independent things, they're going to be a set of interdependent things.

In the first stage of working a problem or trying to master a skill, you find that everything seems totally dependent and that's when you have the least power. The next stage is to make things independent and to break things down and classify them. The whole idea is to segment things into finer and finer individual techniques. The ultimate stage is learning how to reconnect all of those parts that you separated out and now reorganizing them into something where the pieces are interdependent rather than a collection of things that are independent.

I am at the middle stage myself, so I don't quite see how all the pieces fit together. For example, if we don't leave the heater on in the restaurant, then the dough warms up overnight at a different rate than it did a couple of days ago. I think, well, there really doesn't seem to be that much difference but I know there was that two-degree difference, so I'll correct for it. I'll think I'm back where I started, but I am not. And then sometimes you don't even know what's different and then you just literally scratch your head. In a year it will be obvious what was different.

Can you give me an example?

One of the ingredients I had given pretty minimal thought to—and didn't realize how important it was—was oregano. I have a little herb garden in front of my house and I grow some oregano. I didn't like the strain I had. One day I found a better sample in an abandoned herb garden. I dug it up and I put it in my front yard and used it.

So now I'm ready to launch the restaurant and I'm going to all my suppliers looking for oregano. Thirty-three oreganos later, I'm still sitting here saying none of them tastes like the one that I grew in my garden.

You don't realize that there is a difference to be worked on, but that's when you're caught with your guard down. The oregano that I really, really like is a year away from production quantity so now I'm experimenting; maybe there's a better way to dry the oreganos that I have. If I get a fresh one, maybe I can dry it differently and maybe it's the drying process will give me something closer to what I want. So now I've gone down the tunnel trying five, six, or seven ways of drying it; heated drying using a dehydration machine that blows a fan and a little bit of heat over it using dehumidifiers and all these different things.

So it sounds like your method for overcoming this is to try a lot of different things?

It really is, and you know it's funny because I like to say, well, how do you know? I tried everything and a lot of people think, wow, it's amazing you figured this out! People think there is some sort of secret magic, but the problem is that when you get to the end of what's known, when you get to the end of engineering, you're left with hunch and trial and error, but those carry you much farther than people often give them credit for.

Pizza Dough—No-Knead Method

This makes enough dough for one medium-sized pizza with the crust rolled thin. You'll probably want to multiply these quantities by the number of people you're cooking for.

Weigh into a large bowl or plastic container:

1⅓ cups (170g) flour

1 teaspoon (5g) salt

1 tablespoon (10g) instant yeast

Using a spoon, mix together so that the salt is thoroughly distributed. Add:

½ cup (120g) water

Mix in the water using the spoon so that the flour and water are incorporated.

Let rest on counter for at least four hours, preferably longer. You can mix the ingredients together at breakfast time (for example, before running off to that day job at Initech or wherever) and the dough will be ready by the time you get home. It's the same principle as the no-knead bread: the glutenin and gliadin proteins will slowly crosslink on their own.

You can cut and serve pizza directly off the peel. If you don't have a pizza peel, you can use a piece of cardboard to slip a pizza into and out of the oven.

Notes

- *I have a confession to make: when it comes to pizza dough, I'm lazy and don't worry about exact hydration levels, proper kneading method, ideal rest times, and controlling temperature to generate the ideal flavor.*
- *If you want to experiment, order some sourdough yeast culture (which is actually a culture of both the well-known sourdough strain of yeast and the bacteria* lactobacillus*). The ratio of yeast to bacteria in the dough will impact the flavor. You can control that ratio by letting the dough mature for some amount of time in the fridge, where yeast will multiply but bacteria won't; and some amount of time at room temperature, where the bacteria will contribute flavors. If you want to explore these variables, read Jeff Varasano's web page on pizza—see the interview with him on page 236 for details.*

Chemical Leaveners

While yeast allows for the creation of many delicious foods, it has two potential drawbacks: time and flavor. Commercial bakers with high volumes and those of us with limited time to play in the kitchen can't always afford to wait for yeast to do its thing. Then there're the flavors and aromas generated by yeast, which would clash with the flavors in something like a chocolate cake. Chemical leaveners have neither of these problems.

Chemical leaveners are divided into two categories:

Baking soda
: A bicarbonate (HCO_3^-) that's bound with a sodium atom (related compounds use potassium or ammonium to similar effect). When added to water, the bicarbonate dissolves and is able to react with acids to generate CO_2.

Baking powder
: A self-contained leavening system that generates carbon dioxide in the presence of water. Baking powders by definition contain a baking soda and acids for that baking soda to react with.

The idea that these are categories, not single ingredients, is probably foreign to most home cooks, but the chemicals that make up a baking powder or baking soda can vary. Industrial food manufacturers use different compositions and particulate sizes depending upon the food being produced.

Baking Soda

Anyone who's done the third-grade science fair project using vinegar and baking soda to make a volcano can tell you that baking soda can generate a whole lot of gas really quickly. But in the kitchen, baking soda remains one of the bigger mysteries. How is it different from baking powder? And how do you know which one to use?

Always sift dry ingredients together before adding in wet ingredients to make sure any salt, baking soda, or baking powder are truly dispersed. You can use a strainer over a bowl as a sifter or even just mix the ingredients with a wire whisk or a fork.

The quick answer would go something like: "Baking soda reacts with acid, so only use it when your ingredients are acidic." And as simple explanations go, this covers you 99% of the time when cooking. But baking soda is a little more complicated and interesting in a geeky way, so it's worth a brief digression into the chemistry. I promise this'll be short.

The baking soda you buy in the store is a specific chemical: sodium bicarbonate, $NaHCO_3$. Unlike baking powder, which is a blend of chemicals that are self-contained ("just add water and heat!"), when added to a dish, sodium bicarbonate needs something to react with in order to generate gas.

Without something for sodium bicarbonate to dissolve into, it's an inert white powder. Upon getting wet—any moisture in any food will do—the sodium bicarbonate dissolves, meaning that the sodium ions are free to run around separately from the bicarbonate ions.

> The sodium is just there to transport the bicarbonate to your food; we can ignore it once it's dissolved. The sodium does make the food slightly saltier, incidentally, which is why industrial food manufacturers will sometimes use things like potassium bicarbonate: potassium is good for you, and this avoids the sodium for people on a low-sodium diet.

Most of us are familiar with the pH scale (the H stands for hydrogen; it's unclear what the p stands for, "power" and "potential" are the best guesses). The pH scale is a measure of the amount of available hydrogen ions in a solution. Chemicals that affect the number of hydrogen ions can be classified in one of two ways:

Acids (pH below 7)

Proton donors; i.e., chemicals that increase the number of hydronium ions (H_3O^+; the hydrogen binds with a water molecule) in the solution

Bases (pH above 7)

Proton receivers; i.e., chemicals that bind with hydronium ions, reducing their available concentration in a solution

When it comes to pH, a bicarbonate ion has an interesting property that chemists call *amphotericity*: it can react with either an acid or a base. In the kitchen, so few things are actually basic—egg whites, baking soda, maybe the stuff in your fire extinguisher, and that's pretty much it—that you can safely ignore baking soda's ability to react with bases and just think of it as something that reacts with acids. Still, to understand baking soda, it's important to understand that bicarbonates react with other compounds and either raise the pH by reducing the amount of available acids or lower the pH by reducing the amount of available bases.

This phenomenon is called *buffering*: a *buffer* is something that stabilizes the pH level of a solution. Buffers hang out in the solution and, when an acid or base is added, glom on to it and prevent it from affecting the count of available hydronium ions. In a glass of pure water, there's not much for the bicarbonate ions from baking soda to interact with, so they just float around and taste generally nasty. But if you were to add a spoonful of vinegar—which is acetic acid—to that glass, the bicarbonate ions would react with the acetic acid and generate carbon dioxide as part of that reaction.

Depending upon the amount of bicarbonate you started with, after you add the spoonful of vinegar the glass will be in one of three states (none of which involve being half-full or half-empty): bicarbonate ions still available but no acetic acid ions available, no bicarbonate ions available but acetic acid ions still available, or neither bicarbonate nor acetic acid ions freely available. In baking, it's this last state—a neutral balance—that we want to reach. Too much baking soda, and it won't all react with the acids in the food and will leave the food with a soapy, yucky taste. Not enough baking soda, and the food will remain slightly acidic (which is okay) and not have as much lift as possible (which is probably not okay—your food will be flat). To repeat one of my favorite quotes: "Dosage matters!"

The reaction between baking soda and an acid is the key to understanding when you should use baking soda versus baking powder. This balancing act between acids and baking soda isn't a problem with baking powder, of course. This is because the baking powder is already balanced for you—the ratio of acids to bicarbonate is preset by the manufacturer.

If your ingredients aren't very acidic, baking soda won't have much to react with, so use baking powder. On the other hand, if your ingredients are extremely acidic, using baking soda will work, since there will be enough hydronium ions to react with. How much baking soda to use depends on the pH of the ingredients in your dish. Short of testing or calculating the pH, experimentation is the easiest way: take a guess and keep notes. Keep adding baking soda until the additional baking soda no longer helps with lift (or can be tasted). If you're still not getting enough lift at this point, switch to adding baking powder.

Baking soda doesn't need an acid to decompose; heat will do it, too. Try melting some sugar, just as though you were making caramel (see 212 in Chapter 4), and instead of adding cream, add a small spoonful of baking soda and stir. The baking soda will break down and cause the sugar to bubble up.

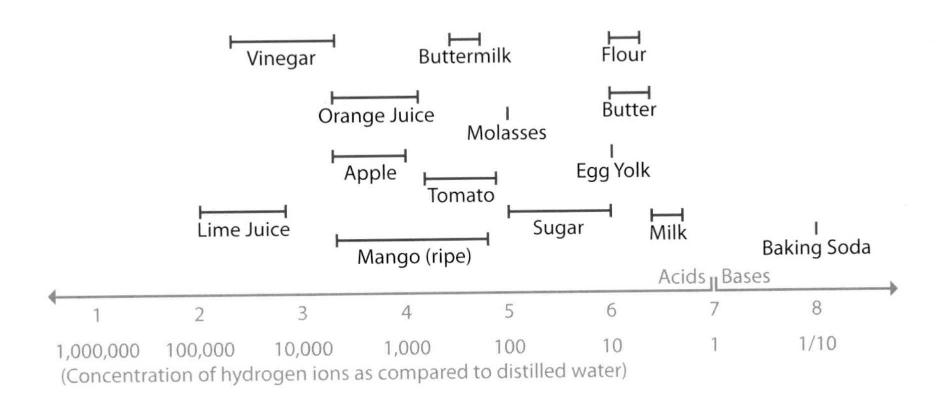

The pH of common ingredients.

Buttermilk Pancakes

Given time, yeast and bacteria generate flavors that we often find pleasant. But what about those times when you're craving that taste right now—or at least, sometime this morning? You can take a shortcut by using buttermilk, which has already been munched on by bacteria.

Whisk together to combine thoroughly:

2 cups (240g) bread flour

5 tablespoons (60g) sugar

1½ teaspoons (7g) baking soda

1 teaspoon (5g) salt

In a separate bowl, melt:

½ cup (115g) melted butter

In the same bowl as the butter, add and whisk together:

2½ cups (610g) buttermilk (lukewarm!)

2 large (120g) eggs

Mix the wet ingredients into the dry, stirring with a whisk or spoon to combine. Cook on a griddle or nonstick frying pan set over medium heat (if you have an IR thermometer, 325–350°F / 160–175°C) until golden brown, about two minutes per side.

Notes

- *You don't need to butter the griddle or pan before cooking these—there is enough butter in the batter that the pancakes are self-lubricating—but if you do feel the need, wipe any excess butter out of the pan before cooking the pancakes. If you have any dots of oil on the surface, they'll interfere with the Maillard browning reactions.*
- *Pull the buttermilk and eggs out of the fridge an hour or so before you're ready to use them, to allow them to come up to room temperature. If you're in a rush, you can double-duty a microwave-safe mixing bowl: melt the butter in it, add the buttermilk, then nuke it for 30 seconds to raise the temperature of the buttermilk.*

Try using this batter for buttermilk fried chicken. Slice cooked chicken into bite-sized pieces, dredge them in cornstarch, dip them in this batter, and then deep-fry them in vegetable oil at 375°F / 190°C. The starch will help the batter adhere to the chicken. (No cornstarch? Use flour.) For the ideal texture, cook the chicken sous vide, as described on page 333 of Chapter 7.

Gingerbread Cookies

Chemical leaveners aren't always used to create light, fluffy foods. Even dense items need some air to keep them enjoyable.

In a bowl, mix together with a wooden spoon or electric beater:

½ cup (100g) sugar

6 tablespoons (80g) butter, softened but not melted

½ cup (170g) molasses

1 tablespoon (17g) minced ginger (or ginger paste)

In a separate bowl, whisk together:

3¼ cups (400g) flour

4 teaspoons (12g) ginger powder

1 teaspoon (5g) baking soda

2 teaspoons (3g) cinnamon

1 teaspoon (1g) allspice

½ teaspoon (2g) salt

½ teaspoon (2g) ground black pepper

Sift the dry ingredients into the bowl with the sugar/butter mixture. (I use a strainer as a sifter.) Work the dry and wet ingredients together using a spoon or, if you don't mind, your hands. The dough will get to a crumbly, sand-like texture. Add ½ cup (120g) water and continue mixing until the dough forms a ball.

Turn out the dough onto a cutting board coated with a few tablespoons of flour. Using a rolling pin, roll out the dough until it is about ¼" (0.6 cm) thick. Cut it into shapes using a cookie cutter or a paring knife and bake them on a cookie sheet in an oven set to 400°F / 200°C until cooked, about eight minutes. The cookies should be slightly puffed up and dry, but not overly dry.

Baking gingerbread cookies is, of course, a great holiday activity with kids.

Gingerbread Cookie Frosting

In a microwave-safe bowl, mix together with a fork or electric beaters:

3 tablespoons (40g) butter, softened but not melted

1 cup (200g) powdered sugar

1 tablespoon (15g) milk

1 teaspoon (4g) vanilla extract

Add food coloring if desired. Microwave the frosting for 15 to 30 seconds—long enough to melt the frosting, but not so long that it boils. This will give you a frosting that you can then quickly dip the cookies into and that will set into a nice, thin coating that adheres well to the cookies.

One-Bowl Chocolate Cake

I have a thing against cake mixes. Sure, commercial mixes produce very consistent results—they use food additives and stabilizers exactly calibrated for the other ingredients in the cake mix—but even for a quick birthday cake, you can make a truly homemade one that actually tastes like chocolate without much more work.

Cakes are commonly made using a *two-stage method*, in which dry ingredients are weighed out and whisked in one bowl, wet ingredients are whisked in a second bowl, and then the two are combined. In the *streamline method*, all ingredients are mixed in the same bowl: first dry (to make sure the baking powder is thoroughly blended), then wet, then eggs.

In a *large* bowl or the *large* bowl of a mixer, measure out:

2¼ cups (450g) sugar

2 cups (240g) pastry or cake flour (all-purpose flour is okay, too)

¾ cup (70g) cocoa powder (unsweetened)

2 teaspoons (10g) baking soda

½ teaspoon (2g) salt

Whisk together the dry ingredients, then add to the same bowl and whisk to combine thoroughly (about a minute):

1½ cups (360g) buttermilk

1 cup (218g) canola oil

1 teaspoon (5g) vanilla extract

Add eggs and whisk to combine:

3 large (180g) eggs

Prepare two 9" / 22 cm or three 8" / 20 cm round cake pans by lining the bottom with parchment paper. Yes, you really need to do this; otherwise, the cakes will stick and tear when you try to remove them. Spray the paper and pan sides with nonstick spray or coat with butter, and then dust with either flour or cocoa powder.

The parchment paper doesn't need to cover every last square millimeter of the bottom of the cake pan. Cut a square of parchment paper, and fold it in half, then in quarters, and then in eighths. Snip the top off the folded paper, unfold your octagon, and place it in the pan.

Divide the batter into the cake pans. Try using a scale to keep the weights of the pans the same; this way the cakes will all be roughly the same height.

Bake in an oven preheated to 350°F / 175°C until a toothpick comes out clean, about 30 minutes. Allow to cool before turning out and frosting. If your cakes sink in the middle, your oven is probably too cold. Check that it is calibrated correctly; see page 42 in Chapter 2 for details.

Even professional bakers use toothpicks to check doneness. For brownies, check that a toothpick inserted 1" / 2.5 cm deep comes out clean; for cakes, push the toothpick in all the way.

Notes

- *When placing the cake pans in the oven, put them on a wire rack in the middle of the oven. If you keep a pizza or baking stone in your oven (which is recommended), don't set the cakes directly on the stone; put them on a rack above the stone.*
- *Like buttermilk, baking cocoa powder is acidic!* Dutch process *cocoa powder, however, is alkalinized—that is, it has had its pH level adjusted, changing it from a pH of around 5.5 to a pH between 6.0 and 8.0, depending upon the manufacturer. Don't just blindly substitute Dutch process cocoa powder for straight-up cocoa powder; some of the baking soda will need to be switched out for baking powder.*
- *You can sometimes make a buttermilk substitute by adding 1 tablespoon (15g) of vinegar or lemon juice to 1 cup (240g) of milk. This will adjust the pH to be roughly the same as that of a cup of buttermilk, but it will not create the same texture or thickness, so don't use that substitute for this recipe. If you don't have buttermilk, use regular milk and substitute baking powder for half of the baking soda.*
- *Note the baking temperature of 350°F / 175°C—as discussed on page 210 of Chapter 4, sugar doesn't begin to noticeably brown until above this temperature. You can tell a lot about what kind of baked good you'll end up with just by looking at the baking temperature.*

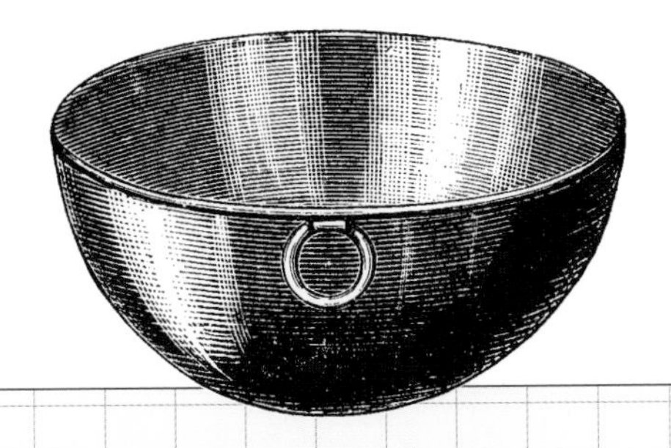

Simple Chocolate Ganache Frosting

In a saucepan over medium heat, heat 1 cup (240g) of heavy cream until it just begins to boil. Remove from heat and add 2 tablespoons (30g) of butter and 11½ oz (325g) of finely chopped bittersweet chocolate. (You can use semi-sweet chocolate if you prefer your cakes on the sweeter side. Most chocolate chips are semi-sweet; try using those.) Allow to rest until the chocolate and butter have melted, about five minutes. Add a pinch of salt and whisk to thoroughly combine.

To frost the cake, you can just pour the still-warm ganache over the top, allowing it to run down the sides. (This can get messy—which can serve as an excuse for eating half the ganache while frosting.) Or, to create a more traditional frosting, allow the ganache to set in the fridge, about 30 minutes, and then use an electric beater or mixer to beat it until it's light and fluffy. Coat the top of each layer of the cake with the whipped ganache and stack them, leaving the sides exposed.

Notes

- *Try spiking the frosting with a tablespoon of espresso or liquor such as rum, port, or Grand Marnier. For a tangier frosting, substitute buttermilk for half of the heavy cream. If you really want to push the boundaries, trying using anything you think would work in a truffle. Cinnamon is easy to imagine, but why not cayenne pepper or lavender? Or infuse the cream with Earl Grey tea.*
- *Make sure your cake is cool before frosting; otherwise, the heat will melt the ganache.*

Meg Hourihan's Mean Chocolate Chip Cookies

PHOTO USED BY PERMISSION OF MEGHAN HOURIHAN

Meg Hourihan co-founded the company that created Blogger.com, one of the Internet's first blogging platforms.

Tell me about yourself and what you do with food.

I started cooking, mostly baking, making very elaborate cakes, when I was eight or nine. I always liked cooking and technology, and went back and forth for a while. I had been working on the Web and doing the startup thing for a long time. I got burned out and decided to take some time off of the Internet, so I got a job working at a restaurant in a kitchen. It was a nice change of pace.

What similarities and differences between developing software and working in the restaurant did you find?

The kitchen has a very specific life cycle, just like a software project does, but it's incredibly compressed. Every morning you come in and do your prep work, almost like your requirements phase. You figure out what you're going to need to make it through the dinner service. You then get in the moment of cooking and then after the last order has gone out you break down your kitchen and clean up your station. The whole life cycle comes to an end and you have a chance to say, "What did we do well? What can we improve on?" You get to learn from your mistakes, and the next day the whole process starts again. It reminds me of web-based stuff: once your product has launched, you can push an update every single day and respond to customer feedback rather than a packaged software cycle where you're disconnected from end users. When you're working in a restaurant, your end users will tell you in five minutes if that dish is no good and it's going to come right back to the kitchen. You find out pretty quickly if you're doing it right and who you're doing it for because they're right on the other side of the wall.

Have there been any real surprises in the learning process of becoming a better cook?

It's one of those things that takes a lot of work. I'm lucky that I like doing it and have been doing it so long; I have this knowledge base to fall back on. My husband was making a soup the other day. The recipe said to cook the vegetables on medium heat for 45 minutes. He sent me a text message: "We're not having the soup tonight, the vegetables are burned to a crisp." I looked in the pot when I got home. The vegetables were just carbonized. I said, "Oh my god, you can't cook these tiny little vegetables for 45 minutes on medium heat. This is exactly what's going to happen!" He was so mad. He said, "But this is what the instructions told me. I was following the recipe!" If I had been making it, I would have known that that couldn't possibly be right; I have enough experience. If you don't have the confidence, the recipe becomes your crutch and you forget to back up and rely on common sense.

Is there a dish that you are particularly fond of?

When I was doing my blog I had asked people for chocolate chip cookie recipes. I was tired of the one I was using and I said, "If you send me your recipe, I'll cook it to discover the best chocolate chip cookie recipe." I probably got 30 or 40 different recipes and I realized, "Holy cow! There's no way I'm going to be able to make and evaluate all these cookie recipes in any reasonable amount of time." In talking with my husband, we decided that we were going to average all of the recipes and then make that cookie, whatever the result was. The recipe is crazy. Heat the oven to 354.17°F / 178.98°C. Use 1⅓ eggs. It's all these impossible measurements because I just averaged across all the ingredients. You think this is going to be a really gross cookie because you just cobbled together all of these things and you can't possibly average together 40 cookie recipes or whatever it was, but it turned out pretty good.

A Mean Chocolate Chip Cookie

Preheat oven to 354.17°F / 178.98°C, or as close as you can get.

In a medium bowl, sift or thoroughly whisk together:

2.04 cups (245g) flour
0.79 teaspoons (3.81g) salt
0.79 teaspoons (3.63g) baking soda

Set dry ingredients aside.

In another bowl, using a hand or stand mixer, cream until incorporated and smooth:

6.44 tablespoons (87.9g) unsalted butter, softened to room temperature
2.1896 tablespoons (29.9g) unsalted butter, cold
4.2504 tablespoons (58g) unsalted butter, melted
0.84 cups (169g) light brown sugar
0.10 cups (20g) dark brown sugar
0.54 cups (109g) white sugar

Add and mix until all ingredients are combined:

1.33 (46g) eggs
0.33 (8g) egg yolk
1.46 teaspoons (6.08g) vanilla extract
0.17 tablespoons (2.51g) water
0.25 tablespoons (3.84g) milk
1.53 cups (257g) semi-sweet chocolate chips

Add dry ingredients and blend until fully incorporated.

Cover and chill dough in the refrigerator for 25 minutes.

Place parchment paper on one-third of cookie sheet, drop dough by rounded tablespoons onto sheet. Some cookies will be on parchment, others off. Cook for 13.04 minutes.

Baking Powder

You did calibrate your oven, yes? If not, see the sidebar "The Two Things You Should Do to Your Oven RIGHT NOW" on page 42 in Chapter 2.

Baking powder solves the "balancing act" problem encountered when using baking soda by including acids alongside the bicarbonates. And since the acids are specifically mixed into baking powder, they can be optimized for baking; you don't have to rely on whatever acids happen to be present in the food being made.

Baking powder, at its simplest, can be made with just one type of bicarbonate and one type of acid. This is why, in a pinch, you can make your own baking powder: 2 parts cream of tartar to 1 part baking soda. Cream of tartar—potassium hydrogen tartrate—will dissolve in water, freeing tartaric acid ($C_4H_6O_6$) to react with the sodium bicarbonate.

Commercial baking powders are a bit fancier than this, though. Different acids have different rates of reaction and reaction temperatures, so using multiple types of acid allows for the creation of a baking powder that's essentially time-released. This isn't just clever marketing: in baked goods, if the CO_2-generating reaction occurs too slowly, you'll end up with a dense, fallen product. And if those reactions happen too quickly, the food won't have time to properly set so as to be able to hold on to the gas, resulting in things like collapsed cakes.

Some people find that baking powder made with sodium aluminum sulfate tastes more bitter than that made with other acids, such as monocalcium phosphate.

Double-acting baking powder—this is the stuff you'll find at the grocery store—uses both slow- and fast-acting acids to help prevent these types of problems. Fast-acting acids, such as tartaric acid (in cream of tartar) and monocalcium phosphate monohydrate, can work at room temperature; slow-acting acids, such as sodium aluminum sulfate, need heat and time to release CO_2. As long as the ratio of ingredients in your baked products is roughly correct and you're baking within an acceptable temperature range, baking powder is unlikely to be the culprit in failed baking experiments.

Still, if you're getting unexpected results with a commercial baking powder, check whether your ingredients are highly acidic. Acidity impacts baking powder; more acidic ingredients in a recipe will require less baking power. If that doesn't turn up any suspects, check how long it has been since the baking powder was opened. Even though commercial baking powders contain cornstarch, which absorbs moisture to extend the shelf life, the chemicals in baking powder will eventually react with each other. Standard shelf life is about six months after being opened.

Pizza Dough—Yeast-Free Method

While baking powder is most commonly used in sweets, it can be used in savory applications, too. Try making a quick-rising pizza dough—especially handy if someone has a yeast allergy.

Whisk 3–4 cups (360–480g) of flour with 1 teaspoon (6g) of salt and 2 teaspoons (10g) of baking powder. Add 1 cup (240g) of water and knead to create a dough that has roughly a 66–75% hydration level. Let rest for 15 minutes and then proceed with par-baking instructions as described on page 235.

Pumpkin Cake

There are two broad types of cake batters: high-ratio cakes*—those that have more sugar and water than flour (or by some definitions, just a lot of sugar)—and* low-ratio cakes*—which tend to have coarser crumbs. For high-ratio cakes, there should be more sugar than flour (by weight) and more eggs than fats (again, by weight), and the liquid mass (eggs, milk, water) should be heavier than the sugar.*

Consider this pumpkin cake, which is a high-ratio cake (245g of pumpkin contains 220g of water—you can look these sorts of things up in the USDA National Nutrient Database, available online at http://www.nal.usda.gov/fnic/foodcomp/search/*).*

In a mixing bowl, measure out and then mix with an electric mixer to thoroughly combine:

1 cup (245g) pumpkin (canned, or roast and puree your own)
1 cup (200g) sugar
3/4 cup (160g) canola oil
2 large (120g) eggs
1 1/2 cups (180g) flour
1/4 cup (40g) raisins
2 teaspoons (5g) cinnamon
1 teaspoon (5g) baking powder
1/2 teaspoon (5g) baking soda
1/2 teaspoon (3g) salt
1/2 teaspoon (2g) vanilla extract

Transfer to a greased cake pan or spring form and bake in an oven preheated to 350°F / 175°C until a toothpick comes out dry, about 20 minutes.

Notes

- *Try adding dried pears soaked in brandy. You can also hold back some of the raisins and sprinkle them on top.*
- *One nice thing about high-ratio cakes is that they don't have much gluten, so they won't turn out like bread, even with excessive beating. With a total weight of 920 grams, of which only roughly 20 grams is gluten, there just isn't enough gluten present in this cake to give it a bread-like texture. There's also a fair amount of both sugar and fats to interfere with gluten development.*

If you're making a quick cake like this pumpkin cake as the finale to an informal dinner party, try serving it directly on a single plate or even a cutting board. Besides lending a pleasant casual feel, this'll mean fewer dishes to wash!

Tim O'Reilly's Scones and Jam

Tim O'Reilly is the founder of O'Reilly Media, which started out as a publisher of technical books and has more recently branched out into offering content in a variety of media, running technical conferences, hosting online and in-person workshops, and creating other ways to spread the knowledge of innovators. (This book is published by O'Reilly Media.) Above, Tim shows his method for drying apples: sliced into rings, arranged on a window screen, and left to dry in the hot and dry California sun.

You say you don't consider yourself a foodie at all?

No. In fact, I kind of make a small number of things that I make repetitively. A lot of what I do is driven by the fact that I hate to waste things. So hence jam because there's all this great fruit. [Tim has numerous fruit trees.] Right now I'm doing dried apples. But let me put these scones in. [Tim had been making scones as we started.] This is something that I figured out a long time ago. I make this big batch and it's too much for two people so I made a batch and then I was like, oh wait, I can just freeze it.

How did the thought of freezing it come to you?

Oh, I don't know, it was just sort of like duh. It's sort of like so obvious. You just make it and freeze it and then I have it and I can throw in a bunch. When somebody visits it literally just takes me a few minutes. The raspberry jam—I have raspberries, but I don't have enough to make jam all at once, but I'll go out and pick them every day and now you can see what I've now got... [Tim holds up a bag of frozen raspberries.] By the time I get two of these bags I'll have enough to make raspberry jam. You don't have to do it all at once.

What's your favorite kitchen tool?

I like things that seem magical. When you see this particular apple peeler-corer-slicer, you'll go, "Oh! That's so cool! It's magical." It just does a fantastic job.

Tim O'Reilly's Jam-Making Tips

Tim says there are two secrets for making jam:

- *Use a low-methoxyl pectin, such as Pomona's Universal Pectin. Unlike standard pectin, which requires sugar to create a gel, Pomona's is activated by calcium. This basically takes one variable out of the picture, in the sense that you don't have to add sugar for both taste and stability, but just for taste.*
- *Throw some spoons in the freezer before you start. When making the jam, drip the hot jam onto the cold spoon to let it cool, and then you can tell whether it has a good gel or not.*

With these two points in mind, you're totally free to experiment with flavor, because that's the only variable left to optimize.

Tim O'Reilly's Scone Recipe

In a bowl, measure out:

2½ to 3 cups (350–400g) flour (experiment to see how much you prefer)

½ cup (115g) butter, chilled

Using a pastry blender or two knives, cut the butter into the flour. When done, the butter and flour should look like small pebbles or peas.

Add and whisk to combine:

3 tablespoons (36g) sugar

4 teaspoons (20g) baking powder

½ teaspoon (3g) salt

(At this point, you can freeze the dough for later use.)

In the center of the dough, make a "well" and add:

½ to 1 cup (50–100g) currants (or raisins, if you prefer)

½ to 1 cup (130–260g) milk (or soy milk; goat milk is also great)

Stir with a knife until you get just shy of a gooey consistency. Start with only ½ cup (130g) of milk, adding more as necessary until the dough begins to hang together. If it gets very sticky, you've put in a bit too much milk. You could add more flour if you've gone in with less flour to begin with. It's better to bake them sticky than to add more than a total of three cups of flour: the stickiness is just a problem for shaping them, since it sticks too much to your fingers; too much flour, and they can become tough.

Prepare a baking sheet by lining it with parchment paper or a Silpat (nonstick silicone baking mat). If you don't have either, lightly grease a baking sheet. (You can just rub it with the paper from the stick of butter.) Using your hands, shape the dough into small lumps spaced evenly on the baking sheet.

Bake at 425°F / 220°C until the tops are browned, about 10 to 12 minutes.

Tim's homemade strawberry jam on "bottom" of currant scones. Tim pointed out that it's easier to flip the scone over and jam the bottom side of it, instead of trying to slice it open.

Serve with jam, and, if you're feeling piggy, with Devonshire cream (whipped cream works, too, from one of those aerosol cans, so you can just put a spot of it on).

Notes

- *You can use a cheese grater to grate the butter into the flour. Chill the butter for a few minutes so it's easier to handle.*
- *Tim freezes the partially mixed dough, adding the milk and currants to the dough after it's pulled out from the freezer. (The frozen dough has an almost sand-like consistency, so you can pull out as much or as little as you want.) The benefit of the frozen dough is that you can bake scones a few at a time, adding just enough milk to bring the cold dough to a sticky consistency. This makes for a great quick treat, especially if you are the type that has unexpected guests occasionally. It's also in the spirit of learning to cook like a pro: nothing goes to waste this way, and it's efficient!*

Mechanical Leaveners

Mechanical leaveners work by trapping air within a liquid—usually by whipping egg whites, egg yolks, or cream—or by generating steam from water present in the food.

Unlike biological or chemical leavening methods, which rely on the chemical makeup of the food to generate air, mechanical rising techniques rely on the physical properties of the food to hold air. Because of this, mechanical leaveners can't just be added to a dish without considering the impact of the moisture or fat that they also add, which can throw off the ratios between ingredients such as flour and water or sugar and fats.

"Cream the butter and sugar" has nearly three million exact-phrase matches on Google, and plenty is written about the microscopic air bubbles that the sugar crystals drag through the butter when creamed. When you see a recipe call for creaming butter and sugar, use room-temperature butter—it needs to be plastic enough to hold on to the air bubbles but soft enough to be workable—and use an electric mixer to thoroughly combine the ingredients until you have a light, creamy texture.

Egg Whites

Whisked egg whites are the Styrofoam of the culinary world: besides acting as space fillers in cakes, waffles, and soufflés and as "insulators" in desserts like lemon meringue pie, when overcooked, they taste about the same as Styrofoam, too. All metaphors aside though, egg whites are much more forgiving than many cooks realize. With a little attention spent on understanding the chemistry and a bit of experimentation, egg-white foams are easy to master.

A foam is a mixture of a solid or liquid surrounding a *dispersion* of gas; that is, the gas (usually air) is dispersed through the liquid or solid, not in a single big cavity. Bread is a solid foam; whipped egg whites are a liquid foam. (See 303 in Chapter 6 for a description of colloids.)

The key to understanding egg whites is to understand how foams themselves work. Whisking egg whites turns them into a light, airy foam by trapping air bubbles in a mesh of denatured proteins. Since regions of the proteins that make up egg whites are *hydrophobic*—literally, water-fearing—they normally curl up and form tight little balls to avoid interacting with the water. But when whisked, those regions of the proteins are slammed against air bubbles and unfold, and as more and more proteins are knocked against an air bubble, they form a layer around the bubble and essentially trap it in the liquid, creating a foam that's stable.

Oils—especially from egg yolks or any trace oils present in the whisking bowl—prevent egg whites from being whisked into a foam because they're also able to interact with the hydrophobic sections of the proteins. Water and sugar don't interfere with the formation of protein-based foams for the same reason.

Once the air bubbles are encapsulated by the proteins in the egg white, it takes quite a bit of effort to get them to break. Exposing the whites to any oil before whisking is a problem; even a trace amount of fat from a small amount of stray egg yolk will interfere with the creation of the foam. But once the eggs are whisked, they're much more resilient. Try this experiment: whisk an egg white to soft peaks, then add ½ teaspoon (5g) olive oil and continue to whisk. It might surprise you how long it takes before the oil starts to noticeably interact with the foam, and even then, that the foam remains mostly stable.

What Should You Whisk Your Egg Whites In?

Definitely not plastic. Copper bowls work best; a clean stainless steel or glass bowl is fine.

Plastic is chemically similar enough to oil that oil molecules stick around on it and are impossible to completely remove. Whisking egg whites in plastic bowls doesn't produce as good a result because there's enough oil lingering on the surface of the bowl to interfere with the development of the foam. (Of course, it's fine to whip cream in a plastic bowl; more fat isn't going to interfere with the fat-based foam structure.)

When you use a copper bowl, trace amounts of copper ions interact with the proteins in the egg whites to make a more stable foam. It's not a subtle effect: egg whites whisked in copper bowls are definitely easier to work with. Copper bowls are expensive, but if you find you're whipping up egg whites a lot, it's probably worth breaking down and spending $40 on one.

For more occasional use, most of us have stainless steel or glass bowls on hand, and those are fine. While these materials won't help with the egg-white foam by adding copper ions, they also won't hold on to problematic fats. Cream of tartar is commonly used as a chemical buffer—any time you see a recipe calling for egg whites to be whisked and you're using stainless steel or glass, "auto-complete" it with a pinch of cream of tartar (⅛ teaspoon [½ g] per egg white). Don't use cream of tartar in a copper bowl, though; it'll interact with the copper.

Try an experiment: whisk three whites in a plastic bowl, three whites in a stainless steel or glass bowl, and, if you should happen to have a copper bowl, three whites in a copper bowl. Take a spoonful of the resulting foams and smear it on a cookie sheet to compare the difference between the foams.

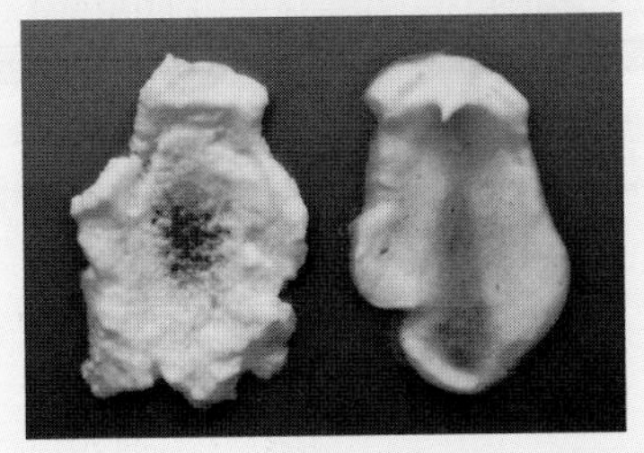

Egg whites whisked to stiff peak in a stainless steel bowl (left) and a copper bowl (right), smeared on a cookie sheet to show the difference in texture.

Air: Baking's Key Variable

Meringues

Egg whites, when whisked and combined with sugar, turn into a sweet, airy mixture suitable for folding into heavier bases, bringing a lightness and sweetness. Of course, sugar and egg whites are pretty good on their own—meringue cookies are nothing more than egg whites and sugar that have spent a little time in the oven. The sugar isn't just for taste, though; it helps stabilize the egg-white foam by increasing the viscosity of the water present in the foam, meaning that the cell walls in the foam remain thicker and are thus less likely to collapse. Net result? The meringue is better able to support the weight of anything you add into the foam.

Stirring and Whisking

When whisking, think about the goal. If you're trying to whisk air into the food to create a foam, such as whipped cream or whipped egg whites, whisk—preferably by hand!—in an up-and-down circular motion, catching and trapping air. If you're trying to mix ingredients together without necessarily adding air, whisk in a flat circular motion. This is especially important for dishes like scrambled eggs, where incorporating air actually reduces the quality.

I prefer whisking things by hand. Why? Electric beaters won't work in as much air before the foam is set because of the motion of the beaters. Also, when whisking, avoid tiny little stirring motions. This is true for stirring almost anything, whether you're holding a whisk or a spoon. Sautéing vegetables? Either get in there like you mean it, or don't touch them; just let them sit so that they brown. Likewise, when whisking foams, get in there like you mean it and whisk some air in there!

Once you've got the motion down, how do you know when it's done? It depends on the recipe. If it calls for soft peaks, the foam should still be supple and pliable, but if it calls for firm or stiff peaks, the foam should hold and set its shape; stiff peaks should be firmer and glossier than firm peaks. (See page 39 in Chapter 1 for more photographs of cream whipped to various states.)

When it comes to folding egg whites into a batter, as long as you're using a whisk or spatula—not an electric mixer—you can be a bit more vigorous than conventional wisdom suggests.

Soft peak stage: the foam stays on the whisk but the peak falls over.

Firm and stiff peak stages: the foam maintains its shape and can be sculpted.

French and Italian Meringue

There are two general forms of meringues: those in which the sugar is directly added as the egg whites are whisked (French Meringue), and those in which the sugar is dissolved before the egg whites are whisked (Swiss and Italian Meringue—we'll cover Italian here, but they're similar). The French version tends to be drier (sugar is hydroscopic, sucking the moisture out of the whites—this is why it increases viscosity) and also grittier; the Italian version has a smoother, almost creamy texture.

French Meringue

In a clean bowl, whisk 3 egg whites to soft peak stage.

Add ¾ cup (150g) of sugar—preferably superfine sugar—one tablespoon at a time, while continuously whisking. If using regular sugar, you'll need to whisk longer to make sure the sugar is entirely dissolved. To check, roll a little bit of the meringue between two fingers (it shouldn't feel gritty).

Italian Meringue

Create a simple syrup by heating in a saucepan ½ cup (100g) sugar and ¼ cup (60g) water to 240°F / 115°C. Set aside.

In a clean bowl, whisk 3 egg whites to soft peak stage. Slowly pour in sugar syrup while whisking continuously.

Meringue Cookies

To make meringue cookies, start with either egg-white meringue recipe. Optionally fold into the meringue whatever ingredients you'd like—ground almonds, chocolate chips, dried fruit, cocoa powder.

Using a spoon or piping bag, portion the meringue onto a cookie sheet lined with parchment paper. Bake in an oven preheated to 200°F / 95°C for a few hours, until they freely come off the parchment paper.

No piping bag? No problem. Put your filling in a large resealable bag and snip off one of the corners. You can use Italian meringue as a topping on desserts as well.

For more on meringues, see page 323 in Chapter 6.

Chocolate Port Cake

One of the great things about this chocolate port cake—besides the chocolate and the port—is the recipe's wide error tolerances. Most foam cakes—those cakes that rely on a foam to provide the air—are very light (think angel food cake). The reason this recipe is so forgiving is that it uses a foam without trying to achieve the same lightness.

You'll need a small saucepan, two clean bowls, a whisk, and a round baking pan or springform pan, 6–8" / 15–20 cm.

In the saucepan (over a burner set to low heat), melt and mix together, but do not boil:

½ cup (125g) port (either tawny or ruby)

½ cup (114g) butter

Once butter is melted, turn off heat, remove pan from burner, and add:

3 oz (85g) bittersweet chocolate, chopped into small pieces to facilitate melting

Leave the chocolate to melt in the port/butter mixture.

In two bowls, separate:

4 large (240g) eggs

Make sure to use a clean glass or metal bowl for the egg whites, and be careful not to get any egg yolk into the whites.

Whisk the egg whites to stiff peaks.

In the bowl with the egg yolks, add:

1 cup (195g) granulated sugar

Whisk the egg yolks and sugar together until thoroughly combined. The yolks and sugar should become a slightly lighter yellow after whisking for a minute or so. Pour the chocolate mixture into the egg yolk/sugar mixture and whisk to thoroughly combine.

Using a flat wooden spoon or flat spatula, add to the chocolate mixture and fold in (but do not overstir!):

¾ cup (100g) all-purpose flour

Then fold in the egg whites in thirds. That is, transfer about a third of the whisked egg whites into the chocolate mixture, mix together, and then repeat twice more. Don't worry about getting the whites perfectly incorporated, although the batter should be relatively well mixed together.

Grease your cake pan with butter and line the bottom with parchment paper, so as to make removing the cake from the pan easier. Transfer the mix to the cake pan and bake in an oven preheated to 350°F / 175°C until a toothpick or knife, when poked into the center, comes out clean, around 30 minutes.

Let cool for at least 10 to 15 minutes, until the edges have pulled away from the sides, then remove from pan. Dust with powdered sugar (you can use a strainer for this: place a few spoonfuls of powdered sugar in the strainer and then jog it with your hand above the cake).

Note

- *When working with chocolate in baking, don't just substitute, say, 80% bittersweet chocolate for a semisweet bar. In addition to differences in sugar, the two types of chocolate have different quantities of cocoa fat, and recipes that rely on the fat level will need to be adjusted accordingly.*

Optimal Cake-Cutting Algorithm for *N* People*

Technically, a suboptimal pie-cutting protocol

If you grew up with a brother or sister, you're undoubtedly familiar with the technique for avoiding fights when splitting food: one person divides it, and the other person chooses. ("You can halve your cake, and eat it, too!") But what to do if you have more than one brother or sister?

There is a solution, but it's a bit more involved. Here's the algorithm for cutting a round cake for *N* people. It's not perfect—don't use this for negotiating land divisions after minor land wars—but when it comes to a table of kids and a large chocolate cake, it'll probably work. (If you find yourself cutting cake for hardcore math geeks, however, I suggest reading up on the literature. Start with An Envy-Free Cake Division Protocol—*http://www.jstor.org/pss/2974850*—and plan to be at it for a while.)

Only one person actually does any cake-cutting, and that person can either be a cake-eater or just a referee. Start with the cake in front of you, along with a knife and *N* plates. Proceed as follows:

1. Make a first cut in the cake, as normal.
2. Explain that you're going to slowly hover the knife above the cake while moving it clockwise around the cake, just like someone thinking about how big the next slice should be. Anyone—including the person cutting the cake—can say "stop" at any point to declare that they want a piece that size, at which point, that's where you'll cut the next slice.
3. Slowly move the knife above the cake until someone calls stop.
4. Slice the cake and hand the person who called stop the new slice. Continue with step 3 with the remaining cake eaters. (To be clear, anyone who calls "stop" is now out of the negotiation and doesn't get to call it again.)
5. When you're down to just one last person, cut the cake wherever he or she likes, which may leave a leftover piece.

One of the nice things about this protocol (a protocol is similar to an algorithm, but allows for accepting user input after being started) is that it allows people who for whatever crazy reason want small slices to do so, and gets them out of the way at the beginning, meaning if somebody else wants a larger slice than an equal *N* division would allow, they get more cake and can eat it, too.

If someone is being greedy and wants a too-big piece, they'll end up getting the last slice—which will normally be the largest slice. If two or more people end up being greedy, though, they could allow the referee to reach the end of the cake by never calling stop, in which case I suggest eating the cake yourself. There's no guarantee that this protocol will satisfy everyone—just that the honest actors are protected from the dishonest ones.

Egg Yolks

If Eskimos have *N* words for describing snow, the French and Italians have *N*+1 words for describing dishes involving egg yolks. A number of these dishes use egg yolks to create light, airy foams by trapping air bubbles.

Egg yolks are much more complex than egg whites: ~51% water, ~16% protein, ~32% fat, and ~1% carbohydrates, while egg whites are only protein (~11%) and water. In their natural state, egg yolks are an emulsion.

An emulsion is a mixture of two liquids that are immiscible—that is, unable to mix (think oil and water). Mayonnaise is the classical culinary example. Egg yolks are an emulsion, too: the fats and water are held in suspension by some of the proteins, which act as emulsifiers—compounds that can hold immiscible liquids in suspension. For more on the chemistry of emulsions, see 304in Chapter 6.

Like egg-white foams, egg-yolk foams trap air with denatured proteins that form a mesh around air bubbles. Unlike whites, though, the only way to denature the proteins in the yolk is with heat; the optimal temperature for egg-yolk foam creation is 162°F / 72°C. Too hot, though, and the proteins coagulate, leading to a loss of air and affecting the texture.

Extra Leavening

Some recipes rely on more than just one method of incorporating air into food. Some English muffins and Chinese pork buns, for example, use both yeast and baking powder. Waffle recipes often call for both whipped egg whites and baking powder. And some mousse recipes call for both whipped egg whites and whipped cream. If you find that a recipe isn't turning out as light as you'd like, look to see if other methods of leavening can be added. If a recipe doesn't rely on chemical leaveners, adding a small amount of baking powder is usually a safe bet. Or, if the recipe has eggs, try separating some of the eggs, whisking the whites, and folding the egg-white foam into the batter.

Simple White Wine and Cheese Sauce

This sauce needs very few ingredients and not much in the way of equipment—a whisk, a bowl, and a stovetop—making it an easy impromptu dish even in an unfamiliar kitchen. (For more on sauces, see pages 116–117 of Chapter 3.)

The only tricky part is preventing the eggs in this sauce from getting too hot and scrambling. If you have a gas burner, this can be done by moving the saucepan on and off a flame set to very low heat. Position yourself so that you can hold the pan with one hand while whisking with the other; you'll need to move the pan to regulate the temperature. If you have an electric burner, use a double-boiler instead: fill a large saucepan with water and place the saucepan with the mixture inside it.

In a saucepan, separate 3 egg yolks, saving the egg whites for some other dish. Add ¼ cup (60g) white wine and whisk to combine.

Once you're ready to start cooking, place the pan over the flame or in the water bowl bath and whisk continuously until the egg yolks have set and you have a frothy foam, about two to three times the volume of the original. This can take 5 to 10 minutes; have patience, it's better to go too slow than too quick.

Add 2 to 3 tablespoons (20–30g) freshly grated Parmesan cheese and whisk until thoroughly combined. Add salt and pepper to taste, and serve on top of an entrée such as fish with asparagus.

Note

- *White wine is quite acidic, with pH levels around 3.4 (Chardonnay) to 2.9 (Riesling). Since acids help prevent egg yolks from coagulating under heat, the wine actually helps protect against coagulation. (Pour yourself a glass; that'll help, too.)*

Zabaglione (Sabayon)

This dish is easy, but it does benefit from a few practice runs. Luckily, the ingredients are cheap!

Zabaglione is the dessert equivalent of white wine and cheese sauce, made by whisking wine, sugar, and egg yolks over low heat; it's essentially a foamy custard, but without the milk. And, like the white wine and cheese sauce, this is a great recipe to have tucked away in the back of your head.

Measure out ¼ cup (60g) Marsala wine and set aside.

Marsala—a white wine fortified with extra alcohol—is traditionally used in zabaglione, but you can use other alcohols, such as Grand Marnier, Prosecco, or port.

In a saucepan, separate out the yolks from 3 eggs, saving the whites for something else (meringues!). Add ¼ cup (50g) sugar to the yolks and whisk to combine.

Place pan over heat, following the directions for the white wine and cheese sauce. Pour in a tablespoon of the Marsala and whisk. Continue adding the Marsala a tablespoon or so at a time, whisking for a minute between each addition. You're looking for the egg yolks to froth up and foam; the heat will eventually set the egg yolks to make a stable foam. If you notice that the egg yolks are scrambling, quickly pour in more of the Marsala to cool the mixture down; it's not ideal, but it'll prevent you from having an entire dish of sweet scrambled eggs on your hands. Once the sauce begins to show soft peaks, remove from heat and serve.

Traditionally, zabaglione is served with fruit: spoon a small portion into a bowl or glass and top with fresh berries. You can also store it in the fridge for a day or two.

Strawberry or Raspberry Soufflé

You're probably wondering what soufflé is doing in the section on egg yolks, right? After all, it's the egg whites that famously give soufflés their rise. I have a confession to make. I make my fruit-based dessert soufflés by making zabaglione. (I am so never going to win a James Beard award—the Oscar of the culinary world.)

Preheat your oven to 375°F / 190°C. Prepare a 1 quart / 1 liter soufflé bowl—which will hold enough soufflé for two to three people—by buttering the inside and then coating it with sugar (toss in a few spoonfuls, then rotate the dish back and forth to coat the side walls).

Prepare the fruit:
Fresh strawberries, raspberries, and white peaches work exceptionally well; wet fruits such as pears can work, but the water may separate while cooking, so start with berries. Rinse and dry the fruit. If using strawberries, hull them; if using peaches or other stone fruits, quarter them and remove the pit. Reserve about ½ cup—a small handful—of the fruit for placing on top of the cooked soufflé. Prepare a second handful of fruit, again about ½ cup, for cooking by slicing it into small pieces; cut strawberries into eighths and peaches into very thin slices. (Raspberries will fall apart on their own.)

Make zabaglione:
Start by making a zabaglione: whisk the 3 egg yolks with ¼ cup (50g) sugar over low heat and add ¼ cup (50g) of kirsch—cherry-flavored brandy—instead of Marsala. (Save the egg whites for whisking.) After adding the kirsch, add the fruit that's been sliced into small pieces and stir, thoroughly mashing in the fruit. You don't need to actually cook the egg yolks until they set; you're just looking to stir and whisk them until you have a frothy, warm, soft foam. Set aside while preparing the egg whites.

Whisk egg whites, fold, and bake.
Whisk the egg whites to soft peak stage, adding a pinch of salt for taste. Fold the egg whites into the fruit base and transfer the mixture to the soufflé bowl. Bake in an oven until the soufflé has risen and the top is browned, about 15 to 20 minutes. Remove and place the soufflé dish on a wooden cutting board. Dust with powdered sugar, place the reserved fruit on top (slice strawberries or peaches into thin slivers), and serve at once. If you're in informal company, it's easiest to just set the soufflé in the center of the table and hand everyone a fork to dig in.

You can use this same technique with the white wine and cheese sauce from the previous page to make a savory soufflé.

Whipped Cream

Unlike eggs, in which proteins provide the structure for foam, cream relies on fats to provide the structure for a foam when whipped. During whisking, fat globules in the cream lose their outer membranes, exposing hydrophobic portions of the molecules. These exposed parts of the fat globules either bind with other fat globules or align themselves to orient the stripped region with an air bubble, forming a stable foam once enough of them have been aggregated together.

When working with whipped cream, keep in mind that the fats provide the structure. If the cream gets too warm, the fats will melt. This is why whipped cream can't be used to provide lift in most baked goods: the cream will melt before the starches and gluten in the flour can trap the air. Be sure to chill your bowl and the cream before whisking.

Whipping high-quality cream increases its volume by about 80%, while whipped egg whites can expand by over 600%!

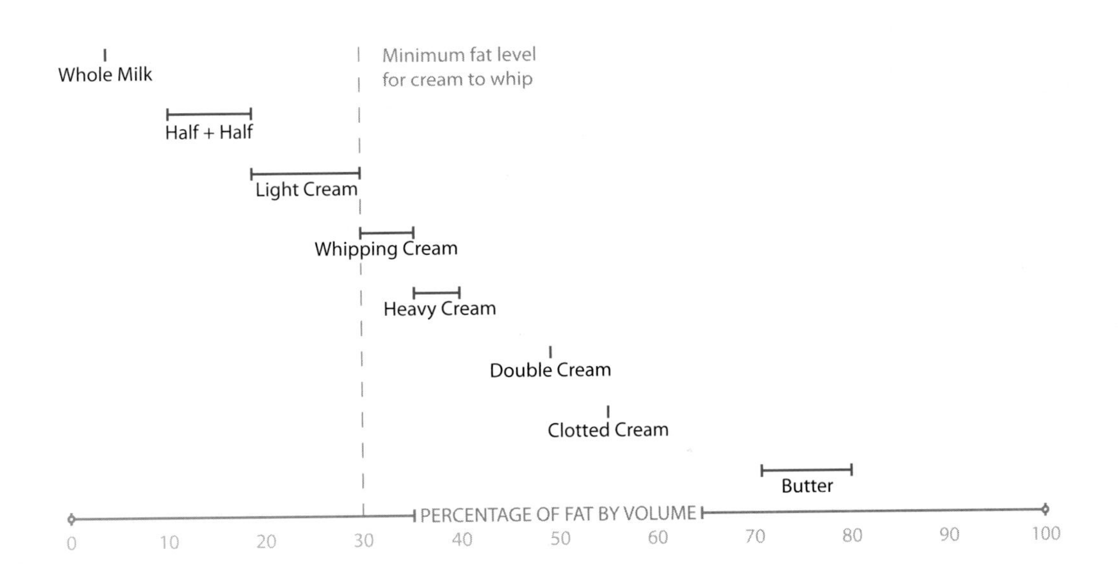

Percentage of fat in dairy products. If the cream doesn't have enough fat, there won't be enough fat globules to create a stable foam.

Michael Chu's Tiramisu

Tell me a little bit about your background and how you started your blog, Cooking for Engineers (at *http://www.cookingforengineers.com*).

I am an engineer in my professional life, but I developed recipes on my own. "Developing" sounds grander than it really is. I would try out recipes and then the ones that I really liked I would keep and then tweak so that I would like them more. Friends would come over to watch TV; at the time we called them "Family Guy Nights" because we would watch the television show *Family Guy*. I would be cooking and writing down my recipes when I found something I liked. I started using Blogger, and at some point I decided I would post a recipe. I took pictures of me making salsa and then I posted that recipe. My friends liked it and so I just thought, "Oh, well, maybe this is a good way to share my recipes." A lot of people started to come and look at the recipes, and it just kind of blew up from there.

One thing I notice about your posts is the number of variations you go through. Do you think people have a fear of trying variations when they go into the kitchen?

I think a lot of times people don't like wasting food. There is a whole culture where wasting food is something that you don't do. I totally agree with that, but when you're trying to learn how to do something it's inevitable that you will make mistakes. There will be some waste; that's something people shouldn't shy away from. When you're trying a new fancy dish for the first time and you've never used the ingredients, you might use too much.

For example, Vietnamese fish sauce is a wonderful ingredient, but if you use a little bit too much it ruins the dish. What do you do at that point? If you eat it

The Original Tiramisu

<table>
<tr><td>4 large (70g) egg yolks</td><td>beat</td><td colspan="2" rowspan="3">beat</td><td rowspan="3">whisk over steam</td><td rowspan="4">beat</td><td rowspan="5">fold</td><td rowspan="8">assemble:
filling
ladyfingers
filling
ladyfingers</td><td rowspan="9">sift on top</td><td rowspan="9">refrigerate 4 hours</td></tr>
<tr><td>½ cup (100g) granulated sugar</td><td rowspan="2"></td></tr>
<tr><td>½ cup (120 mL) sweet Marsala wine</td></tr>
<tr><td>1 lb. (450g) mascarpone cheese</td><td colspan="4">beat</td></tr>
<tr><td>1 cup (240 mL) heavy cream</td><td colspan="5">whip to soft peaks</td></tr>
<tr><td>about 40 ladyfinger cookies</td><td colspan="2"></td><td colspan="4" rowspan="3">soak 2 seconds</td></tr>
<tr><td>12 oz. (355 mL) prepared espresso</td><td colspan="2" rowspan="2">dissolve</td></tr>
<tr><td>2 tsp. (8.5g) granulated sugar</td></tr>
<tr><td>2 Tbs. (11g) cocoa powder</td><td colspan="7"></td></tr>
</table>

Time and activity bar chart for the Original Tiramisu.

you might be turned off from fish sauce for the rest of your life. There is a lot of food waste that goes through my kitchen. There wouldn't be as much if I weren't running Cooking for Engineers, but it's really important to experiment. It's one thing to read a cookbook, but once in a while it's important to deviate and try something completely new.

Sometimes, these mistakes can be expensive; you might ruin your pot. Sometimes they will reveal something awesome. In some recipes, you're told to be sure to not burn the garlic, but then if you experiment and do overcook the garlic, it becomes these little crispy bitter pieces that work really well with certain types of vegetables. People want to get it right the first time. Part of that is due to not wanting to waste the food or the money, but the other part is they haven't gotten to the point where they're enjoying doing it over until they get it right.

Have there been any particular recipes whose success has caught you off-guard?

Tiramisu is the recipe that launched Cooking for Engineers. I posted the tiramisu recipe, and three days later I was getting maybe 100 page views a day on that article. Enough people saw it that I got attention from Slashdot, which wrote an article about this new cooking website geared toward geeky people. Boom, I got a lot of readership. So much so that I had a little trouble keeping up with the number of people who were looking at the web pages on the little server that I was running on.

The tiramisu recipe that we have on Cooking for Engineers is a bit more simplified than many of the other tiramisu recipes. I spent a lot of time developing it. I wanted to come up with something that inexperienced cooks could do without extra steps, so I came up with a method where the cream is mixed in with the Mascarpone cheese to produce the lighter, fluffier texture. I modified the amount of ingredients so that it was well balanced. The tiramisu recipe is probably one of the best we've ever tasted, and very simple to make. It's called "simple tiramisu." After the success of the simple one I included one that was closer to what the original tiramisu was as well, to let people compare them.

For photographs and step-by-step directions, see Michael's site. The two recipes are located at http://www.cookingforengineers.com/recipe/26/Simple-Tiramisu *and* http://www.cookingforengineers.com/recipe/60/The-Classic-Tiramisu-original-recipe.

Simple Tiramisu

about 20 ladyfinger cookies		dip	layer & spread twice	cover
2 shots (2 ounces; 60 mL) prepared espresso	mix & chill			
½ cup (120 mL) prepared coffee				
1 cup (240 mL) heavy whipping cream	whisk to stiff peaks	fold		
1 lb. (455g) mascarpone cheese	mix			
½ cup (100g) granulated sugar				
3 tablespoons (44 mL) rum or brandy				
cocoa powder				
shavings of unsweetened dark chocolate				

Time and activity chart for Simple Tiramisu.

Cream whippers—canisters that can be filled with a liquid and then pressurized with gas (usually nitrous oxide)—are also a form of mechanical leavening. The gas dissolves into the liquid and then, upon spraying, bubbles back out of saturation, foaming up the liquid. From a structural point of view, foams created this way are entirely different from foams created by whisking: instead of a 3D mesh of surfactants holding on to the air bubbles, the air bubbles are essentially just in suspension. This is why hand-whipped cream is more stable than whipped cream from a can. For more on cream whippers, see 370 in Chapter 7.

Chocolate Mousse

Compare the following two methods for making chocolate mousse. The egg-white version creates a creamy, dense mousse, while the whipped cream version creates a stiffer version.

Chocolate Mousse (Whipped Egg White version)	Chocolate Mousse (Whipped Cream version)
In a saucepan, heat ½ cup (120g) of whipping or heavy cream to just below a boil and turn off heat. Add 4 oz (115g) of bittersweet chocolate that's been chopped into small chunks. Separate 4 eggs, putting 2 of the yolks into the saucepan and all the whites into a clean bowl for whisking. Save the other 2 yolks for a different recipe. Whisk the egg whites with 4 tablespoons (50g) of sugar to soft peaks. Whisk the cream, chocolate, and yolks together to combine. Fold the whites into the sauce. Transfer mousse to individual serving glasses and refrigerate for several hours—overnight, preferably. **Note** • *The egg whites in this are uncooked, so there is a chance of salmonella. While it's rare in chicken eggs in the United States, if you are concerned, use pasteurized egg whites.*	Melt 4 oz (115g) of bittersweet chocolate in a microwave-safe bowl. Add 2 tablespoons (28g) of butter and 2 tablespoons (28g) of cream and whisk to combine. Place in fridge to cool. In a chilled bowl, whisk 1 cup (240g) of whipping or heavy cream with 4 tablespoons (50g) of sugar to soft peaks. Make sure the chocolate mixture has cooled down to at least room temperature (~15 minutes in the fridge). Fold the whipped cream into the chocolate mix. Transfer mousse to individual serving glasses and refrigerate for several hours; overnight, preferably. **Note** • *Try replacing the 2 tablespoons of cream with 2 tablespoons of espresso, Grand Marnier, cognac, or another flavoring liquid.*

What About Steam?

While steam doesn't involve mechanically trapping air as the other methods in this section do, it's still a physical process by which air is introduced into food. Most of the recipes given so far also rely on steam generation as part of their leavening; few baked goods truly rely on only one method for providing lift. Try this popover recipe, which is a classic example of a baked good leavened by steam.

Popovers

Traditionally, these are made in specialized popover cups, which are narrow cups with a slight slope to them and that have some heft to them, giving them good heat retention. You can use muffin tins or ramekins instead.

Whisk together in a mixing bowl or blend in a blender:

1½ cups (380g) whole milk
3 large (180g) eggs
1½ cups (180g) flour (try half AP, half bread)
1 tablespoon (15g) melted butter
½ teaspoon (2g) salt

Preheat both the oven and the popover cups or muffin tin at 425°F / 220°C.

Heavily grease the popover cups or muffin tins with butter: melt a few tablespoons of butter and put a teaspoon in the bottom of each cup. Fill each cup about ⅓ to ½ full with batter and bake. After 15 minutes, drop the temperature to 350°F / 175°C and continue baking until the outside is set and golden-dark brown, about another 20 minutes.

Serve at once with jam and butter.

Notes

- *How does gluten affect the inside and crust of the popover? As an experiment, make two batches of the batter, one with either cake or AP flour and the second with a higher-gluten flour. Fill half the cups with one batter and the other half with the second batter and bake them at the same time to eliminate the potential for differences between runs.*
- *Try adding grated cheddar cheese or Parmesan cheese for a savory version, or sugar and cinnamon for a sweet version. You can also pour the popover batter into a large cast iron pan (preheated), top with sliced fruit such as pears or peaches, and bake to make a large, tart-like breakfast pastry.*
- *Don't peek while these are baking! Opening the oven door will drop the air temperature, causing the popovers to drop in temperature and lose some of the steam that's critical to their rise.*

David Lebovitz on American Cooking

PHOTO OF DAVID LEBOVITZ USED BY PERMISSION OF PIA STERN

David Lebovitz was a pastry chef at the renowned Chez Panisse in Berkeley, California for over a decade. Since then, he's written several well-received books on desserts. His blog is at http://www.davidlebovitz.com.

What was working at Alice Waters's Chez Panisse like for you?

Chez Panisse is a great place to work. Money is no object when it comes to sourcing ingredients, and it's a great training ground for cooks. The restaurant really supports the owners and the other cooks, who are very, very interested in producing good food. Once you're in that environment, it's hard to leave. You go somewhere else and you're working with a bunch of line cooks that just care about who won the game last night and how fast they can cook the steaks on the grill so they can get out and go drink beer.

The whole idea of Chez Panisse is to find good ingredients and do as little to them as possible. When we had beautiful fruit, we would often just serve a bowl of fruit or a fruit tart with ice cream; or if we had really good chocolate, we would make a chocolate cake, but it wasn't a cake that was highly decorated, it didn't have a lot of technical swoops and things. Chez Panisse is all about flavor. A lot of the fancy stuff doesn't taste good, so we were more concerned with flavor.

I had dinner last night at a fancy restaurant. They brought this chocolate mousse and there was tapenade on the side. Someone was, like, "Olives: it would be really cool on the plate!" But if someone tasted it? Disgusting. I just wanted to go into the kitchen and say, "Have you guys tasted this food? Because it's stupid."

You had worked at Chez Panisse for years before taking culinary training. What surprises did you run across in that culinary training?

I wasn't expecting things not to taste good. I took a course in making cakes in France, and I thought, "We're going to make cakes that are delicious." It actually was making mousses with gelatin and with fruit purees from the freezer, and everything was like sponge cake, gelatinized fruit puree, and decorations. It was interesting, and I learned something, but those skills don't even translate to what I do. Even if you use fresh fruit, it's just not the best way to use it. I'm an ingredients-based cook.

I did go to chocolate school and that was great; I learned a lot about chocolate, how to work with it, how to manipulate it. Once again, I'm more interested in finding wonderful hazelnuts and in rolling them in chocolate, rather than opening up a can of hazelnut paste and making chocolates out of it.

What would you recommend to somebody who wants to learn how to bake?

The best thing they can do is just bake. The thing about baking is it's very recipe-oriented. If you want to learn to make a pound cake, you just make a recipe, and the longer you go, the more you see how things work, how you can change things. You can add an egg yolk to make things richer or substitute sour cream for the milk in the recipe.

A lot of bakers are very precise, and we do have a reputation, especially in the professional world. A chef once said to me, "Why are you guys all so weird?" There are a lot of strange people in the pastry world, because we are very precise, we do like to go in our own little world, and we're very analytical people in general. We think a lot about things, whereas a line cook, it's a lot of brawn; it's big, bold flavors; it's roasting meat; it's frying vegetables; it's grilling. Those are ways of coaxing flavor out, but pastry is a much more delicate thing, it demands a lot more care, a lot more softer skills.

This is probably sexist, but a lot of women work in pastry for that reason, because a lot of women are very sensitive. I've always worked in women-owned restaurants, except for one, which was interesting. I never was aware of the whole "macho" thing—the way the guys would talk and treat people—until I went to other restaurants. I read these kitchen memoir books about sexual harassment and stuff, but to me, it's about the food.

When you're working on a pastry, how do you go about getting unstuck when it's just not coming out the way you want it to?

If you knew how to get out of that, you wouldn't be in there in the first place. I develop recipes and write books, so I'll be making things, I'll make them over and over again, and if I'm really stumped, I have a decent network of people who can help me. I might write to a friend who is a bakery cooking professor and say, "I'm trying to make persimmon pie; have you ever made it?" and he'll be like, "Oh, persimmons have a chemical in there that prevents this from happening, and try doing this..." Bakers are sharers, so we do have a loose-knit community. Also, a lot of baking is science. If I make a cake and I want it to be moister and higher, I just have to sit down with my calculator and work it out.

How do you know what the formula is for working it out?

There are printed formulas, which some bakers use. But I'm not so good with math. Michael Ruhlman wrote a wonderful book on ratios, but my brain isn't wired to think that way. So I just make things a million times, until I get it right.

So yours is a much more try-it-and-see approach, as opposed to sitting down and trying to figure out the optimal formula?

Yeah.

A lot of people are very analytical about cooking, and they want to know how things work. It's a different method. It's like a lot of Europeans wonder why Americans won't give up their measuring cups and spoons, which is a terrible way to cook. It's inaccurate and leads to people doing all sorts of weird things. Americans like to hold measuring cups and spoons; it makes us feel good, so we're not going to give them up. Cooking is a visceral thing, a lot of people like to overanalyze recipes. They're like, "Can I make this cake without the quarter teaspoon of vanilla extract?" and I'm like, "Okay, well, think about it, what do you think?" A lot of people don't know, because they're overanalyzing the recipe. They're not stupid, it's just that they're not, I don't know what... It's like, "If I let 5% of air out of my tire, can I still drive?" "Yes. Better if it's full."

Why do you think Americans overanalyze recipes?

That's the big question nowadays. Everyone's trying to figure out why are Americans scared to cook? I think that Americans are in this weird space now where they want to be told what to do; they want to have a recipe; they want an authority to tell them that this is the recipe, don't change it. We spent eight years under Bush and nobody questioned what he did for four and a half years. Everyone just wanted to be told what to do rather than say, "Wait a minute, look at the facts!" A recipe might say bake a chicken for an hour, and someone will write and say they baked it for an hour, and it was too dry. Well, your chicken was probably four pounds instead of six. There's only so much stuff you can put in a recipe.

Where do you think this fear of failure comes from?

That's something I haven't been able to figure out, because everybody makes mistakes. A lot of people look at food magazines and the pictures are beautiful, and they're like, "Oh, mine doesn't look like that!" Well, you don't have a team of food stylists and a camera and a photographer lighting it right. The best piece of pie is not supposed to hold together with 2" sides that are perfectly smooth. The best chocolate chip cookies are not the ones that look perfect; they're the ones that are full of oozy chocolate chips that are gushing all over the place.

Why did you start your blog?

The site was started in 1999, when my first book came out, because I thought—famous last words—I thought it would be a good way for people to get in touch with me in case they had problems with the recipes. You don't want people saying the recipes don't work; you'd rather have them write to you and say, "I made this cake and it didn't work; what did I do wrong?" Now it's like, "I made Bill Smith's chocolate cake and it didn't work; what did I do wrong?"

I have a recipe—actually, it's in the oven right now—for a cake that has one egg in the whole cake; that's the only fat in it. Some woman wrote me—she's trying to eat less fat—what could she replace the egg with? I'm like, one egg yolk? That's 5 grams of fat for 12 servings. Somebody actually asked that, and then I wonder how these people go to the bank every day, get their driver's license, pay bills, write a check, and work. What's going through their minds?

I'm not quite sure I follow you there.

Those kinds of things seem common sense to me. Somebody who is concerned about eating an eighth or a twelfth of an egg yolk because they're on a low-fat diet? I don't understand that thinking. If the recipe had six egg yolks or four egg yolks, maybe I could see it, but it's a cake, and it's like saying, "I don't like chocolate; how can I make these chocolate chip cookies without chocolate?" It's like sorry, that's what it is.

I just read that book, *French Women Don't Get Fat,* because somebody had it at their house and I borrowed it. I was reading it, and I was like, "Oh. My. God." They pass on this myth that French women eat a certain way, that they drink half a glass of champagne only once a week. The book sold millions of copies in America based on something I don't consider necessarily to be true. There's a lot of fat women here. [David lives in Paris.] Everyone's asking me what I think about that book, and I'm like, ask the one who wrote it. Do you really think French women don't eat junk food and don't smoke their brains out? Wake up. It's like French people saying, "Don't all Americans carry guns?" I'm like, "Yes, when we're born, they put a gun in our hand. When you're two, everyone gets a gun in America."

I think there's a certain cultural gullibility that we have in both directions, both Americans dealing with people with international backgrounds and being in other countries and talking about Americans. I was at a Thanksgiving that was 18 international students from Harvard's Kennedy School and I started talking about the ghost of Thanksgiving past, present, and future, where the ghost of the turkey that you had previously eaten would show up. These international students just ate it up. They totally believed that this was part of the American "story." I was like, "Really, no, this isn't true guys; I'm completely pulling your leg." It's amazing how much cultural misunderstanding there seems to be about these things. I wonder what cultural differences there are in learning how to cook?

Well, French people, and this is a generalization, because it's not necessarily true for everybody, but they're much more relaxed about how things look. Americans jumped on the fast food wagon in the 50s and 60s, whereas the French jumped on it in the 90s and now; so they're losing that whole connection with homemade things, but they still are closer to it than we are. It's not unusual to go to someone's house for lunch and they made a quiche; whereas, in America, you'd be like, "Oh my God, I can't believe you made that, you made the crust, you made the filling, you chopped the vegetables?" That's changing here [in France]; everyone is eating frozen food now.

Is there any sort of cultural backlash against that from part of the French community?

Not yet. Americans have seen our cuisine decline and are now interested in farmers' markets and all that stuff, whereas the French didn't fall as far as we did. They're like, "Everything at the market is local, and is fresh." I'm like, "Well, everything's from Morocco, look at the box." To which they say, "Oh, well, it's not like America."

But you're saying Europe is actually becoming more like America, in that sense?

Yeah.

What do you think of people who really feel like they need to have the most up-to-date technical equipment and toys?

Well, that's an American thing. I go back to America and everyone has wine refrigerators, and they're filled with Kendall Jackson Chardonnay. If you have good wine, you don't put it in one of those refrigerators, because they have compressors that shake, which is bad for wine. Unless you have a very good wine refrigerator that doesn't shake, you're better off without it. It's funny to see people who have wok burners and wine refrigerators and all that stuff in their house. A lot of people want to have the illusion of cooking; they want to have all these bottles of olive oil wrapped up on the counter in baskets and things, but on the other hand, do they really need all that stuff?

It sounds like one piece of advice you would give to somebody is to not obsess over equipment?

Yes. You don't need every saucepan in the world, you need like three. For me, having a mixer is very important; for me, having an ice cream machine is important. But you don't need a panini grill; you can use your skillet and just put a weight on top of it, something like a can of tomatoes, and there you have it.

PHOTO OF DAVID BY KRISTIN HOHENADEL / APARTMENT THERAPY: THE KITCHN

6

Playing with Chemicals

MANKIND HAS BEEN ADDING CHEMICALS TO FOOD FOR MILLENNIA. Salt is used both as a preservative (curing meats, inhibiting bacterial growth) and as a flavor enhancer (masking bitterness). Acetic acid, a key component in vinegar and a byproduct of some strains of bacteria, turns cucumbers into pickles and cabbage into the Korean dish *kimchi*. And citric acid in lemon juice brightens the smell and taste of fish by neutralizing the amine compounds that can create that "fishy" smell as the tissue breaks down.

In recent history, the food industry—the collection of businesses that farm, distribute, prepare, and package the foods we eat—has developed a number of techniques to help perishable foods last longer. Refrigeration slows down bacterial growth, "modified atmosphere packaging" (MAP) displaces oxygen to reduce oxidation and retard the growth of aerobic bacteria, and chemical food additives extend shelf life, fortify foods, and aid in mass production. These same chemicals are also used to create entirely new types of foods, including many candies, and as key ingredients in some techniques of an entirely new kind of cooking given names such as *molecular gastronomy* or *modernist cuisine*.

By definition, food itself is made up of chemicals, of course. Corn, chicken, and bars of chocolate are just big piles of well-structured chemicals. For our purposes, we'll consider a food additive to be any chemical—a compound with a definable molecular structure—used in food that by itself cannot be harvested directly from nature without further refinement or processing.

In this chapter, we'll take a look at cooking techniques that use food additives, both traditional and modern. Some recent culinary techniques rely on chemical stabilizers, gelling agents, and emulsifiers to create new types of dishes. We'll cover these chemical-based techniques in the second portion of this chapter. Even if you're not the type who wants to use chemicals to make foams, to "spherize" liquids, or to turn liquids into gels, understanding how food additives work and what they do makes recovering from kitchen errors quicker and decoding ingredient lists easier.

One of the largest challenges facing commercial food preparers is extending shelf life while maintaining the taste, texture, and appearance of foods. To reduce costs, speed up manufacturing, and increase the shelf life of products like vegetables and baked cookies from days to months, industry relies on chemicals.

Take a look at the food additives used in a certain popular cream-filled chocolate cookie:

Baking soda (a.k.a. sodium bicarbonate)

Speeds up manufacturing by immediately giving rise to a dough or batter (via chemically reacting to release carbon dioxide) so you don't have to wait for the rising action of yeast.

Cornstarch

A thickener, also used as a stabilizer. (Cornstarch is derived from corn but is sufficiently processed, filtered, centrifuged, dried, and treated with acids that it should be considered a food additive.)

Enriched flour (wheat flour, niacin [B3], reduced iron, thiamin mononitrate [B1], riboflavin [B2], folic acid [B9])

Fortified with micronutrients that are removed during the processing of white flour. The FDA requires that white flour be supplemented with B vitamins (to prevent various deficiencies) and iron (to prevent anemia, a low red blood cell count).

Salt (a.k.a. sodium chloride)

Used to mask bitterness, to improve flavor, and in some cases to act as a preservative.

Soy lecithin

An emulsifier, used to prevent oils and water from separating. If you were following a recipe for a cream filling, it would likely call for egg yolks, which are around 10% lecithin, as an emulsifier.

Vanillin (artificial flavor)

Used as a flavoring agent, vanillin is the primary component of natural vanilla extract and is responsible for the majority of vanilla's flavor. Vanillin has the molecular formula $C_8H_8O_3$, regardless of whether the source from which it is derived is "natural" or "artificial."

Some of these items—baking soda, cornstarch, and salt—might not strike you as food additives, either because of their "natural" origins or their long history in the kitchen. But even baking soda arrived only relatively recently on the food scene, when in 1846 John Dwight and Austin Church figured out a commercial method for manufacturing it.

Food additives are used for the following purposes:

- To preserve nutritional value (preventing the breakdown of food)
- To address dietary needs (via fortification, such as the addition of iodide in table salt)
- To extend shelf life or stability in order to keep the food enjoyable longer (e.g., treating apricots with sulfur dioxide to preserve color)
- To aid in manufacturing, processing, or transportation—that is, to address issues caused by scaling to mass production (i.e., keeping larger volumes of food consistent)

Most commercially prepared food products use food additives for more than one of these purposes. In the cookie example just cited, baking soda speeds up manufacturing, cornstarch and soy lecithin aid in the manufacturing process, salt and vanillin improve flavor for enjoyment, and flour is fortified to address dietary needs.

Food additives have gotten something of a bad rap in recent years. The politics, economics, and trade-offs of a food supply that is necessarily driven by economics are well beyond the scope of this book. For now, keep in mind that food itself is chemical, and to cook is to cause chemical and physical reactions.

Just as there have been food additives that were once thought safe but turned out to be dangerous (e.g., red dye no. 2), there are "natural" items—foods from the earth—that pose their own risks without human processing (e.g., hydrocyanic acid in raw lima beans, which is neutralized by cooking). The source of a chemical—natural versus man-made—should not be your sole distinguisher of safety. No one would argue that hemlock or botulinum toxins—both "all natural"—are things you should be adding to your midnight snack.

Taste tests done by America's Test Kitchen have found that most pastry chefs are unable to discern the difference between natural and artificial vanilla, much to the chefs' embarrassment.

Traditional Cooking Chemicals

Before getting into modern industrial chemicals—chemicals that began to routinely appear in food only after World War II—let's take a look at some traditional food additives and the chemicals and chemistry behind them: salt (sodium chloride), sugar (sucrose), acids and bases (citric acid, lye), and alcohol (ethanol).

Salt

IMAGE COURTESY OF NASA

Salt crystals.

Ahh, salt: responsible for the salvation of many a food (or is that salivation?). The oldest seasoning in use, in small quantities it helps reduce the bitterness of foods and enhances the other flavors in a dish (for a discussion of the gustatory system, see 82 in Chapter 3). In larger quantities, it can be used chemically to preserve food (dry and wet brining) as well as mechanically to alter how foods cook (salt roasting).

From a chemistry perspective, salt is an ionic compound composed of a cation from a metal or ammonium and an anion from an acid. In solid form, salt is a crystal of atoms arranged in an alternating pattern based on charge: cation, anion, cation, anion, arranged in a 3D checkerboard pattern.

Our tongues detect one kind of salt, sodium chloride, as being "salty." Sodium chloride (common table salt) is made up of sodium (a metal, and one that in its pure form happens to react violently when dropped in water) and chloride (chlorine with an extra electron, making it an anion). Other salts can register as different tastes. Monosodium glutamate, for example, triggers our taste receptors for umami. In water, the salts dissolve and the individual ions are freed, and they are then able to react and form bonds with other atoms and molecules.

While at first glance the chemistry of salt may not seem important to everyday cooking, it's helpful to understand the basics of how it works when preparing and cooking food. Here's a quick refresher on a few chemistry definitions that'll pop up throughout this chapter. (Finally, a use for that high school chemistry!)

Atom

Basic building block of matter; these are the elements listed in the periodic table.

Molecule

Two or more atoms bonded together (where "or more" can be millions). H = hydrogen atom, H_2 = dihydrogen molecule.

Cation

Any positively charged atom or molecule (i.e., one that has more protons than electrons).

Anion

Any negatively charged atom or molecule (i.e., one that has more electrons than protons).

Cations and anions can be a single atom (Ca^{2+}) or anything from a small molecule (NO_3) to a really large one, such as alginate (composed of many thousands of atoms).

Osmosis and Salt

Applying salt to the outside of fish causes osmosis, which is the physical process of a solvent passing through a membrane to equalize the concentration of solute on the membrane's other side.

In animal tissue, salt (the solute) is unable to penetrate the cell walls (the membrane) present in the tissue, so water (the solvent) leaves the cells in order to equalize the differences in concentration. (The process of equalizing osmotic pressure is called *diffusion*.) If there's a large enough difference in solute concentrations, at some point plasmolysis occurs—the cell structure collapses—and if enough water leaves the cell, the cell dies.

From a food safety perspective, the amount of salt necessary to cause sufficient plasmolysis to render bacteria nonviable depends on the species of bacteria involved. Salmonella is unable to grow in salt concentrations as low as 3% and *Clostridium botulinum* dies at around 5.5%, while *Staphylococcus* is hardy enough to survive in a salt concentration up to 20%. *Staphylococcus* is not a common concern in fish, according to the FDA, so food safety guidelines consider salt solutions of ~6% sufficiently safe (except for those in an at-risk group) when curing fish.

Dry brining

Beef jerky, salmon gravlax, sausages, hams, prosciutto, and corned beef are all cured using salts, typically sodium chloride (table salt) or sodium nitrate, which gives foods like salami a distinctive flavor and pinkish color. Besides adding flavor, salt preserves these types of foods by creating an inhospitable environment for microorganisms (see the section "Foodborne Illness and Staying Safe*" in Chapter 4).

Salt curing has been used for centuries to preserve fish caught at sea, and it's also something that you can easily do at home. Surrounding it with a sufficient quantity of salt draws moisture out of food; this is called dry brining. But salt doesn't just "dry out" the food (along with any bacteria and parasites). At sufficient concentration, it actively disrupts a cell's ability to function and kills it, rendering bacteria and parasites nonviable.

This killing ability isn't limited to just foods. For an adult human, the lethal dose of table salt is about 80 grams—about the amount in the saltshaker on your typical restaurant table.

Overdosing on salt is reportedly a really painful way to go, as your brain swells up and ruptures. Plus, it's unlikely the ER physicians will correctly diagnose the cause in time. (Paging Dr. House.)

Wet brining

Wet brining—the process of soaking meat in salted water—can be used both to add flavor and to reduce water loss during cooking.

As an experiment, try doing an A/B test with brined and nonbrined pork chops. Does brining change the weight loss during cooking? Using a gram scale, weigh a pork chop pre-brining, post-brining, and after cooking, and compare the percentage weight loss to that of a "control" pork chop that is cooked without having been brined. You may also want to test how brining changes the flavor. If you're cooking for others, enlist them as tasters. Cook both brined and nonbrined pork chops, serve a portion of each to everyone, and see what preferences your tasters have.

Salmon Gravlax

In a bowl, mix together:

5 teaspoons (30g) kosher salt

1 tablespoon (12g) sugar

3 tablespoons (12g) finely chopped fresh dill

1 teaspoon (4g) vodka

1 teaspoon (2g) crushed peppercorns (ideally, use a mortar and pestle)

On a large piece of plastic wrap, place:

1 pound (450g) salmon, washed and bones removed; preferably a center cut so that its shape is rectangular

Sprinkle salt mixture over fish and massage into salmon. Wrap fish in plastic and store in fridge, flipping and massaging twice a day for a day or two.

Store in the fridge and consume within a week.

Notes

- *Note the use of vodka as a solvent. Try substituting other spirits, such as cognac or whiskey. And in place of dill, try using coriander seed, loose tea leaves (e.g., Earl Grey or Lapsang Souchong), shallots, or lemon zest. The Scandinavians traditionally serve salmon gravlax on top of bread with a mustard dill sauce.*
- *You can substitute other fatty fish for the salmon and obtain a similar texture. What happens if you try tuna?*
- *Curing inhibits most common bacterial growth but does not prevent all types of bacteria from growing. Avoid serving this to anyone in an at-risk group. This recipe is a bit heavy on the salt—6% by weight—to err on the side of safety. You can reduce the saltiness by rinsing the finished product in fresh water, followed by recoating it with dill and ground pepper to restore some of the flavor. For food safety issues related to parasites, see the section "How to Prevent Foodborne Illness Caused by Parasites" on page 170 in Chapter 4.*

For an extremely technical guide to curing fish and potential pathological hazards, see *http://www.fda.gov/Food/ScienceResearch/ResearchAreas/SafePracticesforFoodProcesses/ucm094579.htm*; for a more practical guide, see *http://www.cfast.vt.edu/downloads/fstnotes/salting.pdf*.

- *Salt curing—as is done in salmon gravlax—is the first step in making lox. After curing, lox is also cold smoked, which is the process of exposing a food to smoke vapors that have been cooled down. You can approximate the flavor of lox by adding liquid smoke to the rub (see page 328).*

You can remove the skin from a piece of fish by placing it skin-side down on a cutting board and carefully running a knife along the surface between the skin and flesh while using your hand to keep the fish from sliding around.

Pork Chops Stuffed with Cheddar Cheese and Poblano Peppers

Brined pork chops are a good example of wet brining. This is also one of those dishes that's both tasty and easy.

In a container, mix 2 tablespoons (60g) salt with 4 cups (1 liter) of cold water. Stir to dissolve salt. Place 2 to 4 boneless pork chops in the brine and store them in the fridge for an hour. After pork chops have brined, remove from water and pat dry with paper towels. Lay out the pork chops on a clean plate to allow them to come to room temperature.

Create a filling by mixing together in a bowl:

- **¼ cup (40g) poblano pepper, roasted and *then* diced, about 1 pepper (see notes)**
- **¼ cup (40g) cheddar cheese or Monterey Jack cheese, cut into small cubes**
- **½ teaspoon (3g) salt**
- **½ teaspoon (1g) ground black pepper**

Prepare the pork chops for stuffing: using a small paring knife, make a small incision in the side of the pork chop, then push the blade into the center of the pork chop. Create a center cavity, sweeping the blade inside the pork chop, while keeping the "mouth" of the cavity—where you pushed the knife into the meat—as small as possible.

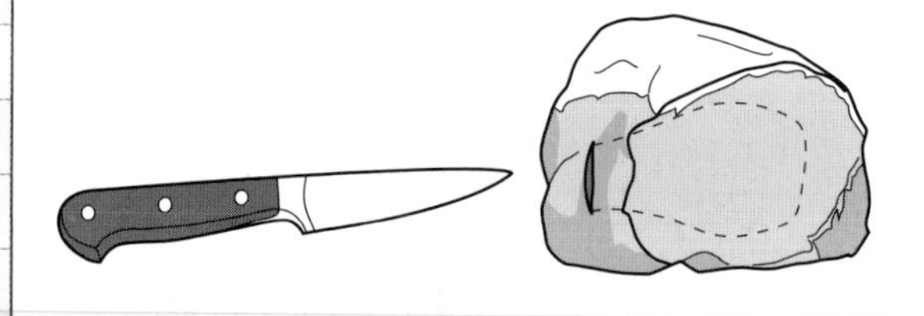

Stuff about a tablespoon of the filling into each pork chop. Rub the outside of the pork chops with oil and season with a pinch of salt.

You'll have leftover filling. It's better to make too much than risk not having enough. Save the extra stuffing for scrambled eggs.

Heat a cast iron pan over medium heat until it is hot (about 400°F / 200°C, the point at which water dropped on the surface sizzles and steams). Place the pork chops in the pan, searing each side until the outside is medium brown, about five to seven minutes per side. Check the internal temperature, cooking until your thermometer registers 145°F / 62.8°C. Then remove the pork chops from the pan and let them rest on a cutting board for five minutes.

You can pull the pork chops from the pan before they reach temperature and let the carryover bring them up to 145°F / 62.8°C, but make sure they do get up to this temperature. You should also verify that your thermometer is calibrated correctly and that you properly probe the coldest part of the meat.

To serve, slice the pork chops in half to reveal the center. Serve on top of rosemary mashed potatoes (see page 201 in Chapter 4).

Notes

- *How do you roast a poblano pepper? If you have a gas stovetop, you can place the pepper directly on top of the burner, using a pair of tongs to rotate it as the skin burns off (expect the skin to char and turn black; this is what you're going for). If you don't have a gas stovetop, place the pepper under a broiler (gas or electric) set to high, rotating it as necessary. Once the skin is burnt on most sides of the pepper, remove from the heat and let it rest on a cutting board until it's cool enough to handle. Using a cloth or paper towel, wipe off the burnt skin and discard. Dice the pepper (discarding the seeds, ribbing, and top) and place into a bowl.*

- *Try other fillings, such as a mixture of sage, dried fruits (cranberries, cherries, apricots), and nuts (pecans, walnuts); or pesto sauce.*

Trichinosis and Pork

145°F / 62.8°C? I thought pork had to be cooked to 165°F / 73.9°C!

Good question; glad you asked. Trichinosis—a parasitic infection from roundworm—has historically been a concern in pork, but this is no longer the case in the United States. The U.S. Code of Federal Regulations requires commercial processors that cook pork to heat it to 140°F / 60°C—well below the well-done temperature of 165°F / 73.9°C—and hold it at that temperature for at least one minute.

To be safe (well, safer—see the discussion on food safety in Chapter 4), give yourself at least a 5°F / 2°C error window. When cooking pork chops, leave the temperature probe in after the chop reaches temperature and check that the temperature *remains at or above* 145°F / 62.8°C for at least one minute. If you see the temperature drop down, transfer the chops back to the pan. The pan itself—even off-heat—should have enough residual heat to keep them at 145°F / 62.8°C.

If you're curious about the history of trichinosis, see the USDA's Parasite Biology and Epidemiology Lab fact sheet at *http://www.aphis.usda.gov/vs/trichinae/docs/fact_sheet.htm*. A century ago, ~1.4% of pork was infected; in 1996, of 221,123 tested animals in the United States, 0 were infected.

Salt-Roasted Fish

Salt can also be used as a "protective outer layer" on food during cooking. By packing foods such as fish, meats, or potatoes in a mound of salt, you ensure that the outer surface of the cooked food doesn't reach the same surface temperatures as it would if uncovered, leading to a less extreme gradient of doneness (see "Temperature gradients" on page 153 in Chapter 4).

Traditionally, the salt is mixed with egg white or water to make a thick paste that will hold its shape and can be packed around something like a fish.

When salt roasting, leave the fish skin on. It'll prevent the fish from getting too salty.

You don't need to bury the fish too deeply. Go for about 1/2" / 1 cm of salt on all sides—enough to take the brunt of the surface temperature but not so much that the center of the fish takes too long to actually reach temperature.

Try this with a whole fish, something medium to large (2 to 5 lbs / 1 to 2 kilos), such as a striped bass or rockfish (check *http://seafoodwatch.com* for suggestions). Rinse the fish thoroughly, and add some herbs (rosemary, bay leaf, etc.) and lemon wedges in the center. Line a baking pan with parchment paper (this'll make cleanup easier), and add a thin layer of salt. Place the fish on top of the salt, and then pack the rest of the salt around the sides and top of the fish.

Bake the fish in an oven set to 400–450°F / 200–230°C, using a probe thermometer set to beep when the internal temperature reaches 125°F / 52°C.

Remove from oven and let rest 5 to 10 minutes (during which carryover will bring the temp up to 130°F / 54°C). Crack open the layer of salt and serve.

Notes

- *Don't have a probe thermometer? The Canadian Department of Marine Fisheries recommends measuring the thickest part of the fish and cooking for 10 minutes per inch. (Add ~10 minutes for the 1" / 2.5 cm of salt around the fish.)*
- *Try this with other foods, such as pork loin (add spices—black pepper, cinnamon, cayenne pepper—to the salt mixture) or even entire standing rib roasts.*
- *Sugar can also work as a "packing material," provided that the oven temperature remains sufficiently low. Salt melts at 1474°F / 801°C, well above oven temperature; sugar melts at 367°F / 186°C. Given this, you should be able to do the equivalent of salt roasting with sugar at temperatures below 367°F / 186°C (try around 325°F / 163°C). However, it takes less water to dissolve the same amount of sugar, so moist items such as fruits will end up giving off too much water for this to work. When I tried "sugar roasting" an apple at 340°F / 170°C, the moisture in the apple was enough to allow the sugar to dissolve into a syrup. It was still delicious, though.*

100 grams of salt (left) and 100 grams of sugar (right), each with 30 grams of water.

Carolyn Jung's Preserved Lemons

PHOTO BY JOANNE HOYOUNG-LEE

Carolyn Jung started as a hard news reporter covering everything from plane crashes to trials. She then transitioned to food, working for the San Jose Mercury News *for over a decade as a food writer and editor. With "the whole journalism media industry imploding," she started her own blog at* http://www.foodgal.com.

What's a day in the life of a food writer like?

It's one of the most fun, most creative, and most enjoyable professions there is. Food is this innocuous way to get strangers talking, and it's a very innocuous way to educate people, and not just about food. It teaches people about culture, about history, about different ethnicities, about different places in the world, about politics, about religion. All of those aspects are what really make it interesting, much more so than people think at the outset.

Where does this recent fascination that people have for cooking come from?

A large impetus has been the Food Network, which has made food such a phenomenon. A lot of people who wouldn't normally cook were attracted to shows like *Iron Chef* because it was almost like watching a boxing match or a football game. Who doesn't dream about being the quarterback on their favorite team? Cooking shows have been the same; you imagine yourself in that contestant's position. "Oh, my God, if I got a box with mushrooms and lemongrass and chicken and avocado, what the heck would I make?"

Why do you think the Food Network took off?

I was watching a documentary on the history of how it all developed. Apparently when it started it was a couple of people sitting at a desk, like regular news shows. As the audience started to grow and funding became available, they got people in a kitchen cooking on sets, but it was very rudimentary. I think what they said really made their number of viewers take off was when they started doing shows where they would go to the nation's barbeque festival or the crawfish festival in the South, things where they showed food as a participatory event.

What's been the most unexpected difference between your experience in the print world and your blog?

As a newspaper reporter, I was used to writing some very long, involved pieces. On the Web, people don't have that kind of attention span. You have a shorter window of time to attract a reader online, but you're also able to build a very loyal audience. If someone likes what you're doing, they will stay with you.

Are there any particular blog posts that have had much stronger reactions than you expected?

I wrote about how to make preserved lemons, and how I got, as my husband calls it, almost obsessed with watching my lemons. It's the simplest thing ever. All you do is make slashes into fresh lemons, fill the cavities with salt, and then pack these lemons in a sterilized glass jar. You top it with a little bit of fresh lemon juice and put the cap on it. As the days go by, the lemons start breaking down and getting softer, exuding more of their juice, and it brines itself in this mixture of lemon juice and salt. I remember the first time I made this, I would wake up every day and go look at my jar of lemons to see what they looked like. It was almost like a science experiment. The fun part is discovering all the uses there are once you have this jar.

Preserved Lemons

All you need are washed and preferably organic lemons (either Eurekas or Meyers), kosher salt, and a glass jar with a tight lid that has been sterilized by running it through the dishwasher.

Make two cuts (lengthwise) in each lemon so that the quarters created remain attached. Stuff kosher salt into the crevices of the lemons. Then place the salted lemons tightly into the glass jar. If I have one or two leftover lemons, I'll often squeeze the juice into the jar before closing it. But you don't have to. This just gives the lemons a little bit of a head start.

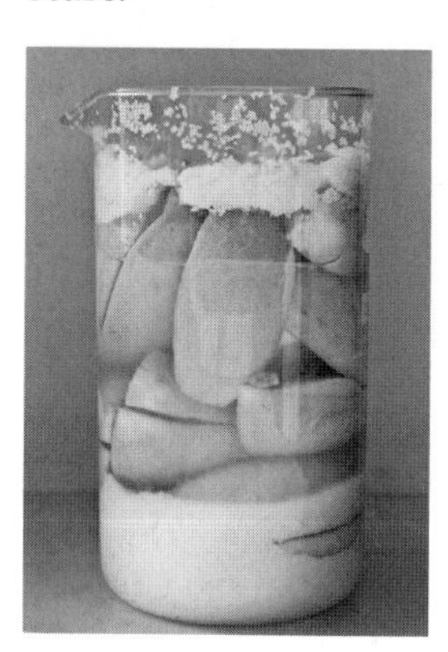

Place the jar on a countertop and then just watch and wait. Over the next few days, more and more juice will exude from the lemons, filling the jar. You can give it a shake now and then—or not—to keep the salt blended well in the liquid. In about three weeks, the lemons will get very soft and the brining liquid thick and cloudy. Once that happens, you can store the jar in the refrigerator. As long as the brine covers the lemons, they'll keep for about a year refrigerated.

To use, pick a lemon or part of one out of the jar with a clean fork. Give the lemon a quick rinse. Remove any seeds. Then, use the peel however you like—chopped or sliced in thin slivers. Some people discard the flesh, but others consider that wasteful. I always add some of the chopped flesh in with the rind in whatever I'm making.

Use preserved lemons in your favorite Moroccan chicken tagine recipes. Or stir it into tuna salad for sandwiches, pasta salad, bean salad, vinaigrettes, marinades for fish or Cornish game hens, or in couscous topped with toasted pine nuts.

For a fast and easy example of how to use these, try making quinoa in a rice cooker. Use kitchen shears to snip one of the lemon slices up into small pieces and mix in with the quinoa before cooking.

RECIPE USED BY PERMISSION OF CAROLYN JUNG; ORIGINALLY INSPIRED BY KITTY MORSE.

Sugar

Sugar, like salt, can be used as a preservative, and it works for the same reasons. The sugar changes the osmotic pressure of the environment, leading to cellular plasmolysis and inhibiting the growth of microbial cells. This is why sugary foods such as candies and jams don't require refrigeration to prevent bacterial spoilage: their water activity is low enough that there's just not any free water for the bacteria.

Sugar's osmotic properties can be used for more than just preserving food. Researchers in the UK have found that sugar can be used as a dressing for wounds, essentially as cheap bactericidal. They used sugar (sterilized, please), glycol, and hydrogen peroxide (0.15% final concentration) to create a paste with high osmotic pressure and low water activity, creating something that dries out the wound while preventing bacteria from being able to grow. Clearly, whoever said "pouring salt on an open wound" didn't try sugar!

Sugar Swizzle Sticks

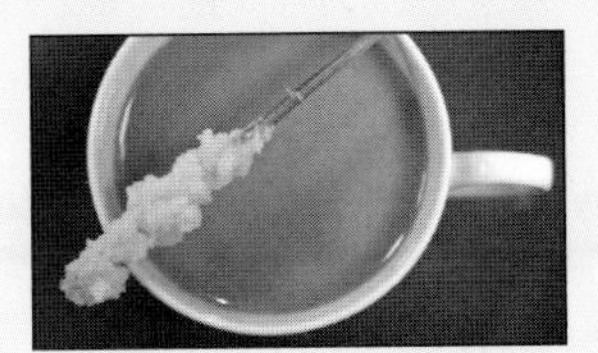

This is just plain fun. You can make fancy sugar sticks for sweetening your coffee or tea with very little effort. While probably not something you'd use on a daily basis, it's a fun project to do with kids.

In a saucepan, boil until completely dissolved:

2 cups (430g) sugar
1 cup (240g) water

Allow the sugar syrup to cool. While waiting, fetch the following:

1 narrow drinking glass
1 small wooden cooking skewer
Tape, such as masking tape
Plastic wrap

Dip the first two or three inches of the skewer into the sugar syrup and then into dry sugar to create seed crystals on the stick.

Stretch a piece of tape across the top of the drinking glass and poke the skewer through the tape so that it's dangling in the center of the glass but not touching the bottom. You might need to use an extra piece of tape around the skewer to keep it from dropping down.

Once the sugar syrup has cooled (to avoid thermal shock breaking the glass), pour it into the glass. Cover with plastic wrap. Set the glass someplace where it won't be disturbed and check it every day as the sugar crystals grow. Remove the skewer when the sugar crystals have reached the desired size.

Note

- *You can add food coloring to the water to make colored sugar crystals. Note that some food colorings are not suitable for vegetarians, such as red food coloring (cochineal or carminic acid), which is derived from the scales of an insect.*

Playing with Chemicals

Simple Lime Marmalade

Marmalade is made by boiling sliced citrus fruits in sugar water and then adding pectin to cause the liquid to gel. For an intensely bitter marmalade—whether you like this style is a matter of personal preference—use Seville oranges. These can be hard to come by, which is why I suggest using limes here. Try this with other citrus fruits, or try a blend!

In a saucepan, bring to a boil and then simmer for half an hour or so, until the rinds are soft:

1 pound (400–500g) limes, cut in half lengthwise, then sliced thinly (about 6 to 8 limes' worth)

2 cups (500g) water, at least enough to cover limes

1.5 cups (300g) sugar

Once the fruit has softened, remove from heat. The marmalade should be intensely bitter at this point; you can add a bit more sugar if you find it overwhelming. Add pectin, following the directions on the box. If you're using a high-methoxyl (HM) pectin, keep in mind that some amount of acid is needed for it to set; in contrast, low-methoxyl (LM) pectin requires a sufficient amount of calcium to set. If your marmalade or jams aren't setting, you'll need to either add something acidic for HM pectin (e.g., lemon juice) or calcium for LM pectin.

Try making your own pectin! See Chapter 4 for details. Once you have the liquid pectin, just add it into the marmalade, simmering to reduce the liquid if necessary.

Cool and store in fridge.

Candied Orange Rind

In a pot, bring to a boil:

2 cups (475g) water

2 cups (430g) sugar

Orange rind from 3 to 6 oranges, cut into strips of width around 0.5 cm / 1/4"

Simmer for 20 to 30 minutes, until the rind is tender. Remove rind from pot, dry on paper towels, and transfer to a container. Add more sugar to container to help pull out moisture in the rind.

Notes

- *The bitter compound in citrus pith (or as a biologist would call it, the mesocarp) is* limonin, *which can be neutralized either by heat or by steeping in a base. Sugar is used for its preservative qualities that prevent bacterial growth, not for counteracting the bitterness of the raw pith.*
- *Try other citrus fruits, such as grapefruit, lemon, lime, or tangerines; or fruits such as cherries, peaches, or apples. You can add spices such as cinnamon to the water as well, or substitute liquors such as Grand Marnier or dark rum for part of the water.*
- *You can chop up candied rind and use it in baked goods, or try dipping the candied rind in chocolate and serving it as a simple candy.*

Hervé This on Molecular Gastronomy

PHOTOS USED BY PERMISSION OF HERVÉ THIS

Hervé This (pronounced "teess") is a researcher at the Institut National de la Recherche Agronomique in Paris known for his studies of chemical changes that occur in the process of cooking. Along with Nicholas Kurti and others, he started a series of workshops entitled "International Workshop on Molecular and Physical Gastronomy," first held in 1992 at Erice in Sicily, Italy.

What was the original reason you and Dr. Kurti had for picking the name "molecular and physical gastronomy"?

Nicholas Kurti was a retired professor of physics. He loved cooking, and he wanted to apply new technology in the kitchen, ideas from the physical lab, mostly vacuum and cold, low temperatures. For myself, the idea was different: I wanted to collect and test the old wives' tales of cooking. Also, I wanted to use some tools in the kitchen that were already in chemistry labs.

For many years, when I was doing an experiment in Paris, he was repeating it in Oxford, and what he was doing in Oxford, I was repeating in Paris. It was great fun. In 1988, I proposed to Nicholas to create an international association of the kind of thing that we were doing. Nicholas said to me that it was too early but, probably, it would be a good idea to make a workshop with friends meeting together. This is why we needed a name, so I proposed molecular gastronomy, and at that time, Nicholas, who was a physicist, had the feeling that that would put too much emphasis on chemistry, so he proposed molecular and physical gastronomy. I accepted the idea only because Nicholas was a great friend of mine, not because I was convinced scientifically.

In the beginning, I published a paper in a main journal in organic chemistry, and in this paper I made the confusion between technology and science. In 1999, I realized that a clear distinction should be made between engineering and science because it is different.

How does the work that you do with molecular gastronomy differ from what a food scientist does who publishes in journals such as the *Journal of Food Science*?

It is a question of history. At that time [1988], food science was more the science of food ingredients or food technology. You had papers on, let's say, the chemical composition of carrots. Nicholas and I were not interested at all in the chemical composition of carrots, in the chemistry of ingredients.

We wanted to do science, to explore the phenomena that you observe when you cook, and cooking was completely forgotten at that time. In the previous centuries, Lavoisier and others studied how to cook meat broth. This was exactly what we are doing. Food science had drifted; cooking was completely forgotten. Recently, I took the 1988 edition of *Food Chemistry* by Belitz and Grosch—a very important book in food science—and looked at the chapters on meat and wine. There is almost nothing about cooking wine or cooking meat; it is very strange.

It seems like there is much confusion about what you mean with the term "molecular gastronomy."

Molecular gastronomy means looking for the mechanism of phenomena that you observe during cooking processes. Food science in general is not exactly that. If you look at the table of contents of the *Journal of Agricultural and Food Chemistry*, you will see very little material referring to molecular gastronomy.

So, molecular gastronomy is a subset of food science that deals specifically with transformation of food?

Exactly, it is a subset. In 2002, I introduced a new formalism in order to describe the physical organization of colloidal matter and of the dishes. This formalism can apply to food and also to any formulated products: drugs, coatings, paintings, dyes, cosmetics. It has something to do with physical chemistry and, of course, it has something to do with molecular gastronomy. So it's true that molecular gastronomy is a particular kind of food science, but also it's a particular kind of a physical chemistry.

It's fascinating to see how easy it is to make inventions or applications from science. Every month I give an invention to Pierre Gagnaire. I should not, because it is invention, not discovery, but I can tell you that I just have to snap the finger and the invention is there. I take one idea of science, I ask myself, "What can I do with that?" and then I find a new application. It is very, very easy. The relationship is of use, and this is probably the reason why there is so much confusion between science and technology. We've been studying carrot stocks. We were studying what is going out of carrot roots into the water and how is it going out. One day, I came to the lab. I was looking at two carrot stocks made from the same carrot. One stock was brown, the other was orange. It was the same carrot, same water, same temperature, same time of cooking, and one stock was brown; the other was orange. I stopped everybody in the lab saying, "We have to focus on this, because we don't understand anything."

We focused on this story, and it was due to the fact that one preparation was made in front of light, and the other was in the darkness, and, indeed, we discovered that if you shine some light on the carrot stock, it will turn brown. So we explored the mechanism, how it turned brown. It was a discovery, not an invention, and thus it was science. At the same time, the application is of use, because cooks want to get a beautiful golden color to stocks, and in order to get the brown color, they grill onions and they put them in the stock. I can tell cooks now: avoid the onions and just add some light. So you see, the discovery is leading to invention immediately.

Tell me more about your collaboration with Chef Pierre Gagnaire.

I don't know if it is a collaboration, it's a friendship. Pierre's wife told Pierre more than 10 years ago, "You're crazy and Hervé is crazy, so you probably could play together."

The real story is that, in 1998, Pierre opened a new restaurant in Paris. He was launching the restaurant with lunches for the press, for the media, for politics, etc., and I was invited. I did not know him, except from reputation at the time. One year passed, and I was asked by the newspaper *Libération* for recipes for Christmas—scientific recipes. I told them I'm not a chef, and that I should not give recipes. I proposed, instead, that I would invite two wonderful chefs to do recipes from ideas that I would give to them, and Pierre Gagnaire would be one of the two chefs.

When I was in the cab driving to the restaurant for the interview and the picture, I realized that beer can make a foam. It means that you have proteins that are surfactants that can wrap the air bubbles. If the proteins can wrap the air bubbles, it means that they can wrap oil. When I arrived in the restaurant, Pierre was there; immediately I asked him, "Do you have some beer, and some oil, one whisk and one bowl?" He looked at me, and he asked for the ingredients and the hardware, and I told him, "Please, put some beer and then whisk the oil into the beer; I can predict that you will get an emulsion." And he got it. He tasted the emulsion, and he found it very interesting, and he decided to make the dish after this wonderful emulsion.

One year later I was invited to lecture at the Academy of Sciences. I proposed to them to make the lecture with a dinner from Pierre. We worked for three months, meeting every Monday morning between 7 and 10. It was so fun that we decided that we had to play on and we never stopped. It's not collaboration, it's just playing together, where we are children.

It seems like some of the more novel cuisines that places like elBulli or Alinea do are removed from the normal dining experience. How much of that experience is created by taking scientific discoveries and applying them to a meal, as opposed to a chef having a concept and coming to a scientist and asking is there a way to make this?

Well, there are many questions in that one. I have the feeling that we don't cook the way we should. For example, we are still roasting chicken. Is it a good idea? I don't know. We ask the question, "Should we go on as we always have?" Many chefs are changing their ways. Many of my inventions are free on Pierre Gagnaire's website (*http://www.pierre-gagnaire.com/francais/cdthis.htm*), and I know that chefs go there to get ideas for the kitchen. I publish the ideas for free; there are no patents, there is no money involved. It is all for free because I want to rationalize the way we cook. We don't cook in a rational way. We are still roasting chicken.

For one of the books that I published, the title was translated as *Cooking: A Quintessential Art*, but in French it was *Cooking: Love, Art and Technique*. The idea that cooking is an art was not even admitted some years ago: "Real art is painting or music or sculpture or literature." I remember talking with a minister of public education in France. He was saying, "No, no, no, it's not art. You're just joking; it's cooking." It's love first, then art, then technique. Of course, technology can be useful only for the technical part, not for the art, and not for the love component. Nowadays, Ferran of elBulli and Alinea's Grant Achatz are using the technique, but there are a lot of possibilities for improvement. They will make their own interpretation, and then science has nothing to do with that. It is personal interpretation; it is feeling.

Do you think that elBulli and Alinea, or restaurants like them, are able to sufficiently use all three components: love, art, and technique?

The love component of cooking is not really formalized. The science needed is still not there. I have the idea that we need to do some science on the love component. Because I'm a physical chemist, it's not very easy for me to make this study. It's still very primitive. Currently, the chef behaves intuitively with the love component. If someone is friendly, he will greet you at the entrance of the restaurant, "Ah, here you are, very happy to have you," and you are happy because you're greeted as kind of a friend. But this is intuition. What I'm saying is that we need to scientifically study the mechanism of phenomena of this friendship. We don't have this mechanism currently.

It almost sounds like psychology or sociology.

It is, exactly. My way of doing molecular gastronomy is to do physical chemistry, daily, at the lab, but I'm producing the concepts so that other people can pursue them in their own way. Their own way can be psychology, sociology, history, geography; we need the knowledge to understand the mechanism of phenomena that we observe in cooking. It is a very foolish idea to think that we cannot investigate all the phenomena. It can be done. Imagine that I discover, or someone discovers, a way to give more love to a dish. It means that the guest will be happier. But imagine that you give this knowledge to a dishonest guy, then the guy would use the knowledge dishonestly, and this will increase the power of dishonest people. If you give the same knowledge to kind people, they will do their best. This is the same question as with nuclear physics. If you are acting poorly, you will make a bomb; if you try to act for the good of humankind, you will make electricity. Science is not responsible for the application; you are responsible for the application.

I asked Dr. This if he had a favorite experiment that could be done at home to learn more about food. His reply:

The most exciting discovery that I did was to put fruits like plums in various glasses full of water but with different quantities of sugar dissolved. In light syrups the fruits sink, but in concentrated syrups they float. This is, of course, linked with density, but when you wait, the fruits in light syrups swell (by osmosis) and explode, whereas they shrink in concentrated syrups.

This experiment is useful to know how to make a syrup of the exact concentration for preserving fruits: put them in concentrated syrup and slowly add water until they begin sinking. The osmotic pressure is then nil so that they will keep their shape and consistency.

Left: cherries in water only; center: cherries in a light sugar syrup; right: cherries in a heavy sugar syrup.

Acids and Bases

We've already discussed chemical reactions that generate air via acid neutralization with baking powder and baking soda in Chapter 5 (see 239), but there's more to pH in cooking than just that. Acids and bases are commonly used to adjust the pH level for two reasons: to cook foods and to prevent foodborne illness.

When it comes to cooking with acids, ingredients such as lime juice can be used to essentially "cook" the proteins in items like shrimp and fish, resulting in similar changes to those that happen when applying heat. On the molecular level, a protein in its native state is structured so as to balance the various attracting and repulsing charges between both its internal regions and the surrounding environment. Portions of proteins are nonpolar—flip ahead a few pages to the section on "When a Molecule Meets a Molecule" to read about polarity—while water is polar. Because of this, proteins normally contort and fold themselves up so that the polar regions of their structures are arranged in a stable shape. Adding an acid or base denatures a protein by knocking its charges out of balance. The ions from an acid or base are able to slip into the protein's structure and change the electrical charges, causing the protein to change its shape. For dishes like ceviche (citrus-marinated seafood), the acid from the lime or lemon juice literally causes a change on the molecular level akin to cooking. And this change doesn't just happen on the surface—given sufficient time, acidic and basic solutions will fully penetrate a food.

When it comes to food safety, adjusting the pH level of the environment can both destroy any existing bacteria or parasites and also prohibit their growth. Ceviche is a classic example of this. *Vibrio cholerae*—a common seafood-borne pathogen—rapidly dies in environments with a pH level below 4.5, even at room temperature. With sufficient lime juice, *V. cholerae* will not survive. Or consider cooked white rice. Left out at room temperature, it becomes a perfect breeding ground for *Bacillus cereus*: it's moist, at an ideal temperature, and has plenty of nutrients for bacteria to munch away on. (Uncooked rice is dry, and since bacteria need moisture to reproduce, they remain dormant. See the FAT TOM variables from 162 for more.) But drop the pH level of the rice by adding enough rice vinegar—down to about 4.0—and the rice falls well outside a hospitable range for bacteria to grow. This is why proper preparation of sushi rice is so critical in restaurants: failure to correctly adjust the pH level can result in sickening diners.

> Spores for *B. cereus* are highly prevalent in soil and water; they're essentially impossible to get rid of. They're heat-stable, too—you can't boil them away. Whoever joked about cockroaches being the only thing to survive a nuclear blast clearly hadn't read up on these things.

Scallop Ceviche

This scallop ceviche is a simple dish to prepare, and surprisingly refreshing on a warm summer day. It's also a good example of how acids—in this case, the lime and lemon juices—can be used in cooking.

In a bowl, mix:

½ cup (130g) lime juice

¼ cup (60g) lemon juice

1 small (70g) red onion, sliced as thinly as possible

2 tablespoons (20g or 1 bulb) shallot, thinly sliced

2 tablespoons (18g) olive oil

1 tablespoon (15g) ketchup

1 clove (7g) garlic, chopped or run through a garlic press

1 teaspoon (4g) balsamic vinegar

Add and toss to coat:

1 lb (500g) bay scallops, rinsed and patted dry

Store in fridge, toss again after two hours, and store overnight to give sufficient time for acid to penetrate scallops. Add salt and pepper to taste.

Notes

- *Try slicing one of the scallops in half after two hours. You should see a white outer ring and a translucent center. The outer ring is the portion that has had time to react with the citric acid, changing color as the proteins denature (just as they would with heat applied). Likewise, after marinating for a day or two, a sliced scallop should show a cross-section that's entirely white.*
- *Keep in mind that the pH of the marinade is important! At least 15% of the dish should be lime or lemon juice, assuming the remaining ingredients are not extremely basic. Lime juice is more acidic than lemon juice (pH of 2.0–2.35 versus 2.0–2.6).*
- *Try adding minor quantities of herbs like oregano to the marinade or adding cherry tomatoes and cilantro to the final dish (after marinating).*

What, you're worried that the scallops are still "raw" and full of bacteria? To quote from the literature: "In the face of an epidemic of cholera, consumption of ceviche prepared with lime juice would be one of the safest ways to avoid infection with [Vibrio] cholerae." (L. Mata, M. Vives, and G. Vicente (1994), "Extinction of Vibrio cholerae in acidic substrata: contaminated fish marinated with lime juice (ceviche)," Revista de Biologiá Tropical 42(3): 472–485.)

Still, since some types of bacteria can withstand more extreme environments, if you really want to play it safe, avoid serving this to anyone in an at-risk group.

Extra credit for using water buffalo milk or milking the cows yourself.

Mozzarella Cheese

Making your own cheese is neither a time-saver nor a money-saver, but it's a great experiment to see how closely related two seemingly different things can be. Cheese is made from curds*—coagulated casein proteins—in milk. The whey is separated out via an enzymatic reaction, allowing the curds to be cooked and then kneaded, stretched, and folded to create that characteristic structure found in string cheese.*

American string cheese is really mozzarella cheese that's been formed into long, skinny logs. Other countries make string cheese using goat or sheep's milk, sometimes adding in cumin seeds and other spices, and often braid several thin strands together.

You'll need to order a few chemicals to do this. (See the upcoming notes and the sidebar "Buying Food Additives" on page 303 for how to do so.) In two small bowls or glasses, measure out and set aside:

½ teaspoon (1.4g) calcium chloride dissolved in 2 tablespoons distilled water

¼ tablet rennet, dissolved in 4 tablespoons distilled water (adjust quantity per your rennet manufacturer's directions)

In a stock pot, mix and slowly heat to 88°F / 31°C:

1 gallon (4 liters) whole milk, *but not* ultra-pasteurized or homogenized

1½ teaspoon (12.3g) citric acid

¼ teaspoon (0.7g) lipase powder

Where it says "not homogenized," it really means not homogenized. (The milk can, and probably should, be pasteurized, though.) If you use homogenized milk, you'll end up with a squeaky mess that vaguely resembles cottage cheese but doesn't melt together. The homogenization process disrupts the protein structures such that they can no longer bind together.

Once the liquid is at 88°F / 31°C, add the calcium chloride and rennet mixtures and continue to *slowly* heat to 105°F / 40.5°C, stirring every few minutes. At this point, you should begin to see curds separating from whey.

Once the liquid is at 105°F / 40.5°C, remove from heat, cover the pot, and wait 20 minutes. At this point, the curds should be fully separated from the whey; if not, wait a while longer.

Transfer the curds to a microwave-safe bowl using a slotted spoon, or strain out the whey and transfer it from your strainer. Squeeze as much of the whey out of the curd as possible, tipping the bowl to drain the liquid. Microwave on high for one minute. Squeeze more of the whey out. The cheese should now be sticky; if not, continue to microwave in 15-second increments until it is warm and sticky (but not too hot to handle).

Add ½ teaspoon flaked salt to the cheese and knead. Microwave for one more minute on high until the cheese is around 130°F / 54.4°C. Remove and stretch, working it just like playing with silly putty: stretch, fold in half, twist, and stretch again, over and over, until you've achieved a stringy texture.

Notes

- *The addition of acid denatures proteins in the milk, helping curd formation. Citric acid is commonly used. For similar reasons, many cheeses use rennet—traditionally derived from calf stomach—because it has a number of enzymes that break down proteins in the milk.*
- *The lipase powder is not* chemically *required, especially given that animal-based rennet contains lipase. Your rennet source might not contain it, however, and the lipase enzyme is responsible for the characteristic flavor of mozzarella because of the way it cleaves the fats in milk. For a lacto-vegetarian mozzarella cheese, use vegetable-based rennet and skip the lipase powder, but note that the cheese will not have the traditional flavor. For a source of lipase, try* http://www.dairyconnection.com *or* http://thecheesemaker.com/cultures.htm.
- *For a good writeup on making mozzarella following a more traditional, more authentic, and much more involved approach, see* http://fiascofarm.com/dairy/mozzarella.htm.

Green Olives

If you are lucky enough to have access to an olive tree during the fall, when the unripe fruit is available, try your hand at making green olives.

Unlike the mature black fruit of the olive tree, olives in their green form can be soaked in lye (sodium hydroxide, a.k.a. caustic soda) to remove the bitter compound oleuropein that is present in the unripe flesh.

Obtaining food-grade sodium hydroxide might require some searching online; http://www.lyedepot.com *carries food-grade micro beads. Do not use industrial products such as Drano, because they contain other chemicals! Also, because lye is extremely corrosive, take great care not to come into direct contact with it. Use rubber gloves and eye protection, and consider finding an outdoor space that is more forgiving of accidental spills than your kitchen.*

In a large plastic bucket or glass jar, place:

Green olives of a consistent size, with any of the fruit that is bruised or soft discarded

Room-temperature water (add enough to completely cover olives)

Transfer the water to a second plastic bucket or glass jar, and measure how much water you used. (Adding it to the first container was just to determine the amount needed.) Add to the water:

1 tablespoon of lye per quart of water for a 1.5% solution

Stir carefully to combine, and gently pour over olives into the first container. Soak for one day.

After a day, discard the lye/water solution and refresh with a new batch of water and lye. Soak for an additional day.

After the second day, take an olive out and cut into it, exposing the pit. If there is any whiteness near the center, discard the lye/water solution again and refresh with a new batch. Repeat until the entire olive flesh is a consistent color.

Once the olives are cured to the center, drain the liquid and soak in fresh water. A day later, replace the fresh water with salt water, using 1 tablespoon of salt per quart. Replace the salt water daily for three to five days. Transfer to a jar, fill with salt water, and store in the fridge.

Notes

- *As with the scallops in ceviche, you can see the flesh of the olive change over time as the sodium hydroxide penetrates to the center of the olive.*
- *Try adding vinegar and spices (e.g., bay leaves, rosemary) to the final saltwater brine to impart those flavors into the olives. For suggestions on further variations, including other curing techniques, see UC Davis's excellent writeup at* http://anrcatalog.ucdavis.edu/pdf/8267.pdf.
- *Be careful with the lye! No, really, I mean it. Pretend that it's oil at 400°F / 200°C. Also, avoid metal containers or utensils, because lye reacts with metal, especially aluminum. Wood, plastic, stainless steel, and glass are okay.*
- *If any of the olives float above the surface, you can place a glass or plate on top of the bowl to immerse them fully. (Floating olives will oxidize.)*

Alcohol

A number of organic compounds that provide aromas in food are readily dissolved in ethanol but not in water. You will invariably encounter dishes where alcohol is used for its chemical properties, either as a medium to carry flavors or as a tool for making flavors in the food available in sufficient quantity for your olfactory system to notice.

Ethanol can react with carboxylic acids in acid-catalyzed conditions, forming compounds that then react with more ethanol to generate water and the ester compounds that help carry aromas up into the nasal cavity.

Alcohol is often added to sauces or stews to aid in releasing aromatic compounds "locked up" in the ingredients. Try adding red wine to a tomato sauce or dribbling a bit of Pernod (anise liqueur) on top of a piece of pan-seared cod served with roasted fennel and rice.

You can also make your own flavor-infused vodkas by adding diced fruit, berries, herbs, or other spices to straight vodka. And since your concoction doesn't have to be shelf-stable like commercial varieties, you can generate better-tasting infusions. Don't limit yourself to just vodkas, either; try adding mint and a small quantity of sugar syrup to bourbon whiskey and storing it in the freezer.

Does Alcohol "Burn Off" in Cooking?

No, not entirely. Even though the boiling point of pure ethanol (C_2H_5OH) is lower than that of water at atmospheric pressure (173°F / 78°C), the intermolecular bonding between ethanol and other compounds in the food is strong enough that its boiling point varies based on the concentration of ethanol in the food and how the other chemicals in the food hold on to it.

The table to the right shows the percentage of alcohol remaining after various cooking methods according to a paper published by researchers at the University of Idaho.

Cooking method	% remaining
Alcohol added to boiling liquid and removed from heat	85%
Alcohol flamed	75%
No heat, stored overnight	70%
Baked, 25 minutes, alcohol not stirred into mixture	45%
Baked/simmered, alcohol stirred into mixture:	
...for 15 minutes	40%
...for 30 minutes	35%
...for 1 hour	25%
...for 2 hours	10%

Fat-Washing Alcohols: Butter-Infused Rum, Bacon-Infused Bourbon

The term "fat washing" comes from the process of using fat to "wash out" undesirable molecules, but it is more useful in the home kitchen (and in molecular mixology) as a way of infusing oil-soluble compounds into alcohol. If you use a non-neutral flavored fat—a fat that has other molecules mixed in—some of the flavorful molecules will bind with the alcohol molecules (it is a solvent, after all) and remain behind in the drink.

Why do this? Because you can create infused alcohols with flavors that might not come out in traditional infusing. The flavors can either be native to the fat (butter, bacon) or fat-soluble compounds bloomed in the fat before fat washing.

Create an infusion of 3–5% fat and 95–97% alcohol. Try 2 teaspoons (10g) of melted butter with 1 cup (200g) of rum or 2 teaspoons (10g) of bacon fat (filtered!) with 1 cup (200g) of bourbon. Let rest at room temperature for 12+ hours. Longer times and higher temperatures will yield a stronger infusion, so you'll want to experiment.

Try using an immersion blender to kick-start the infusion.

After infusing, place infusion in freezer until fats have solidified, and then filter through a coffee filter or other ~20-micron filter (see the filtration section on page 361 in Chapter 7).

Unfiltered.

100 micron filter.

~10–20 micron filter.

Notes

- *Try this with blue cheese, nut butters, and other fats.*
- *A key step in refining alcohol is the removal of undesirable compounds. It's impossible to remove every last "bad" molecule, but the more that are removed, the better tasting the beverage will be. This is why "the good stuff" costs more: refiners are able to remove more of the off-tasting compounds by increasing the number of steps in processing or giving the alcohol longer to age, which allows for better yield of the chemical reactions that remove the compounds. Fat washing can be used as a DIY way to further refine an alcohol: the compounds will bind with some of the fat molecules, which can then be removed by simple filtration. Try using a neutral-flavored fat, such as lard, for refining without altering the flavor.*
- *Incidental advice if you ever find yourself writing a book involving alcoholic drinks: don't write after drinking your experiments.*

Vanilla Extract

In a small glass jar with a tight-fitting lid, put:

1 vanilla bean, sliced open lengthwise and chopped into strips to fit jar

1 oz (30g) vodka (use enough to cover vanilla bean)

Screw lid on jar or place plastic wrap over top and store in a cool, dark place (e.g., pantry) for at least a few days. For better flavor, allow the extract to steep for a few weeks.

Notes

- *The vanilla bean can be left over from some other recipe. If you cook with vanilla frequently, consider keeping the jar of vanilla constantly topped off. Whenever you use a vanilla bean, add it to the jar, removing an old one when space requires it. And as you use the extract, occasionally top off the jar with a bit more vodka or other liquor such as rum.*
- *Play with other variations: instead of vodka, which is used for its high ethanol content and general lack of flavor, you can use other spirits such as rum, brandy, or a blend of these.*
- *The ethanol dissolves a number of compounds present in the vanilla bean, including the compound vanillin, which gives vanilla its characteristic flavor.*
- *Instead of vanilla beans, try using star anise, cloves, or cinnamon sticks. Or try varying both solvent and substance (e.g., orange rind with Grand Marnier).*
- *Flavored alcoholic drinks can be made with this same technique. Instead of a large quantity of the solute (e.g., vanilla bean) and a minor amount of solvent (e.g., vodka), place a small bit of the solute into a bottle of the solvent. For an example, search online for* nocino, *an Italian walnut liqueur made with* unripe *walnuts, aromatic spices, and ethanol.*

Sage Rush: Gin, Sage, and Grapefruit Juice

This is a simple cocktail and a darn good one. And having a simple, darn-good cocktail in your repertoire can be handy. It only takes knowing one good drink to impress that romantic potential.

Put two or three sage leaves (fresh!) in a shaker and muddle with the back side of a spoon. Add 1 part gin and 1 part pink grapefruit juice—say, 2 oz (50 ml) of each—and add several ice cubes. Shake vigorously. Strain into a martini glass.

Note

- *If you have fresh pink grapefruit, use that. Squeeze the juice from half a grapefruit and add gin to taste. You can muddle the sage leaf post-shaker directly in the glass as well.*

When a Molecule Meets a Molecule...

Alcohol isn't the only solvent in the kitchen. The same chemical interactions that give alcohol its magic apply to oil and water, which is why recipes call for steps such as toasting caraway seeds in oil: the oil captures the molecules responsible for the characteristic nutty flavors developed and released by heating the seeds.

But *how* does a solvent work? What happens when one molecule bumps into another molecule? Will they form a bond (called an *intermolecular bond*) or repel each other? It depends on a number of forces that stem from differences in the electrical charges and charge distributions of the two molecules.

Of the four types of bonds defined in chemistry, two are of culinary interest: polar and nonpolar.

A molecule that has an uneven electrical field around it or that has an uneven arrangement of electrons is *polar*. The simplest arrangement, where two sides of a molecule have opposite electrical charges, is called a *dipole*. Water is polar because the two hydrogen atoms attach themselves to the oxygen atom such that the molecule as a whole has a negatively charged side. When two polar molecules bump into each other, a strong bond forms between the first molecule's positive side and the second molecule's negative side, just like when two magnets are lined up. On the atomic level, the side of the first molecule that has a negative charge is balancing out the side of the second molecule that has a positive charge.

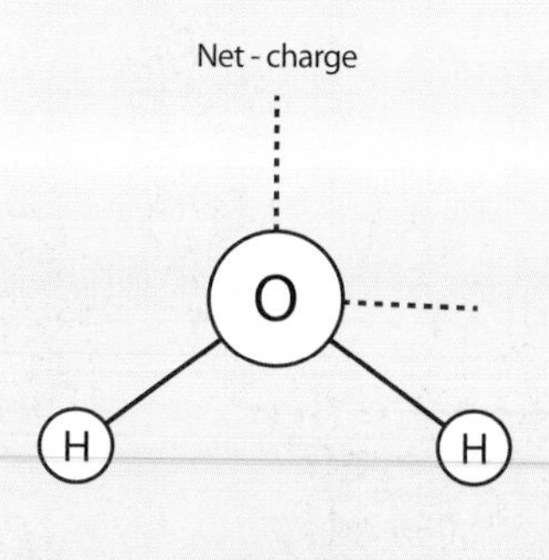

A water molecule is polar because the electrostatic field around the molecule is asymmetric, due to the oxygen atom being more electronegative than the hydrogen atoms and the resulting differences in how the two hydrogen atoms share their electrons with the oxygen atom. (Electron sharing is another type of bond, a covalent bond.)

A molecule that has a spherically symmetric electrostatic field—that is, there is no dipole, and the molecule doesn't have a "side" that has a different electrical charge—is *nonpolar*. Oil is nonpolar because of the shape in which the carbon, oxygen, and hydrogen atoms arrange themselves.

In most cases, when a polar molecule bumps into a nonpolar molecule, the polar molecule is unlikely to find an electron to balance out its electrical field. It's a bit like trying to stick a magnet to a piece of wood: the magnet and wood aren't actively repelled by each other, but they're also not actually attracted. It's the same for polar-nonpolar interaction: the molecules might bounce into each other, but they won't stick and will end up drifting off and continuing to bounce around into other molecules.

This is why oil and water do not mix. The water molecules are polar and form strong intermolecular bonds with other polar molecules, which are able to balance out their electrical charges. At an atomic level, the oil doesn't provide a sufficiently strong bonding opportunity for the negatively charged side of the water molecule.

Water and sugar (sucrose), however, get along fine. Sucrose is also polar, so the electrical fields of the two molecules are able to line up to some degree. The strength of the intermolecular bond depends on how well the two different compounds line up, which is why some things dissolve together well while others only dissolve together to a certain point.

Extracts for drinks

Bitters are to bartenders what extracts and spices are to chefs: they provide flavor with minimal impact on texture, volume, or other variables. *Bitters* refers to any extract that includes a bittering agent, such as gentian, quinine, or citrus rind. Angostura bitters is the "generic" bitter—one of the few to have survived through the Prohibition era—and is what most people think of when a recipe calls for bitters. Campari is also a bitter, although not commonly described this way. Bitters come in a range of flavors: from the complex and spicy (clove, anise, cinnamon) to the bright and clean (orange, grapefruit, mint).

This collection of bitters shows just a small selection of the flavors available.

Bitters can be used as flavorings in things besides alcoholic drinks. Try a dash of bitters in soda water, along with a slice of lime. Since they are a subset of extracts, you can use them in any place where a bitter extract would work. You can balance out bitterness with the addition of sugar, just as is done in an old-fashioned cocktail. Bitters as an accent flavor in a chocolate truffle? As part of a dressing? Try it!

Bitters recipes can be quite complicated, requiring exotic ingredients and involving dozens of steps taking upward of a month. If you want to try your hand at one of the more involved recipes, try the one that follows here. For additional recipes, pick up Gary Regan's book, *The Joy of Mixology* (Clarkson Potter), from which the recipe on the following page is adapted with permission. His recipe uses both ethanol and water as solvents. The ethanol at the beginning dissolves one set of organic compounds present in the spices. Later, the water dissolves a different set. Notice that the ethanol that contains the first set of alcohol-solvent organic compounds is never subjected to heat!

Regan's Orange Bitters No. 5

Combine in a large jar:

2 cups (450g) grain alcohol such as Everclear or vodka

½ cup (160g) water

8 oz (250g) dried orange peel

½ teaspoon (3g) caraway seeds

1 teaspoon (2g) quassia chips

1 teaspoon (2g) cardamom seeds

½ teaspoon (0.75g) cinchona bark, powdered

1 teaspoon (0.50g) coriander seeds

¼ teaspoon (0.25g) gentian

Make sure that the liquid covers the dry ingredients, adding more grain alcohol if necessary, and screw on the lid. Shake vigorously to mix, about 20 seconds, once a day for two weeks.

After two weeks, remove the solids, boil them in water, and then add them to the alcohol again. You can separate the liquid from the solids by straining it with a cheesecloth or fine sieve, returning the liquid to the jar and placing the solids into a saucepan. Muddle the solids with a pestle so that the seeds are broken open. In the saucepan with the solids, add:

3½ cups (800g) water

Bring to a boil and then simmer with lid on for 10 minutes. Turn off heat and allow to return to room temperature for about an hour. Once cool, recombine the solids and water with the alcoholic liquid in the jar.

Shake vigorously for 30 seconds once a day for at least a week. Then strain out the solids and discard them.

Next, we'll make a sugar syrup to add to the liquid. In an empty saucepan, bring to medium heat:

1 cup (200g) sugar

Once the sugar starts to melt, stir constantly until the sugar caramelizes to a dark brown color.

Allow to cool for a few minutes. Add the liquid to the sugar, stirring it until entirely dissolved. Transfer liquid to jar and let rest for a week.

After a week, remove any solids that are floating and decant the clear liquid into another container, leaving behind the sediment.

You should have about 12 fluid ounces (350 ml) of liquid at this point. Add 6 ounces (180 ml) of water, shake thoroughly, and transfer to a bitters bottle (amber or other opaque bottle to prevent light from breaking down some of the organic compounds).

Notes

- *Some of the harder-to-find ingredients can be procured from* http://www.kalustyans.com *and* http://www.starwest-botanicals.com.
- *When using bitters, a "dash" is a solid pour from a bottle with a dash cap: bottle right-side up, rotate 180 degrees, and back. It's not a side trickle. Your "dashes" will be larger as the bottle gets emptier due to the change in air volume in the bottle, but for practical purposes at home, it's probably not worth breaking out the milligram scale. (But if you do, a quick check with my scale shows roughly 6 dashes to the gram.)*
- *A number of online sites exist for ordering bitters, in case you get taken with them but don't want to spend the time making them. Try searching* http://www.kegworks.com *for the word "bitters" or search the Internet for "Fee Brothers" and "Bitter Truth," two makers of specialized bitters.*

Linda Anctil on Inspiration

Linda Anctil is a private chef in Connecticut who blogs about her work at http://www.playingwithfireandwater.com.

How do you think about the visual experience of food?

I approach food as a designer, but because it is food, it also has to function. Ultimately, it has to taste good. Sometimes, I'm inspired by an ingredient, a season, a shape, or a color, but inspiration can come from anywhere. I always try to include an element of surprise, whether it's visual or stimulating to other senses.

Sometimes, I'm inspired by a serving piece. I found this great votive holder with a cup suspended inside of another cup with an empty chamber underneath. I've served brandies and bourbons in it and, actually, one with smoked cedar. I infused the flavor into a bourbon, then filled the bottom empty chamber with cedar smoke. The person drinking it will lift the glass to take a sip, and the smoke just pours out from it. It heightens the whole experience because you're bringing the sense of smell into it.

Votive holder on table with bourbon and cedar smoke.

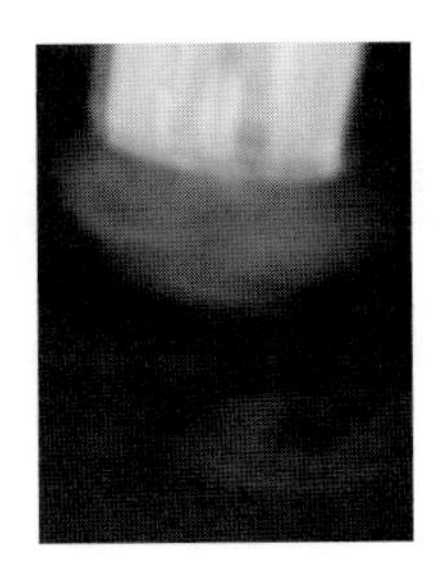

Votive holder being picked up with smoke wisps coming out.

I inverted that same votive candle holder in another post. I put clam chowder in it with small potatoes that were hollowed out and filled with bacon and clams, all the flavors you expect to find in clam chowder. It almost looked like the potatoes were floating in the broth. It should be playful. It should be whimsical as well as delicious.

Nature is a constant source of inspiration to me. I went out to the garden to pick some sage last winter, and I had the scent of conifers from my Christmas tree on my gloves. The smells became intermingled in my mind and, suddenly, conifers became something I could use as an herb. It inspired a whole series of dishes that I put together using the flavor of conifer. I did one where I layered lots of different textures and flavors together, culminating with my video "The Winter Garden" (at *http://www.youtube.com/watch?v=2bYvapNDIJw*). I think it was probably my most abstract or conceptual dish that I've put together, but it really captured that whole feeling of being outside on that one day with the ice and the snow and the frost and the smell of conifers. I was the only one who ate the dish in the end. I enjoyed it a lot. It was a very personal expression to me.

Do you have any suggestions about how to think about presenting food?

Keep an open mind. Pick up a piece of fruit and imagine that you were an alien on this planet who had never seen it before, and experience it through that lens. How does it look to you? What does it smell like? What does it taste like? What can you do with it? Think outside of the box and enjoy the journey!

It sounds like it's really about personal expression.

It absolutely is. You can look at any artist or chef's food, and you'll realize it's a personal expression of who they are. It's telling a story about that person's experiences. That's a wonderful aspect of cooking.

Modern Industrial Chemicals

Over the past century, the food industry has developed or repurposed a number of chemicals to address the issues of scaling created when producing food in large quantities. Preventing illness, maintaining freshness, controlling costs, and meeting changing consumer demands have all presented challenges. Producing larger quantities of food increases the time between harvest and consumption, increasing the chances of spoilage and the amount of time foodborne pathogens have to develop. And aggregating ingredients from a larger number of producers increases the impact that a single contaminated item can have.

Hard on the heels of World War II, when advances in food science had been applied to address these problems in the military's meal rations ("an army travels on its stomach"), the food industry found a new market in the American consumer. Convenience foods and prepared meals burst onto the scene at the same time that freezers went into mass production and television sets became the "must have" item for the American family. Instant food and instant entertainment have been married ever since.

The same family of chemicals that enabled the creation of the TV dinner (mmm, Kraft Macaroni & Cheese) also allowed for a new set of dishes to be created by *haute cuisine* chefs, sometimes called *molecular gastronomy* or *modernist cuisine* (we'll use the latter term). These chefs use industrial chemicals to create entirely different ways of conveying flavors and exciting the senses. When done well, the dishes are not about additives at all, but about the perceptions and emotions that all good meals strive to evoke. No one is suggesting that vegetables and whole foods should be replaced with white powders.

Molecular Gastronomy

What is molecular gastronomy? It depends on whom you ask. To the purists, it's the use of scientific investigative techniques to understand the chemical and physical changes that occur during cooking, a pure science. Under this definition, anything done in the kitchen is merely an application of general physical and chemical principles ("molecular cooking"). To others, molecular gastronomy is the use of unusual processes and chemicals to create sometimes alien experimental dishes that have a reputation for being sensory gimmicks. And somewhere in the middle are a large number of modern foodies: it's the stuff that Marcel did on the reality TV show *Top Chef*.

Regardless of the exact definition, one thing is clear: modern techniques for manipulating food can expand the toolbox from which a talented chef can create new experiences, whether subtle or over-the-top. As with all art, some experiments turn out great, stirring an emotional response and creating (hopefully positive) feelings, while others fail and are politely ignored (or not).

Either way, the diner must be a willing participant in these types of experiences. If you're lusting after a classic cheeseburger, you'd probably be unhappy being served a "deconstructed" burger, never mind how well executed: beef tartar on one corner of the plate, micro-green salad on the opposite corner, a smear of tomato reduction between the two, and toasted sesame bread on the side. As with most such things, frame of mind, expectations, and being open to the experience are key.

P.S. See John Crace's review of "In Search of Perfection" in the Guardian, at http://www.guardian.co.uk/books/2006/nov/21/digestedread.johncrace.

The demand for innovative foods at the high end of the culinary world should not be surprising. Luxury restaurants now have to compete with the enthusiastic hobbyist chef, who has been able to better approximate traditional restaurant fare as the quality of consumer gear and produce has improved. The same technological advances that have enabled the production of convenience foods have also enabled the agro-industrial food complex to deliver an ever-widening—sometimes maddeningly so—variety of food, and also to make those foods available for a longer window of time each year.

Turning to food additives for new dishes is a logical progression in the process of creating something new. Sometimes, the results are amazing; other times, they fall flat. Compare the culinary iconoclasts to the fashions that show up on the Paris runways: while it might not be "everyday" wear or cuisine, the better concepts and ideas that start out at the high end eventually make their way into the clothing shops and onto the general restaurant scene.

Many of the techniques that rely on food additives originated in Europe. Chef Ferran Adrià's restaurant elBulli, in Spain, is considered by many to be the originator of much of modern haute cuisine. Chef Heston Blumenthal's restaurant The Fat Duck, in the UK, has also established an international reputation for pushing the boundaries of food.

By some accounts, one had a better chance of getting into Harvard than getting a reservation at elBulli.

Should you have the opportunity and inclination to dine at them, both Alinea (Chef Grant Achatz's restaurant in Chicago) and wd~50 (Chef Wylie Dufresne's restaurant in New York City) are highly regarded and happen to use food additives in creating some of their dining experiences.

Fortunately, you do not need to eat at these places to understand what this style of cooking offers. With willingness and a certain amount of determination, you can duplicate, or at least roughly approximate, a number of the techniques in use at these restaurants in your own home.

Be forewarned: while the techniques are generally not difficult, the time and costs involved and the resulting product might not leave you clamoring to use these methods in your daily routine; in fact, you might even think they should be classified as a form of culinary terrorism. Still, even if the use of some of these chemicals remains limited to the "fun party trick" category because of their novelty, isn't a part of geeking out understanding how things work? Before jumping into the techniques, however, let's take a slight detour to examine a chemical taxonomy and the chemistry of colloids to help explain the science behind the techniques.

E Numbers: The Dewey Decimal System of Food Additives

It's easy enough to write eggplant on the grocery list, but how does one go about writing up a grocery list for food additives? The *Codex Alimentarius Commission*—established by the United Nations and the World Health Organization—has created a taxonomy of food additives called "E numbers." Like the Dewey Decimal classification system, it establishes a hierarchical tree: a unique E number is assigned for each chemical compound, grouped by functional categories, with the numbering of chemicals determined by each chemical's primary usage.

E100–E199:	**Coloring agents (i.e., food coloring, like those found in the grocery store)** E120: Cochineal or carminic acid ("red 4," in common use)
E200–E299:	**Preservatives** E251: Sodium nitrate (used in curing items like sausages) E290: Carbon dioxide
E300–E399:	**Antioxidants, acidity regulators** E300: Ascorbic acid (vitamin C) E322: Lecithin (emulsifier, typically from soy) E330: Citric acid (in lemons, limes, etc.) E327: Calcium lactate
E400–E499:	**Thickeners, emulsifiers, and stabilizers** E401: Sodium alginate E406: Agar E441: Gelatin E461: Methylcellulose
E500–E599:	**Acidity regulators, anti-caking agents** E500: Sodium bicarbonate (baking soda) E509: Calcium chloride E524: Sodium hydroxide (lye)
E600–E699:	**Flavor enhancers** E621: Monosodium glutamate (MSG)
E700–E799:	**Antibiotics**
E900–E999:	**Miscellaneous** E941: Nitrogen (used in food storage) E953: Isomalt (also known as Isomaltitol)
E1000–E1999:	**Additional chemicals** E1510: Ethanol (alcohol)

An abbreviated table of E numbers including common food additives.

Not everything has an E number; for example, neither common salt (sodium chloride) nor transglutaminase (discussed later in this chapter) is currently included. Which additive to use for a particular effect, such as gelling, depends upon the properties of the food with which you're working and your goals. Most food additives used in modernist cuisine come from the E400–E499 range, which consists of the following:

Thickeners (e.g., cornstarch, methylcellulose, agar, carrageenan)
: Provide structure to items such as gels (Jell-O), traditional French dishes (aspics and terrines), and confections (gummy candies). Food preparers also use them to prevent both water and sugar crystallization in foods such as ice creams, because thickeners inhibit the development of molecular lattices.

Emulsifiers (e.g., lecithin and glycerin)
: Prevent two liquids from separating, as with oil and water in mayonnaise. The food industry uses lecithin in chocolate for similar reasons, to prevent the cocoa solids and fats from separating and to increase the viscosity of the melted chocolate during manufacturing.

Stabilizers (e.g., guar and xanthan gums)
: Lend a smooth "mouth-feel" to a liquid and can also act as emulsifiers by preventing aggregates from separating. Think of how oregano stays suspended in a commercial salad dressing, instead of precipitating out and settling to the bottom.

You will also see compounds from the E300–E399 and E500–E599 ranges used, but usually as secondary additives that help the E400–E499 compounds function. A number of the E400–E499 additives require either certain pH ranges or secondary compounds to react with, such as calcium when working with sodium alginate.

Some additives work in a broad range of pHs and temperatures but have other properties that may prohibit their use, depending upon the recipe. For example, while agar is a strong gelling agent, in some gels it also exhibits *syneresis* (when a gel expels a portion of its liquid—think of the liquid whey that separates out in some yogurts). Carrageenan does not undergo syneresis but cannot handle an environment as acidic as agar can. For example, if you attempt to use carrageenan to gel lime juice, which has a pH between 2.0 and 2.35, you will also need to add an acidity regulator to raise the pH.

Buying Food Additives

As you've probably guessed by now, your local grocery store is unlikely to stock many of the chemicals needed for these techniques. Sourcing food-grade versions will undoubtedly get easier over time, but expect to place some Internet orders before attempting the recipes in the rest of this chapter.

First, a warning: *you should not use "technical grade" or other non-food-grade substances*. While technical-grade powders might be 99.9% pure, there are no guarantees as to what substances comprise the remaining portion. Who knows what carcinogens might be hanging out in that remaining 0.1%!

The following sites can supply most of the food additives used in this chapter in food-grade qualities:

- Check here first, because I will maintain this list with current information: *http://www.cookingforgeeks.com/book/additives/*.
- If you're in the U.S., *http://www.shopchefrubber.com*, *http://www.lepicerie.com*, and *http://www.le-sanctuaire.com* are all good starting points.
- In Europe? Look at *http://www.cuisine-innovation.fr*, *http://www.bienmanger.com*, and *http://www.creamsupplies.co.uk*.

Commercial food preparers have to balance additional variables in their recipes. In the lime gel example, if the pH is raised too much, the food becomes hospitable to bacterial activity, depending on other parameters in the food (e.g., water availability). Balancing all of this can require multiple chemicals, which is why prepared foods can have quite a number of chemicals on their ingredient labels!

Colloids

One of the more common uses of industrial chemicals in food is to form colloids. A colloid is any mixture of two substances—gas, liquid, or solid—where one is uniformly dispersed in the other, but they are not actually dissolved together. That is, the two compounds in the mixture don't form chemical bonds, but the overall structure appears uniform to the naked eye.

Common colloids in the kitchen are whole milk and chocolate. In milk, solid particles of fat are dispersed throughout a water-based solution. In chocolate, particles of cocoa solids are dispersed throughout a solid medium of cocoa fat and other ingredients.

The following table shows the different combinations of particles and media, along with examples of foods for each colloid type. The medium of a colloid is called the *continuous phase* (it's the watery liquid in milk); the particles are known as the *dispersed phase* (for milk, the fat droplets).

	Gas particles	Liquid particles	Solid particles
Gas medium	*(N/A: gas molecules don't have a collective structure, so gas/gas combinations either mix to create a solution or separate out due to gravity)*	**Liquid aerosols** - Mist sprays	**Solid aerosols** - Smoke (convertible to a solid-in-liquid colloid via liquid smoke) - Aerosolized chocolate
Liquid medium	**Foams** - Whipped cream	**Emulsions** - Milk - Mayonnaise	**Sols and suspensions** - Commercial salad dressings
Solid medium	**Solid foams** - Meringue cookies - Soufflés	**Gel** - Gelatin - Jell-O	**Solid sols** - Chocolate

Some of these colloid types might remind you of various dishes served at more experimental restaurants.

One of the surprises of this table is the relatively broad swath of techniques that it captures. Foams, spherifications, and gelled foods are all colloids. Even some of the more recent novel dishes are colloids from the gas medium category. Chef Grant Achatz (Alinea, in Chicago) has used solid aerosols by infusing a pillow with smoke and then placing the dish on top of the pillow, forcing the air containing the aerosol to leave the pillow and diffuse into the diner's environment.

Chef Achatz uses smoke-infused "pillows" to present a pleasant olfactory experience while avoiding the taste sensation for items such as mace and lavender.

Other luxury restaurants have created courses that involve liquid aerosols (by spraying a perfume), and one company (Le Whif) is working on a kitchen gadget that creates solid aerosols from foods such as chocolates.

Some food additives can be used in more than one type of colloid. For example, guar gum can act as an emulsifier (by preventing droplets of oil from coalescing) and as a stabilizer (by preventing solids from settling). Methylcellulose is both a gelling agent and an emulsifier. Don't think of food additives as directly mapping onto the colloids they create, but it's a handy framework for thinking about the types of effects you can achieve.

Making Gels: Starches, Carrageenan, Agar, and Sodium Alginate

The food industry uses gels to thicken liquids, to emulsify sauces, to modify texture ("improve mouth-feel," as they say), and to prevent crystal formation in products such as candies (sugar crystals) and ice cream (ice crystals and sugar crystals). Gels are also found in traditional home cooking: both gelatin (see the section on "Filtration" in Chapter 7) and pectin (see the sidebar "Make Your Own Pectin" on page 197 in Chapter 4) are used in many dishes to improve mouth-feel, and they also help preserve items such as jams.

From the perspective of modernist cuisine, thickeners and gels are used primarily to create dishes in which foods that are typically liquid are converted into something that is thick enough to hold its shape (this is what pectin does in jam), or even completely solid.

Gels can also be formed "around" liquids to create a gelatinous surface in a technique known as *spherification*, originally discovered by Unilever in the 1950s and brought to the modernist cuisine movement by Chef Ferran Adrià of elBulli. For our purposes, gels in foods can be classified into two general types: soft gels and brittle gels (true gels).

You can think of a *soft gel* as a thicker version of the original liquid: it has increased viscosity (it's "thicker"), but it retains its ability to flow. Soft gels can exhibit a phenomenon termed *shear thinning*, which is when a substance holds its shape but will flow and change shape when pressure is applied. Substances like ketchup and toothpaste exhibit shear thinning: squeeze the bottle or tube, and it flows easily, but let go, and it holds its shape.

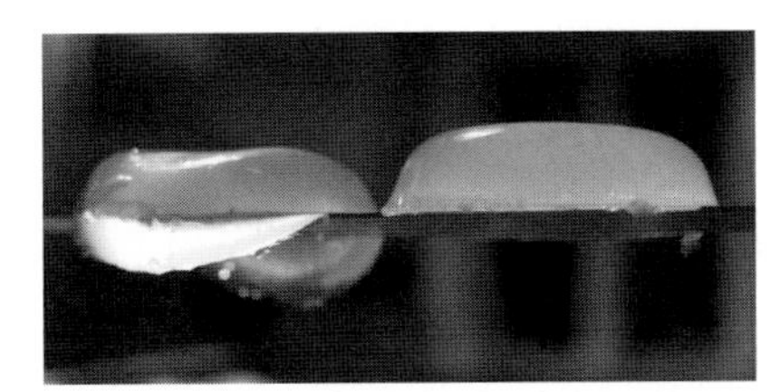

Iota carrageenan (left, 2% concentration) creates a flexible brittle gel, while kappa carrageenan (right, 2% concentration) creates a firm brittle gel. These two samples are resting on top of a narrow bar.

While a soft gel can be described as a "thicker" version of the original liquid, a *brittle gel* can be thought of as a solid. Brittle gels—foods like cooked egg whites and Jell-O—have a tightly interconnected lattice that prevents them from flowing at all. With sufficient quantities of the gelling agent, this type can form a block or sheet that you can pick up, slice into blocks or strips, and stack as a component in a dish, and it has a "memory" of its cast shape, meaning that it will revert to that shape when no other forces are in play.

In the consumer kitchen, cornstarch is the standard traditional gelling agent. In industrial cooking, carrageenan is commonly used in gelling applications. (Try finding cream cheese that doesn't have carrageenan in it.) Iota carrageenan is used when a thickening agent is needed, while kappa carrageenan and agar yield firm, brittle gels. While the gelling agents used to create flexible and rigid gels are generally different, you can create a flexible gel with a gelling agent typically used in rigid, brittle applications by carefully controlling the quantity of gelling agent used.

Making gels: Starches

Starches are used as thickeners in everything from simple roux to pie filling. They're easy, plentiful, and exist in almost all of the world's cuisines: cornstarch, wheat flour, tapioca starch, and potato "flour" (not actually a flour) being the most common. While there are differences among these starches—size of the starch granules, length of the molecular structure, and variations on the crystalline structure—they all act essentially the same. Expose to water, heat up, then cool down, and they thicken up.

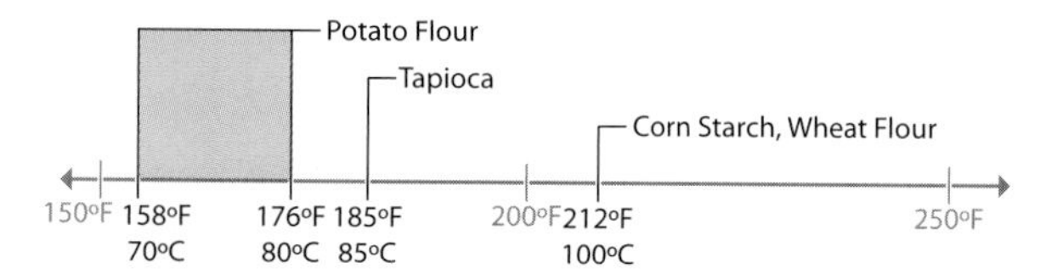

Gelatinization temperature of common starches.

Starch is composed of repeating units of amylopectin and amylose that form crystalline structures. The gelatinization temperature—the temperature at which these crystalline structures melt and then absorb water and swell—can vary, depending upon the ratio of amylopectin and amylose groups. We'll examine cornstarch here, but as you play with the others, keep in mind that the gelatinization temperature can vary.

Instructions for use. To use cornstarch (called "corn flour" in the UK) to make a gel, mix it with a small amount of cold liquid such as water to create a slurry. Adding cornstarch directly to a hot liquid will result in clumps. Add the slurry to the desired dish and bring to a simmer.

Uses. Cornstarch is used as a thickener and has about twice the thickening ability of flour. When a recipe calls for a teaspoon of flour, use half a teaspoon of cornstarch. Cornstarch is gluten-free, making it a good thickening substitute for those with gluten allergies. (Flour isn't as good a thickener because it contains other stuff in addition to starch, such as gluten, fat, fiber, and minerals.)

Origin and chemistry. Derived from corn (shocker, I know). Like other starches used in cooking (e.g., potato, tapioca, wheat), cornstarch is a carbohydrate composed of repeating units of amylopectin and amylose that form crystalline structures. On heating, these structures swell up and break down. Upon cooling, the leached amylose molecules can link together to create a 3D mesh, trapping other molecules into the network. For more on the chemistry of starches, see *http://www1.lsbu.ac.uk/water/hysta.html*.

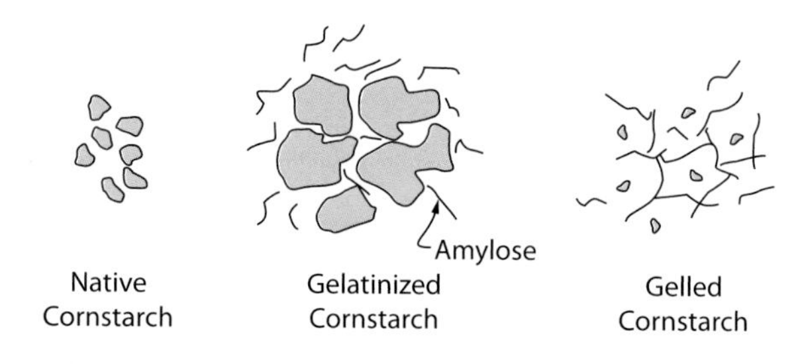

Technical notes	
Gelatinization temperature	203°F / 95°C; maximum thickness at 212°F / 100°C.
Gel type	Thixotropic. (This means it becomes less viscous when pressure is applied. Think ketchup: it holds its shape, but flows under pressure.)
Syneresis ("weeping")	Extensive if frozen and then thawed.
Thermoreversible	No—after gelatinizing, the amylose is leached out from the original starch molecules.

Lemon Meringue Pie

Like many savory foods in which multiple discrete components are combined to create the dish, lemon meringue pie is the combination of three separate components: pie dough, a meringue, and a custard-like filling. We've already covered pie dough (page 223 in Chapter 5) and meringues (page 255 in Chapter 5), so the only thing left for making a lemon meringue pie is the filling itself. Flip to those recipes for instructions on how to make the pie dough and meringue topping.

To make the lemon custard, place in a saucepan off heat and whisk together:

2½ cups (500g) sugar

¾ cup (100g) cornstarch

½ teaspoon (5g) salt

Add 3 cups (700g) of water, whisk together, and place over medium heat. Stir until boiling and the cornstarch has set. Remove from heat.

In a separate bowl, whisk together:

6 egg yolks

Save the whites for making the meringue. Make sure not to get any egg yolk in the whites! The fats in the yolk (nonpolar) will prevent the whites from being able to form a foam when whisked.

Slowly add about a quarter of the cornstarch mixture to the egg yolks while whisking continuously. This will mix the yolks into the cornstarch mixture without cooking the egg yolks (tempering). Transfer the entire egg mixture back into the saucepan, whisk in the following ingredients, and return to medium heat and cook until the eggs are set, about a minute:

1 cup (240g) lemon juice (juice of about 4 lemons)

Zest from the lemons (optional; skip if using bottled lemon juice)

Transfer the filling to a prebaked pie shell. Cover with Italian meringue made using the six egg whites (double the recipe on page 255 in Chapter 5, which is for three whites), and bake in a preheated oven at 375°F / 190°C for 10 to 15 minutes, until the meringue begins to turn brown on top. Remove and let cool for at least four hours—unless you want to serve it in soup bowls with spoons—so that the cornstarch has time to gel.

To create decorative peaks on the meringue, use the back of a spoon: touch the surface of the unbaked meringue and pull upward. The meringue will stick to the back of the spoon and form peaks.

Gelling agents typically come as a powdered substance that is added to water or whatever other liquid you are working with. Upon mixing with the liquid, and typically after heating, the gelling agent rehydrates and as it cools forms a three-dimensional lattice that "traps" the rest of the liquid in suspension. By default, add your gelling agent to a cold liquid and heat that up. Adding gelling agents to hot liquid usually results in clumps because the outer layer of the powder will gel up around the rest of the powder.

Making gels: Carrageenan

Carrageenan has been used in food as far back as the 15th century for thickening dairy products. Commercial mass production of carrageenan gums became feasible after World War II, and now it shows up in everything from cream cheese to dog food, where it acts as a thickener. Modernist cuisine dishes use it for the same reason, although typically to thicken liquids into gels in ways that we might not think of at first glance (beer gel, anyone?).

Instructions for use. Mix 0.5% to 1.5% carrageenan into room-temperature liquid. Gently stir liquid to avoid trapping air bubbles into the gel; lumps are okay at this stage. (They're hard to get out unless you have a vacuum system.) Allow to rest for an hour or so; carrageenan takes a while to rehydrate. To set carrageenan, bring to a simmer either on a stovetop or in an oven. If you are working with a liquid that can't be heated, create a thicker concentration using just water, heat that, and then mix it into your dish.

Uses. Carrageenan is used to thicken foods and to control crystal growth (e.g., in ice cream, keeping ice crystals small prevents a gritty texture). Carrageenan is commonly used in dairy (check the ingredients on your container of heavy whipping cream!) and water-based products, such as fast-food shakes (keeps ingredients in suspension and enhances mouth-feel) and ice creams (prevents aggregation of ice crystals and syneresis, the expulsion of liquid from a gel).

Origin and chemistry. Derived from seaweed (such as *Chondrus crispus*—common name Irish moss), carrageenan refers to a family of molecules that all share a common shape (a linear polymer that alternates between two types of sugars). The seaweed is sun-dried, treated with lye, washed, and refined into a powder. Variations in the molecular structure of carrageenan cause different levels of gelification, so different effects can be achieved by using different types of carrageenan (which, helpfully, grow in different varieties of red seaweed). Kappa carrageenan (k-carrageenan) forms a stronger brittle gel, and iota carrageenan (i-carrageenan) forms a softer brittle gel.

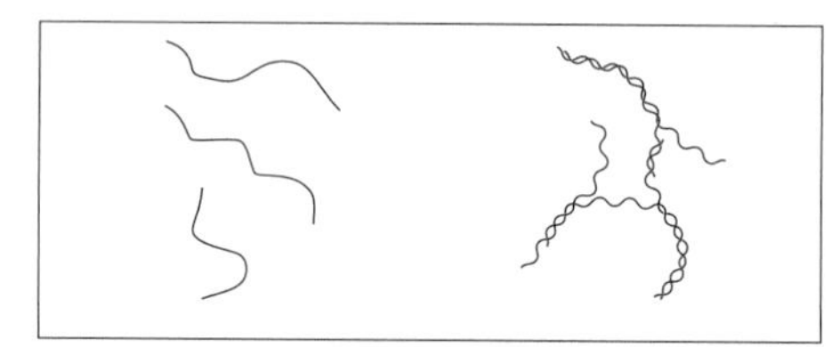

On the molecular level, carrageenan, when heated, untangles and loses its helical structure (left); when cooled, it reforms helices that wrap around each other and form small clusters (right). The small clusters can then form a giant three-dimensional net that traps other molecules.

Technical notes

	i-carrageenan	k-carrageenan
Gelling temperature	95–149°F / 35–65°C	95–149°F / 35–65°C
Melting temperature	131–185°F / 55–85°C	131–185°F / 55–85°C
Gel type	Soft gel: gels in the presence of calcium ions	Firm gel: gels in the presence of potassium ions
Syneresis	No	Yes
Working concentrations	0.3% to 2%	0.3% to 2%
Notes	Poor solubility in sugary solutions Interacts well with starches	Insoluble in salty solutions Interacts well with nongelling polysaccharides (e.g., gums like locust bean gum)
Thermoreversible	Yes	Yes

Gelled Milk with Iota and Kappa Carrageenan

This isn't, in and of itself, a tasty recipe (add some chocolate, though, and you've got something close to commercial prepackaged food). Still, it will give you a good sense of what adding a gelling agent does to a liquid and provides a good comparison between soft and brittle gels.

Soft gel version

In a saucepan, whisk to combine and then bring to a boil:

1 teaspoon (1.5g) iota carrageenan
3.5 oz (100 ml) milk

Pour into a glass, ice cube tray, or mold and chill in the fridge until set (about 10 minutes).

Brittle gel version

Again in a saucepan, whisk to combine and then bring to a boil:

1 teaspoon (1.5g) kappa carrageenan
3.5 oz (100 ml) milk

Pour into a second glass, ice cube tray, or mold and chill in the fridge until set.

Notes

- *Try modifying the recipe by adding 1 teaspoon (4g) of sugar, substituting some cream for a portion of the milk, popping the mixture into a microwave for a minute to set it, and pouring it into a ramekin that has a thin layer of jam or jelly and toasted sliced almonds on the bottom. Once gelled, invert the set gel onto a plate for something roughly approximating a flan-style custard.*
- *Since the carrageenan is thermoreversible (once gelled, it can still be melted), you can take a block of food gelled with kappa carrageenan, slice it into cubes, and do silly things like serve it with coffee or tea (one lump or two?).*
- *You can take a firm brittle gel and break up the structure using a whisk to create things like thick chocolate pudding.*

Making gels: Agar

Agar—sometimes called *agar-agar*—is perhaps the oldest of all the food additives commonly used in industry, but has only recently become known in western cuisines, mostly as a vegetarian substitute for gelatin. First used by the Japanese in the firm, jelly-type desserts that they're known for, such as *mizuyokan*, agar has a history stretching back many centuries.

When it comes to playing with food additives, agar is one of the simplest to work with. You can add it to just about any liquid to create a firm gel—a 2% concentration in, say, a cup of Earl Grey tea will make it firmer than Jell-O—and it sets quickly at room temperature. It comes in two general varieties: flakes or powder. The powdered form is easier to work with (just add to liquid and heat). When working with the flake variety, presoak it for at least five minutes and make sure to cook long enough so that it breaks down fully.

Instructions for use. Dissolve 0.5% to 2% agar by weight in cold liquid and whisk to combine. Bring liquid to a boil. As with carrageenan, you can create a thicker concentrate and add that to a target liquid if the target liquid can't be boiled. Compared to carrageenan, agar has a broader range of substances in which it will work, but it requires a higher temperature to set.

Use. Agar is a gelling agent, used in industry in lieu of gelatin in products such as jellies, candies, cheeses, and glazes. Since agar is vegetarian, it's a good substitute in dishes that traditionally call for gelatin, which is derived from animal skins and bones. Agar has a slight taste, though, so it works best with strongly flavored dishes.

Origin and chemistry. Derived from seaweed. Like carrageenan, agar is a seaweed-derived polysaccharide used to thicken foods and create gels. When heated above 185°F / 85°C, the galactose in agar melts, and upon cooling below 90–104°F / 32–40°C it forms a double-helix structure. (The exact gelling temperature depends on the concentration of agar.)

During gelling, the endpoints of the double helices are able to bond to each other. Agar has a large hysteresis; that is, the temperature at which it converts back to a gel is much lower than the temperature at which that gel melts back to a liquid, which means that you can warm the set gel up to a moderately warm temperature and have it remain solid. For more information on the chemistry of agar, see *http://www.cybercolloids.net/library/agar/properties.php*.

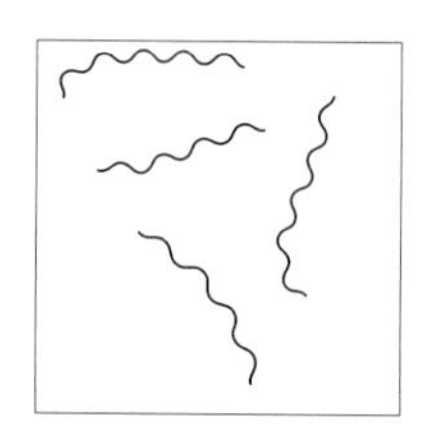

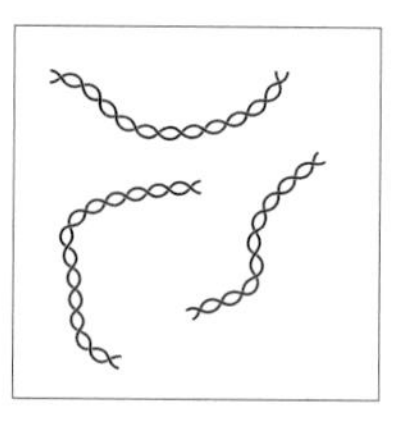

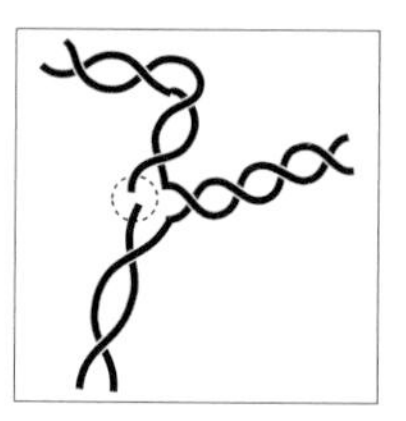

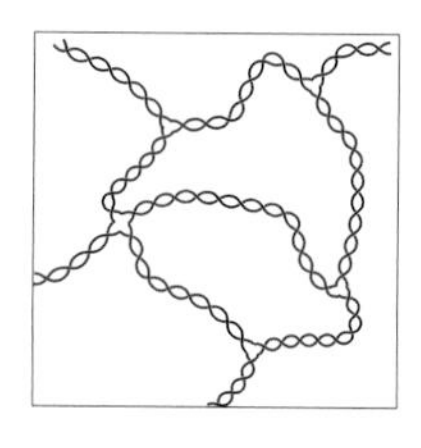

Agar at the molecular level. When heated, the molecule relaxes into a relatively straight molecule (upper left) that upon cooling forms a double helix with another agar molecule (center). The ends of these double helices can bond with other agar double helices (upper right), forming a 3D mesh (left).

Technical notes	
Gelling temperature	90–104°F / 32–40°C
Melting temperature	185°F / 85°C
Hysteresis	140°F / 60°C
Gel Type	Brittle
Syneresis	Yes
Concentrations	0.5%–2%
Synergisms	Works well with sucrose
Notes	Tannic acid inhibits gel formation (tannic acid is what causes over-brewed tea to taste bad; berries also contain tannins)
Thermoreversible	Yes

Chocolate Panna Cotta

Agar can be used to provide firmness, as this example shows. In a saucepan, whisk together and gently simmer (below boiling—just until small bubbles form on surface) for one minute:

3½ oz (100g) milk

3½ oz (100g) heavy cream

½ pod vanilla bean, sliced lengthwise and scraped

8 teaspoons (20g) powdered sugar

1 teaspoon (2g) agar powder

Turn off heat, remove vanilla bean pod, and add, briefly stir, and let rest:

3.5 oz (100g) bittersweet chocolate, chopped into fine pieces to assist in rapid melting

After a minute, add and whisk to thoroughly combine:

2 eggs yolks (reserve whites for some other recipe)

Pour mixture into glasses, bowl, or molds and store in fridge. The gel will set in as little as 15 minutes, depending upon the size of the mold and how long it takes the mousse to drop below agar's setting point (around 90°F / 32°C).

Notes

- *The agar provides a firmness that creates a stronger mousse than that created when using gelatin, so you should plan to use this mousse in applications where firmness is a desired trait.*
- *This chocolate mousse, while good by itself, really works better as a component in a dish. Example uses: roll a ball of the mousse in toasted nuts to create a truffle-like confection, spread a layer of the mousse into a prebaked pie crust and top with raspberries and whipped cream, or smear a thin layer of the mousse in the bottom of a bowl and place a small scoop of vanilla ice cream and some fresh fruit on top.*
- *When working with a vanilla bean, use a spoon or the edge of a knife to scrape the seeds from the pod, and add both pod and seeds to your mixture. Scraping the bean helps get the vanilla into the mixture more quickly.*

Rum Screwdriver Gel

In a small mixing bowl, measure out:

8 teaspoons (40g) rum

In a saucepan, whisk to combine, and bring to a boil and hold for an additional minute:

10 teaspoons (50g) orange juice

¼ cup (40g) sugar

1 teaspoon (2g) agar powder

Pour the hot liquid into the small mixing bowl, and stir thoroughly to combine. Transfer mixture to a glass, ice cube tray, or other food mold and store in fridge for 30 minutes or until set.

Notes

- *Yes, these are basically rapid-setting Jell-O shots. Using agar allows for a higher percentage of alcohol—you can gel rum by itself if careful—but make sure to leave enough juice in for flavor.*
- *Play with substitutions. You can replace the rum and orange juice with fluids such as Malibu and coconut milk.*

Making gels: Sodium alginate

The gels covered so far are all *homogenous*, in the sense that they are incorporated into the entire liquid and then set with heat. Alginate, however, sets via a chemical reaction with calcium, not heat, which allows for an interesting application: setting just part of the liquid by localized exposure to calcium.

This is done by adding sodium alginate to one liquid and calcium to a second liquid and then exposing the two liquids to each other. The sodium alginate dissolves in water, freeing up the alginate, which sets in the presence of calcium ions, which will only occur where the two liquids touch. Imagine a large drop of sodium alginate–filled liquid: the outside of the drop sets once it has a chance to gel with the assistance of the calcium ions, while the center of the drop remains liquid. It's from this application that the technique called *spherification* is derived.

Instructions for use. Add 1.0% to 1.5% sodium alginate into your liquid (use water for your first attempt). Let the liquid rest for two hours or so to hydrate fully. It will be lumpy at first; don't stir or agitate the liquid, as doing so will trap air bubbles in the mixture.

It's probably easiest to add the sodium alginate a day in advance and let it hydrate in the fridge overnight.

In a separate water bath, dissolve calcium chloride to create a 0.67% solution (about 1g calcium chloride to 150g water).

Carefully drip or spoon some of your sodium alginate liquid into the calcium bath and let it rest for 30 seconds or so. (You can use a large "syringe" dripper or turkey baster to create uniformly sized drops.) If your shape floats, use a fork or spoon to flip it over, so that all sides of it are exposed to the calcium bath. Remove from bath, dip into another bowl of just water to rinse off any remaining calcium, and play.

Sodium alginate gels firm up over the span of a few hours, so you'll need to make these near when you intend to serve them.

If your sodium alginate sets without exposure to the calcium bath, use filtered or distilled water. Hard water is high in calcium, which can trigger the gelling reaction.

Uses. The food industry uses alginate as a thickener and emulsifier. Since it readily absorbs water, it easily thickens fillings and drinks and is used to stabilize ice creams. It's also used in manufacturing assembled foods; for example, some pimento-stuffed olives are actually stuffed with a pimento paste that contains sodium alginate. The olives are pitted, injected with the paste, and then set in a bath with calcium ions to gel the paste.

Origin and chemistry. Derived from the cell walls of brown algae, which are made of cellulose and algin. Alginates are block copolymers composed of repeating units of mannopyranosyluronic and gulopyranosyluronic acids. Based on the sequence of the two acids, different regions of an alginate molecule can take on one of three shapes: ribbon-line, buckled shape, and irregular coils. Of the three shapes, the buckled shape regions can bind together via any divalent cation. (A cation is just an ion that's positively charged, i.e., missing electrons. *Divalent* simply means having a valence of two, so a *divalent cation* is any ion or molecule that is missing two electrons.)

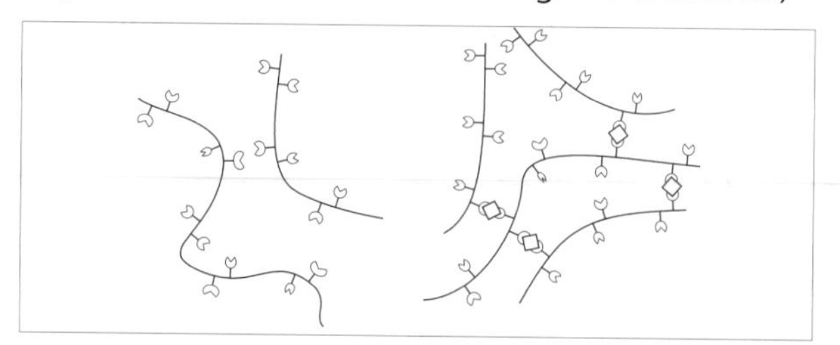

Alginate does not normally bind together (left), but with the assistance of calcium ions is able to form a 3D mesh (right).

Gel "Noodles" and Dots

This is really just a quick experiment to illustrate how to work with sodium alginate.

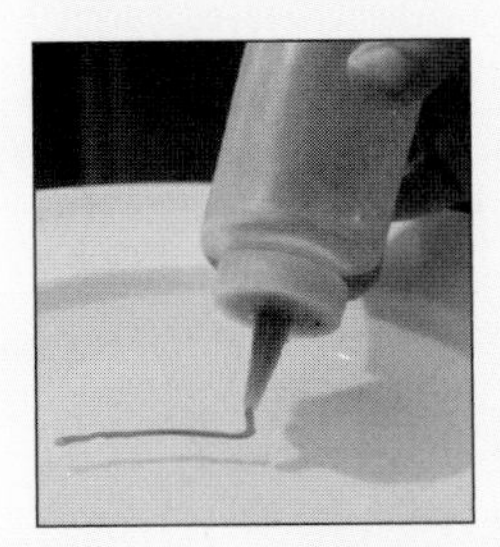

Create a 1% solution of sodium alginate and water. Add food coloring so that you can see the mixture as you work with it. Using a squeeze bottle, pipe out a strand into a bowl containing a 0.67% solution of calcium chloride in water.

Try making drops and other shapes as well. One food trend that's still making the rounds is mini "caviar." The small drops of set sodium alginate liquids have a similar texture and feel as caviar but with the flavor of whatever liquid you use.

Once you've played with this using water, try using other liquids. Jolt Cola? Cherry juice? Keep in mind that liquids that are high in calcium or very acidic will cause the alginate solution to gel up on its own.

Spherification in shapes

Since sodium alginate sets via a chemical reaction, not a thermal one, you can freeze a liquid into a mold and then thaw it in a calcium bath to cause it to partially maintain its shape. The final shape won't retain the crisp edges of the original frozen shape—it'll swell and bloat out slightly—but you'll still get a distinctive shape.

Note that straight-up lime juice won't work, because the alginate will precipitate out in the presence of strong acids. If you're willing to experiment further, try using sodium citrate to adjust the pH.

Try freezing the liquid in a mold before setting the sodium alginate to get more complicated shapes.

Mozzarella spheres

What happens if you want to use sodium alginate in a food that already contains calcium? Depending upon the amount of calcium in the food, adding the sodium alginate straight to it would cause the liquid to set, giving you something similar to a brittle gel.

Swapping the chemicals—adding the calcium chloride to the food and setting it in a sodium alginate bath—doesn't work; calcium chloride is nasty-tasting. Luckily, it's the calcium that's needed for the gelling reaction, not the offensive-tasting chloride, so any compound that's food-safe and able to donate calcium ions will work; calcium lactate happens to fit the bill. This technique is called *reverse spherification*.

To create mozzarella spheres, mix 2 parts mozzarella cheese with 1 part heavy cream under low heat. Add around 1.0% of calcium lactate to this liquid and then set it in a water/sodium alginate solution of 0.5% to 0.67% concentration.

Ann Barrett on Texture

PHOTO USED BY PERMISSION OF ANN BARRETT

Ann Barrett is a food engineer specializing in food textures. She works for the Combat Feeding Directorate of the U.S. Army Natick Soldier Research, Development and Engineering Center (NSRDEC).

What does a food engineer do?

It's like applied chemical engineering but for food. The training focuses on how to process food and how to preserve food, looking at food as material. I happen to have a specialty in food texture or food rheology; rheology means how something flows or deforms. My PhD topic was on the fracturability of crunchy food. How do you measure crunchiness or fracturability, and how do you quantitatively describe the way a food fails? When you chew a food and it breaks apart, can you describe that quantitatively and then relate that to the physical structure of the food?

Tell me a little bit about the NSRDEC.

There are several RDECs (research, development, and engineering centers) throughout the country. NSRDEC is focused on everything the soldier needs for survival or sustenance, aside from weaponry: food, clothing, shelters, and parachutes. The food part is largely driven by the fact that the military is potentially deployed in every kind of physical environment, so we need a wide range of foods to support soldiers operating in a wide range of situations. They have large depots of rations, and that drives a very long shelf-life requirement. Most of the food that we make needs to be shelf-stable for three years at 80°F / 26.7°C. That is not to say that the soldier will always eat something that's three years old, but they definitely might. That drives a lot of the research here; foods that are shelf-stable but also good, that the soldiers will want to eat.

It must be really interesting to work with the constraints that you've got while trying to preserve flavor and texture. How do you go about doing that?

Well, it's often one part experience or knowledge combined with two parts trial and error. There's a lot of bench-top development here. Most of my experience has been in processing and engineering analysis of food, but I do have a project now where we're trying to develop flavors for sandwich fillings. All flavors are chemicals, so you can replicate a natural flavor by knowing what the chemistry is.

For example, we're working on a peanut butter filling for sandwiches. We're trying to make a chocolate peanut butter flavor, a bit like Nutella. We have the peanut butter formula, and we've been looking at adding cocoa and at different chocolate flavors. We put three into storage to see how they would work, and two of them came out just okay, and one of them was delicious. When you're developing something, you have to look at a number of different ingredients to see what works. There will be changes in both the flavor and texture of a food during long-term storage. Flavors tend to become less intense, or off-flavors might develop. Texture can degrade by moisture equilibrating, say in a sandwich, or by staling. There are a multitude of flavors that are commercially available, and also a multitude of ingredients that will adjust texture—for example, starches and gums for liquid or semi-solid foods, enzymes and dough conditioners for breads. So during development you need to optimize a formula to make sure the food is good after you make it and also good after storage.

Even with all the hard science, you still have some degree of, well, "Let's just try it and see what happens?"

Oh, absolutely. You make a product up for a project, sample it, store it, and then sample it again. Everything is actually tasted here, and as a matter of fact, part of our duties is to go over and participate in the sensory panels because the food scientists here, the nutritionists, the dieticians, are all considered expert tasters. The first thing we do is make our product on the bench and then put it into a box that's 120°F / 49°C for four weeks. Those conditions approximate a longer period of time at a lower temperature; it's just a quick test to see if quality holds up. If the product holds up, next is 100°F / 38°C for six months; that's supposed to approximate the quality you would get at three years at 80°F / 26.7°C. Then you have to check that it's microbiologically stable, so it goes to the microbiologist for clearance, and then you can ask people to come and evaluate it. We rate the appearance, aroma, flavor, texture, and the overall quality.

How does the science of food texture work into enjoyment of food?

There are expected textural properties of whatever food category you're dealing with. Sauces are supposed to be creamy; meats are supposed to be at least somewhat fibrous; bread and cake are supposed to be soft and spongy; cereals and crackers are supposed to be crunchy. When texture deviates from what's expected, the food quality is poor. If you are going to measure and to optimize the texture of a product, you need to pinpoint the exact sensory properties you want.

For example, for liquids, flowability or viscosity is the defining physical and measurable characteristic. There are "thin" liquids and "thick" liquids, and you can often change thin to thick by adding hydrocolloids or thermal treatment. Solid foods come in many different textural types. There are elastic solids that spring back after deformation (Jell-O); there are plastic solids that don't (peanut butter). Then besides "solid" solids there are also porous solids—think bread, cake, puffed cereal, extruded snacks such as cheese puffs. Porous foods have the structure of sponges, and like a wet versus a dry sponge, they can be elastic or brittle.

Somebody cooking in the kitchen is actually manipulating these things both physically and chemically?

Yes, that's exactly what cooking is. Take cooking an egg. The protein albumin will denature with heat, causing molecular crosslinking and solidification. Another example is kneading bread dough, which is a mechanical rather than a thermal process that makes the gluten molecules link up; that gluten network is what allows the bread to rise because a structure is developed that will hold gas liberated by yeast. And of course, every time you use cornstarch or flour to thicken a gravy or sauce, you're employing a physico-chemical process. Heat and moisture will make the starch granules absorb water and swell and then bleed out individual starch polymers, which are like threads attached to the granules. The starch polymers then entangle, creating an interconnected structure that builds viscosity. That's why your gravy gets thick.

Gravy

Flour (roux method)

Create a simple roux by melting 2 tablespoons (25g) of butter in a saucepan and then adding 2 tablespoons (17g) of flour. Stir while cooking over low heat until the roux sets and begins to turn light brown, about two to three minutes.

Add 1 to 1½ cups broth or stock; whisk to combine. Simmer over low heat for several minutes, until the gravy has reached your desired thickness. If the gravy remains too thin, add more flour. (To prevent clumping, create a slurry by mixing flour with cold water, and add that.) If the gravy becomes too thick, add more liquid.

Cornstarch

Create a cornstarch paste by mixing 2 tablespoons (16g) cornstarch with ¼ cup (60g) cold water.

In a saucepan, heat up 1 to 1 ½ cups broth or stock. Add the cornstarch and simmer for 8 to 10 minutes to cook the cornstarch. If the gravy remains too thin, add more cornstarch paste. If the gravy becomes too thick, add more liquid.

Notes

- *You can use the drippings from roasted meats, such as turkey or chicken, to bring more flavor to the gravy. If using flour, substitute the fat in the drippings for the butter. If you're searing a piece of meat, use the same pan for making the gravy, deglazing it with a few tablespoons of wine, vermouth, Madeira, or port to loosen up the fond that will have formed on the surface of the pan.*
- *Try sautéing some mushrooms and adding them into the gravy as well. Or, if you're cooking a turkey, slow-cook the turkey neck a day in advance and pull off the meat and add it into the gravy as well.*

Making Things Melt in Weird Ways: Methylcellulose and Maltodextrin

At a high level, making gels is about transforming liquids into solids. In addition to creating gels, though, modern food additives can be used to alter other properties of foods, and another area of play in the modernist kitchen is that of melting. How can we make things change state in unexpected ways?

"Melts" as it cools: Methylcellulose

Methylcellulose has the unusual property of getting thicker when heated (*thermo-gelling* in chem-speak). Take jam: when heated, it loses its gel structure (the pectin melts), causing it to flow out of things like jam-filled pastries. Adding methylcellulose prevents this by causing the jam to "gel" into a solid under heat. And since methylcellulose is thermoreversible, upon cooling after baking, the jam returns to its normal consistency.

Hollywood uses methylcellulose to make slime. Add a bit of yellow and green food dye, and you've got yourself *Ghostbusters*-style slime. To get good consistency, whisk it vigorously to trap air bubbles into the mixture.

Methylcellulose has been used in some modernist cuisine dishes for its thermo-gelling effects. One famous example is "Hot Ice Cream" in which the "ice" cream is actually hot cream that's been set with methylcellulose. As it cools to room temperature, it melts.

Instructions for use. Dissolve methylcellulose into hot water (above 122°F / 50°C) and then whisk while cooling down. Mixing it directly in cold water can be difficult because the powder will clump up as it comes into contact with water. In hot water, though, the powder doesn't absorb any water, allowing it to be uniformly mixed. It's easiest to stir in 1.0% to 2.0% (by weight) into your liquid and let it rest overnight in the fridge to dissolve fully. You can then experiment with setting the liquid. Try baking a small dollop of it, or dropping it by the ice cream scoopful into a pan of simmering water.

Uses. Commercial applications use it to prevent "bake-out" of fillings in baked goods. Methylcellulose also has high surface activity, meaning that it acts as an emulsifier by keeping oil and water from separating, so it is also used in low-oil and no-oil dressings and to lower oil absorption in fried foods.

Methylcellulose increases surface tension—well, actually, "interfacial tension" because "surface" refers to a two-dimensional shape—which is why it works as an emulsifier.

Origin and chemistry. Methylcellulose is made by chemically modifying cellulose (via etherification of the hydroxyl groups). There can be great variation between types and derivatives of methylcellulose, in terms of thickness (viscosity), gelling temperature (122–194°F / 50–90°C), and strength of gel (ranging from firm to soft). If you're having problems getting your methylcellulose to set, check the specifications of the type you have. See Linda Anctil's primer at *http://www.playingwithfireandwater.com/foodplay/2008/03/methylcellulose.html* for additional details.

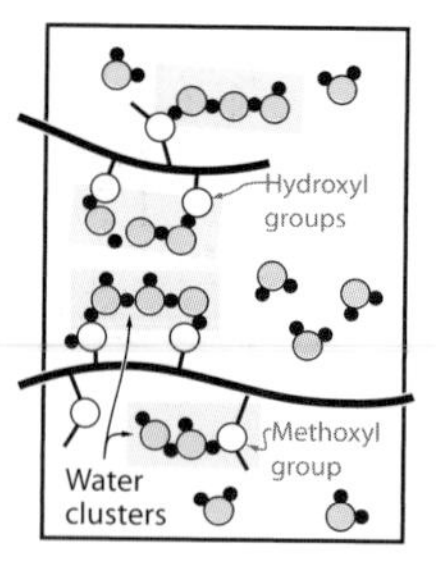

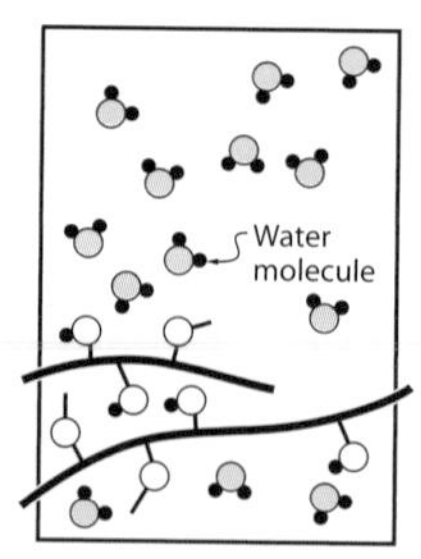

When cold (on left), water molecules are able to form water clusters around the methylcellulose molecule. With sufficient heat—around 122°F / 50°C—the water clusters are destroyed and the methylcellulose is able to form crosslinks, resulting in a stable gel at higher temperatures.

Hot Marshmallows

*These marshmallows remain firm when hot, but melt as they cool. This recipe is adapted from a recipe by Linda Anctil (*http://www.playingwithfireandwater.com*).*

In a saucepan, bring to a boil:

2⅛ cups (500g) water

1 cup (200g) sugar

Let cool, and then whisk in:

10g methylcellulose (use a scale to ensure an accurate measurement)

1 teaspoon (5g) vanilla extract

Let rest in fridge until thick, around two hours. Once thick, whisk until light and foamy. Transfer to a 9″ × 9″ / 20 cm × 20 cm baking pan lined with parchment paper. Bake for five to eight minutes at 300°F / 150°C, until set. The marshmallows should feel dry to the touch and not at all sticky. Remove from oven, cut into desired shapes, and coat with powdered sugar.

Two marshmallows on a plate of powdered sugar.

Two marshmallows after being coated with powdered sugar while still hot.

Same marshmallows after cooling for a few minutes.

When working with gels, you can quickly cool the hot liquid by whisking it while running cold water over the outside of the pan. The water will flow along the bottom of the pan.

Playing with Chemicals

"Melts" in your mouth: Maltodextrin

Maltodextrin—a starch—dissolves in water, but not fat. In manufacturing, it's spray-dried and agglomerated, which creates a powder that's very porous on the microscopic level. Because of this structure, maltodextrin is able soak up fatty substances (they won't cause it to dissolve), making maltodextrin useful for working with fats when designing food. It also absorbs water, so is used as an emulsifier and thickener, as well as a fat substitute: once hydrated, it literally sticks around, mimicking the viscosity and texture of fats.

Since it comes as a white powder, you can also use maltodextrin to turn fatty liquids and solids such as olive oil and peanut butter into powder. Because maltodextrin traps oils but dissolves in water, the resulting powder dissolves in your mouth, effectively "melting" back into the original ingredient and releasing its flavor. Since maltodextrin itself is generally flavorless (only slightly sweet), it does not substantially alter the flavor of the product that is being "powderized."

In addition to the novelty and surprise of, say, a powder dusting on top of fish melting into olive oil in your mouth, powders can carry flavors over into applications that require the ingredients to be effectively "dry." Think of chocolate truffles rolled in chopped nuts: in addition to providing flavor and texture contrast, the chopped nuts provide a convenient "wrapper" around the chocolate to allow you to pick up the truffle and eat it, without the chocolate ganache melting on to your fingers. Powdered products can be used to coat the outsides of foods in much the same way that chopped nuts are used to coat the outside of truffles.

Instructions for use. Add powder slowly to your liquid fat for a ratio of about 60% fat, 40% maltodextrin by weight. You can pass the results through a sieve to change the texture from breadcrumb-like to a finer powder.

Uses. Industry commonly uses maltodextrin as a filler to thicken liquids (e.g., the liquid in canned fruits) and as a way to carry flavors in prepackaged foods such as flavored chips and crackers. Since it traps fats, any fat-soluble substances can be "wicked up" with maltodextrin and then more easily incorporated into a product. For experimental dishes, you can use maltodextrin to create powders that can be sprinkled on the plate as garnish or as a way of transforming something that's normally liquid into a solid.

Origin and chemistry. Derived from starches such as corn, wheat, or tapioca. Tapioca maltodextrin seems to be most commonly used in modernist cooking. Maltodextrin is made by cooking down the starches and running the resulting hydrolyzed starches through a spray-dryer. Chemically, maltodextrin is a sweet polysaccharide composed of typically between 3 and 20 glucose units linked together.

When it comes to understanding how maltodextrin soaks up oils, imagine it being like sand at the beach. The sand doesn't actually bond with the water, but it's still wicking up the liquid in the space between the granules due to capillary action. When working with either sand or maltodextrin, with the right amount of liquid, the powder clumps up and becomes workable. Because maltodextrin is water soluble, however, water would dissolve the starch granules. And, luckily, maltodextrin can soak up a lot more oil per volume than sand can soak up water, making it useful for conveying flavors in a nonliquid form.

Powdered Brown Butter

Whisking any fat such as browned butter (upper left) with maltodextrin (center right) creates a powdered form (bottom) that can be used to create a surprising texture as the powder "melts" back into browned butter when placed in the mouth. Try using this browned butter powder as a garnish on top of or alongside fish, or making a version with peanut butter and sprinkling on desserts.

In a skillet, melt:

4 tablespoons (60g) salted butter

Once melted, continue to heat until all the water has boiled off. The butter solids will start to brown. Once the butter has completely browned and achieved a nutty, toasted aroma, remove from heat and allow to cool for a minute or two.

In a small mixing bowl, measure out:

½ cup (40g) maltodextrin

While whisking the maltodextrin, slowly dribble in the browned butter until a wet sand-like consistency is reached.

Notes

- *Stir slowly at the beginning because maltodextrin is light and will easily aerosolize. The ratio between maltodextrin and the food will vary. If your result is more like toothpaste, add more maltodextrin.*
- *If the resulting powder is still too clumpy, you might be able to dry it carefully by transferring the powder to a frying pan and applying low heat for a few minutes. This will help dry out any dampness present from room humidity. It will also partially cook the food item, which might not work for powders containing items such as white chocolate.*
- *For a finer texture, try passing the powder through a sieve or strainer using the back of a spoon.*
- *Try adding a bit of lemon juice to the brown butter, after it has cooled but before mixing it with the maltodextrin.*
- *Additional flavors to try: peanut butter, almond butter, coconut oil (virgin/unrefined), caramel, white chocolate, Nutella, olive oil, foie gras, bacon fat (cook some bacon and save the fat drippings—this is called* rendering*). You don't need to heat the fats first, but it might take a bit of working to get the maltodextrin to combine. For liquid fats (olive oil), you will need to use roughly 2 parts maltodextrin to 1 part fat: 50g olive oil, 100g maltodextrin.*

Making Foams: Lecithin

Foams are another area of play in modernist cuisine. If you ever happen to be served a dish that has a "foam" component—say, cod served on a bed of rice with a "carrot" foam or *uni* (sea urchin) in a shell with green apple foam, it was probably created by adding a stabilizer such as lecithin or methylcellulose to a liquid and then whipping or puréeing it. (Foams can also be created using cream whippers as described on page 370 in Chapter 7.) While perhaps a little too trendy, it's a clever way to introduce a flavor to a dish without adding much body.

Instructions for use. Add about 1% to 2% lecithin to your liquid (by weight, i.e., 1g lecithin per 100g liquid) and use an immersion blender to froth the liquid. Hold the immersion blender up and at a slight angle so that the blades are in contact with both the liquid and air.

Uses. As an emulsifier, lecithin can be used to create stable flavored foams. It's also used as an antispattering agent in margarines, an emulsifier in chocolate (to reduce the viscosity of the melted chocolate during manufacturing), and as an active ingredient in nonstick food sprays.

Origin and chemistry. Lecithin is typically derived from soya beans as a byproduct of creating soy-based vegetable oil. Lecithin is extracted from hulled, cooked soya beans by crushing the beans and then mechanically separating out (via extraction, filtration, and washing) crude lecithin. The crude lecithin is then either enzymatically modified or extracted with solvents (e.g., de-oiling with acetone or fractionating via alcohol). Lecithin can also be derived from animal sources, such as eggs and animal proteins, but animal-derived lecithin is much more expensive than plant-derived lecithin, so it is less common.

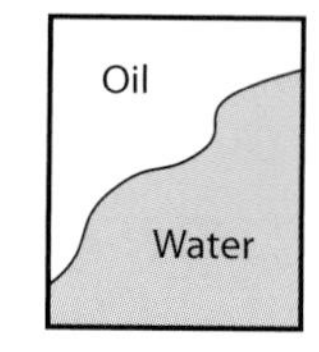

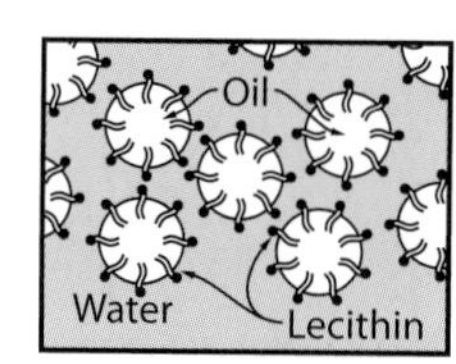

Lecithin molecules have polar and nonpolar regions that are most stable when one side is exposed to a polar substance and the other side to a nonpolar substance. See the sidebar "The Chemistry of Emulsifiers" for a description of how lecithin stabilizes foams.

Fruit Juice Foam

In a large mixing bowl or other similarly large and flat container, blend with an immersion blender:

½ cup (100g) water

½ cup (100g) juice, such as carrot, lime, or cranberry

1 teaspoon (3g) lecithin (powder)

Notes

- *Hold the immersion blender such that it is partly out of the liquid. You want to allow it to siphon air into the mixture.*
- *Allow foam to rest for a minute after blending, so that the resulting foam that you spoon off is more stable.*
- *Try other liquids, such as coffee or beet juice. Lecithin works best at around a 1–2% concentration (2g lecithin per 100g of liquid).*

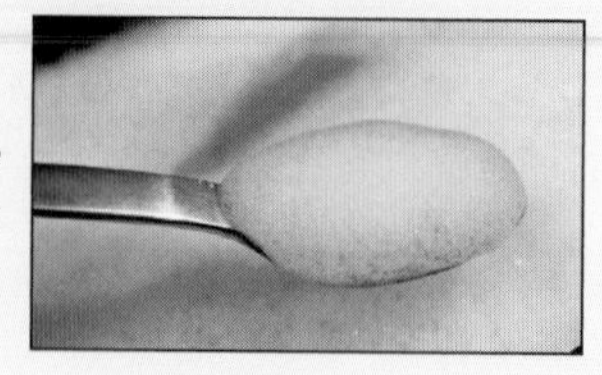

Lecithin can be used to make a large-bubble foam that is surprisingly stable for long periods of time.

The Chemistry of Emulsifiers

You might be wondering why oil and water are able to "mix" in the presence of an emulsifying agent, after the earlier discussion about polar (e.g., water) versus nonpolar (e.g., oil) molecules not being able to mix. An emulsifier has a hydrophilic/lipophilic structure: part of the molecule is polar and thus "likes" the water, and part of the molecule is nonpolar and "likes" the oil. Emulsifiers concentrate at the boundary between water and oil because of the charge structure of the molecules.

Adding an emulsifier keeps foods from separating by providing a barrier between droplets of oil. Think of it like a skin around the oil droplets that prevents different droplets from touching and coalescing. Emulsifiers reduce the chance that oil droplets will aggregate by increasing what chemists call *interfacial tension*. The oil and water don't actually mix; they're just held apart at the microscopic level.

Emulsifiers stabilize foams by increasing their kinetic stability—i.e., the amount of energy needed to get the foam to transition from one state to another is higher. Take the foam of a bubble bath as an example: the soap acts as an emulsifier, creating a foam of air and water. Water doesn't normally hold on to air bubbles, but with the soap (the emulsifier), the interfacial tension between the air and water goes way, way up, so it takes more energy to disrupt the system. The more energy it takes, the more kinetically stable the foam is, and the longer it'll last.

Take a look at the following two photographs to see what a difference an emulsifier can make (and see *http://www.cookingforgeeks.com/book/lecithin/* for a video demonstration).

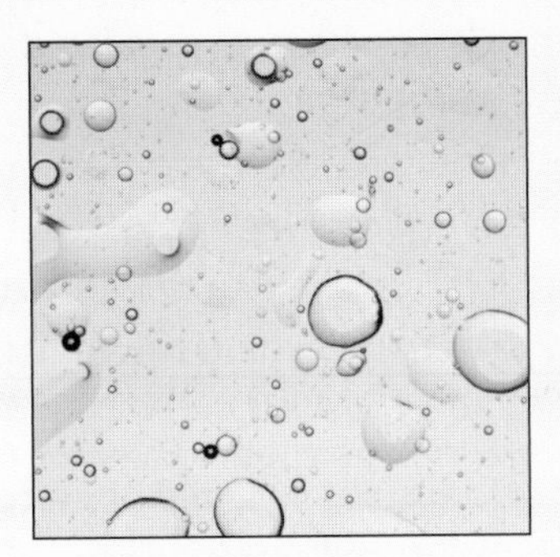

A photo under a light microscope of a half-water, half-oil solution. (The slide is pressing the oil droplets flat.)

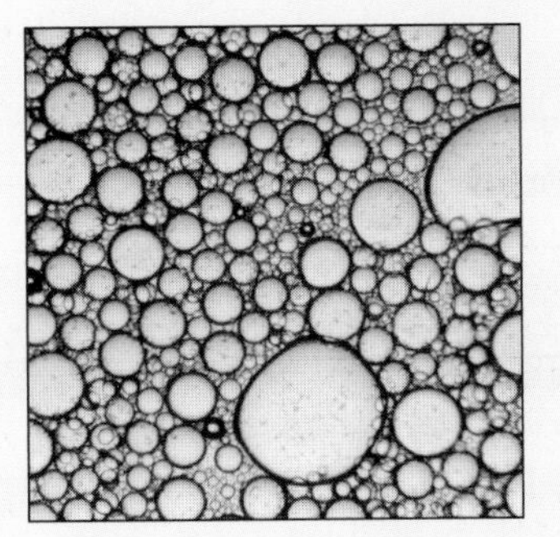

The same mixture with 1% lecithin added. The droplets are stable and do not coalesce into larger drops.

Anti-Sugar: Lactisole

This one is unusual. Unlike the modern additives covered so far, which have essentially focused on either trapping liquids in a gel structure or changing the physical state of food, "anti-sugar" is an additive used to modify a flavor sensation: it reduces the sensation of sweetness. (And no, mixing sugar and anti-sugar does not result in more energy being released than eating just plain sugar.)

One of the challenges facing the food industry is the need to maximize shelf stability and storage potential while maintaining acceptable flavor and texture. Sugar is used in confections and sweets not just for its sweetness, but also as a preservative: because sugar "latches" on to water, it reduces the amount of water available in a food product for bacterial growth. Think back to the FAT TOM rule from Chapter 4: bacterial growth is inhibited by reducing the water activity (the "M" in FAT TOM is for moisture), and because sugar is hygroscopic, adding sugar reduces the freely available water. But more sugar means increased sweetness, so the other flavors in foods can end up being masked with a cloying, overly sweet taste.

In the mid-1990s, Domino Sugar researched chemical modifiers that would reduce the perception of sweetness. The compound lactisole—a carboxylic acid salt—happens to do just that: add it to your foods at a concentration of around 100 parts per million (ppm), and goodbye sweet sensation, as it interferes with your taste buds (the TAS1R3 sweet protein receptor, for you bio geeks). Unlike traditional methods of dampening sweetness in a dish (i.e., adding bitter or sour ingredients), lactisole works by inhibiting the sensation of sweetness on the tongue, so it does not impact perception of saltiness, bitterness, or sourness. Sadly, you can't add it to foods to remove the calories from sugar.

Domino sells a product called Super Envision® that is a blend of mostly sucrose, some maltodextrin, and "artificial flavor" at 10,000 ppm. It's meant to be used at around a 1% concentration in the final product, so the 10,000 ppm becomes 100 ppm. (Gee, I wonder if that "artificial flavor" could be lactisole?)

Try tasting "anti-sugar" in caramel sauce (see page 212 in Chapter 4). Add a small quantity of Domino Super Envision to one bowl of caramel sauce, leaving a second bowl of caramel unmodified for comparison's sake. The taste of the burnt compounds in the caramel sauce will be stronger in the adulterated bowl, because the sweet sensations won't be masking them.

With lactisole, what was once perishable can be mass manufactured without the same worries about spoilage by increasing the amount of sugar and then canceling out the additional perceived sweetness. Some jams and jellies, for example, need a certain level of sugar to remain shelf-stable. Super Envision also shows up in products such as salad dressings, in which sweetness from stabilizers or thickeners would be undesirable, and in some mass-manufactured breads. Pizza dough, when baked, is more visually appealing if it turns golden brown. Adding sugar is an easy way to get a browning reaction, but sweet pizza dough isn't so appealing.

For a list of industrial-style recipes—cereal coatings, instant chocolate milk mix, marshmallows, meringue toppings—see Domino's Envision Applications page at *http://www.dominospecialtyingredients.com/recipes/envision_more.html.*

Savory French Meringues

Without sugar, meringues—well, egg whites—bake into a dry brittle foam that resembles Cheetos (but without the flavor): it's extremely crunchy and, without any flavorings, not particularly pleasant. When sugar is added, the meringue turns into something light, slightly chewy, and delightful.

Try the following experiment to understand how sugar helps stabilize meringues and how lactisole masks the sweetness.

Start by separating six egg whites into a bowl and whipping to stiff peaks. Using a scale, weigh out into three small glass or metal bowls:

Standard meringues

50g whisked egg white

20g granulated sugar

Meringues sans sugar

50g whisked egg white

Meringues with "anti-sugar"

50g whisked egg white

20g granulated sugar

1g Domino Super Envision

Transfer each batch into a piping bag. (A plastic bag with a small cut in the corner works well.) Pipe onto a Silpat or a cookie sheet lined with parchment paper.

Add in chopped nuts, dried fruit, and/or chocolate chips to extend it. Try dipping the baked meringue in tempered chocolate as well.

Bake the meringues in an oven set to 200°F / 95°C for several hours, until dry. (Trying this in the evening? You can set your oven to ~300°F / 150°C, pop the cookies in, and then turn the oven off and come back the next morning.)

Don't try baking meringues directly on the cookie sheet. Proteins are very sticky and will bind to it, making it hard to remove them without breaking them. Because it's flexible, the Silpat or parchment paper can easily be peeled off the back of the cooked meringues.

You can see the difference instantly in the "meringue" made without sugar: the egg white doesn't flow as smoothly out of the piping bag. The standard and anti-sugar meringues have the same texture, but the taste of the standard one is, as expected, sweet. The anti-sugar one tastes pretty much like nothing, as egg white doesn't carry a strong flavor of its own.

Meringues sans sugar.

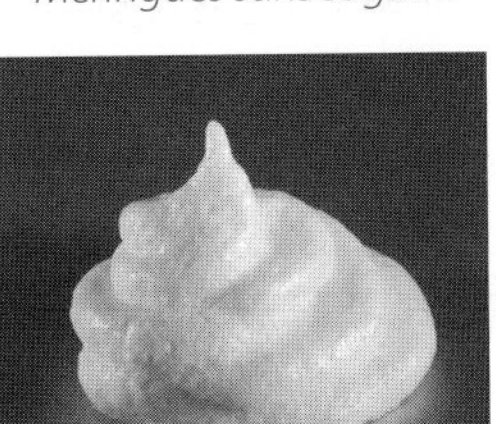

Standard meringues.

Anti-sugar meringues.

Meat Glue: Transglutaminase

One of the more unexpected food additives is *transglutaminase*, a protein that has the ability to bond glutamine with compounds such as lysine, both of which are present in animal tissue. In plain English, transglutaminase is "glue" for proteins.

Transglutaminase isn't used to change the texture of foods or to modify sensations of flavor. Rather, the food industry uses it to re-form scrap meats into large pieces (McNuggets!). You didn't actually think that gorgeous hunk of ham at the deli counter was one piece of meat, did you? From the rare boneless pig?

Transglutaminase is also used to thicken milk and yogurts by making their proteins longer in the same way that adding longer polysaccharides in gelling applications makes things thicker. Additionally, it is used to firm up pastas, to make breads more elastic (able to stretch without tearing), and to improve gluten-free breads for those with celiac disease.

For food hackers, though, the compelling opportunities for transglutaminase reside primarily in meat-binding applications. Food hackers have, of course, seized the opportunity to use it to make Frankenstein meats (all in the name of fun). You can "glue" white fish to red fish, make a turducken (a turkey-duck-chicken dish) that holds together, and make a heat-stable aspic, relying on transglutaminase instead of heat-sensitive gelatins or aspics.

The recipes that follow will give you some starting ideas, but really the concept of "meat welding" can apply to any meats that you want to stay together, including fish and poultry. You can glue scallops together in a long chain, wrap chicken around fillings (binding the chicken to the other end of itself), and wrap bacon around scallops.

The reaction occurs at room temperature and takes around two hours to set, so plan ahead. Use about 1% transglutaminase for the total weight of your food. You can sprinkle it dry on the food item or create a slurry (2 parts water to 1 part transglutaminase) and brush it onto the surfaces to be glued. Once adhered together, let the join rest for at least two hours; otherwise, you will shear and break the bonds as they're setting.

Chicken and steak bonded together with transglutaminase. Mmm, Doublemeat Palace!

Keep in mind that, because *you're* made of protein, you should take care to not get it on your skin or inhale the powder. Unlike real glue, transglutaminase is actually a chemical catalyst that literally bonds the two sides together at a molecular level. Gloves and a respirator mask are good insurance. Since transglutaminase is a protein itself and has the same structures as the amino acids it binds, it's also capable of binding to itself. After a few hours at room temperature, though, it loses its enzymatic properties, so it's not a huge deal if you spill a bit on your work surface. Once opened, store it in your freezer to slow the rate of the binding reaction.

Instructions for use. Create a slurry of water and transglutaminase and brush it onto the surfaces that you want to join. Press them together and wrap with plastic wrap. Store in fridge for two hours or longer.

Try vacuum-packing the food. This will improve the fit between the two pieces of meat.

Uses. Protein binder. Used by the food industry to take scraps of meats and form them into a larger shape, such as deli-style sliced turkey, and to thicken dairy products such as yogurt.

Origin. Manufactured using the bacteria *Streptomyces mobaraensis*. The main producer of transglutaminase is a Japanese company, Ajinomoto, which sells it under the name Activa. (This is the same company that originally formed to manufacture and sell MSG.)

Chemistry. Transglutaminase is an enzyme that binds the amino acid glutamine with a variety of primary amines. Any place where glutamine and a suitable amine are present, transglutaminase can be used to crosslink the two. Transglutaminase is itself digestible (it's a protein) and the enzymatic reaction ceases after a few hours, so there's no danger of it "gluing" your insides together (once it has set, that is, which would happen during cooking anyway).

Transglutaminase acts as a catalyst on glutamine and lysine, causing the atoms composing the two groups to line up so that they form covalent bonds.

A covalent bond is one in which two atoms share an electron, resulting in a lower energy state. Electrons are "lazy" in the sense that they prefer states that take less energy to maintain.

To visualize the reaction, imagine spreading apart the fingers of your left and right hands and touching the tips together, left thumb to right thumb, left pinky to right pinky, etc. Without some amount of coordination, getting the atomic "fingers" to line up just doesn't happen. Transglutaminase helps by providing the necessary atomic-level guidance for the two groups to touch. And once they touch, they can form covalent bonds and stick. Continuing the finger analogy, it's a bit like having superglue on your fingers: once they are lined up and are touched together, they stay together.

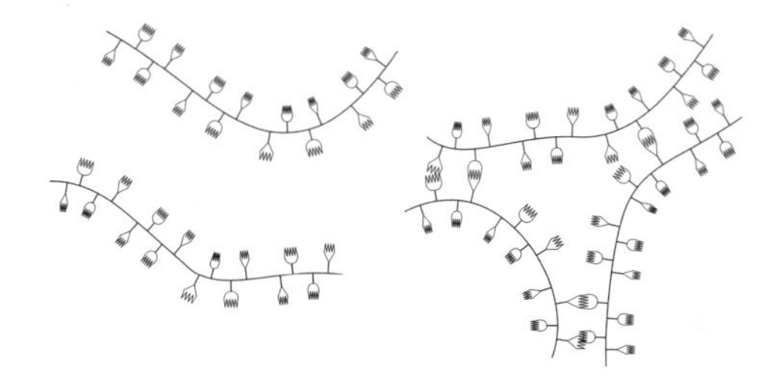

Before interaction, strands of proteins with glutamine and lysine groups are unattached (left); after interaction, the glutamine and lysine groups are covalently bonded wherever transglutaminase has a chance to catalyze. Note that transglutaminase itself does not remain as part of the bond after the reaction.

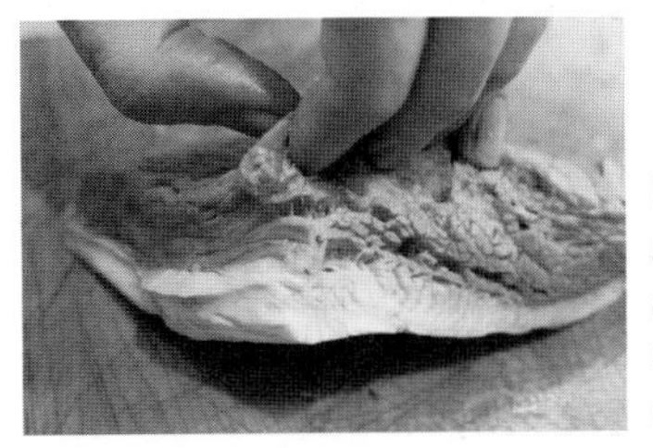

While you can pull apart items joined with transglutaminase, the individual meats themselves may be weaker than the join.

Technical notes	
Concentration	~0.5% to 1% of meat weight.
Notes	Cold-set for at least two hours—that is, apply to meat and let rest in fridge for two hours. Reaction time is correlated with temperature, so it takes longer to set at colder temperatures.
Temperature	Heat-stable once set.

Playing with Chemicals

Bacon-Wrapped Scallops

It's cool to see bacon-wrapped scallops where the bacon just sticks to the scallop. It's also a good example of how to work with transglutaminase.

In a small bowl, mix roughly 2 parts water to 1 part transglutaminase to create a slurry.

On a small plate that will fit in your fridge, lay out:

8 scallops as large and as cylindrical as possible, patted dry

8 slices bacon, cut in half so that they can wrap around a scallop one time

Using a brush, coat one side of each piece of bacon with the slurry. Place a scallop on the bacon and roll the bacon around the scallop. Repeat for each scallop and transfer to the fridge for at least two hours to allow the transglutaminase to set.

After resting, the bacon should be well adhered to the scallops.

Preheat your oven to 400°F / 200°C.

Place the scallops in a hot frying pan lightly coated with oil or a small amount of butter, with one of the "exposed" ends down. This will cause a Maillard reaction and develop a nice layer of flavor on the scallops. After a minute or so, flip the scallops over so that the other exposed side is in contact with the pan and immediately transfer your frying pan to the oven.

Finish in the oven for about five to eight minutes, until the bacon is done and the scallops are cooked.

Use a brush to coat one side of a strip of bacon with transglutaminase. (If you don't already have a pastry brush, consider getting one made with plastic bristles, because that type will not leave strands behind.)

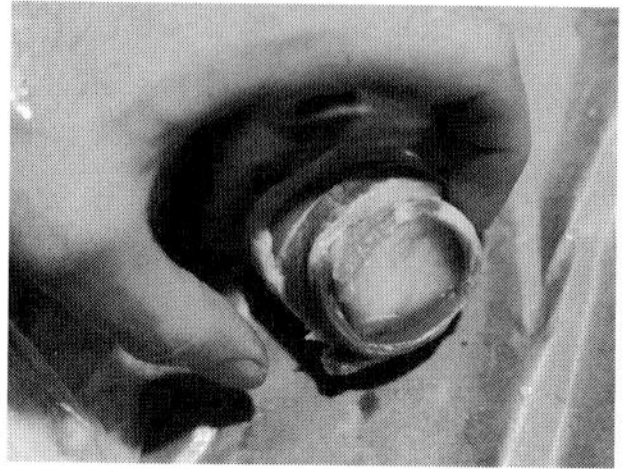

Carefully roll the bacon around the scallop. The transglutaminase will not bond instantly, so you will need to pinch and press the items together. Let set for two hours in the fridge.

Pan sear the scallops on high heat, flip to sear on both sides, and transfer to the oven to finish.

A cross-sectional slice of the finished product shows the joined surface of the bacon and the scallop. (Shown on top of a leaf of bok choy.)

Note

- *Use only an oven-safe frying pan in your oven. Some commercial frying pans have silicone handles—typically blue—that are oven-safe.*

Shrimp Roll-up

Since transglutaminase binds proteins at the molecular level, you can also use it as a binder to form ground meats into a solid form (Spam!). Imagine taking wood glue and, instead of gluing two boards together, using the glue to re-form a piece of wood from sawdust. Yes, just like particleboard or chipboard. The next recipe demonstrates this concept.

Purée with an immersion blender or food processor:

175g shrimp, raw, peeled, and deveined

50g water

10g transglutaminase

Transfer the purée to the center of a large sheet of parchment paper. Using a spatula, fan out the purée so that it's flat enough to place a second piece of parchment paper on top of it. Using a rolling pin, roll the purée out to a thickness of 1/8" / 0.3 cm, just as you would for a pie crust dough. Transfer the "sandwiched" purée to the fridge and let rest for a minimum two hours, preferably overnight.

In a large pot, bring salted water to a rolling boil.

Fill a large bowl with ice water.

Using a sharp chef's knife, cut a portion of the sandwiched purée down to a size that will fit in your pot. Carefully slip the sheet into the boiling water. The parchment paper should detach from the shrimp purée; this is expected. After 30 seconds to a minute, use a spider (or slotted spoon and tongs, if you're careful) to fish out the shrimp sheet from the boiling water, and transfer it to the ice water bath to stop the cooking process.

You should now have a "sheet" of shrimp that you can slice into noodles or use as a wrap around food items. To make shrimp noodles, slice the shrimp sheet into thin slices, which can then be floated in seafood broth or tossed with seared tofu, sesame seeds, sautéed green onions, and soy sauce. Or, try making "reverse sushi," using the shrimp sheet as the wrapper for the rice in place of the customary nori seaweed wrapper.

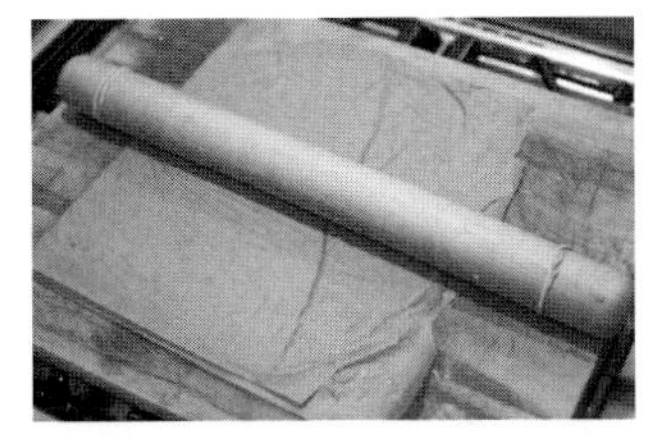

You can use rubber bands on the edges of your rolling pin to achieve a consistent thickness. This photo shows the shrimp and transglutaminase purée sandwiched in parchment paper.

A spider is a handy tool for fetching delicate items from boiling water. Note that the shrimp noodles and parchment paper have separated in the boiling water.

Liquid Smoke: Distilled Smoke Vapor

Smoking—burning wood chips and directing either the hot or cold smoke vapors to come in contact with items such as meats—is a method for curing and preserving foods. Smoking also deposits a number of flavors onto the food that are generated as byproducts of the chemical reactions that occur when wood is combusted. The commercial food industry uses liquid smoke to infuse smoke flavor into foods that are traditionally smoked, such as bacon, and into foods for which the flavor is enhanced by smoke essence, such as "smoked" tofu.

The simplest way of creating a smoked flavor in your cooking—besides actually smoking it—is to include ingredients that are already smoked and contain those chemical compounds. You can infuse smoke flavors into your dish by adding spices such as chipotle peppers or smoked paprika, or by using dry rubs with smoked teas such as Lapsang Souchong. Tobacco, too, can be used to similar effect; some novel restaurant dishes include components like tobacco-infused crème anglaise. However, including smoked ingredients will also bring along the other flavors of the substance being used. Some dishes can use smoked salts, for example, but for many applications, this will contribute too much salt. This is where liquid smoke comes in.

A cook can use liquid smoke to highlight the smoky "toasted" flavors of foods, especially those that have similar molecular compounds to smoke, such as coffee, peanut butter, or Scotch whisky. You can also use it to impart smoke flavor in those situations where grilling isn't an option—say, on the 27th floor of your apartment complex.

When buying liquid smoke, look for an ingredient list that reads "smoke, water." Try to avoid products that have molasses or other additives.

Some of the more unusual uses allow for bringing smoked flavor to foods that can't normally be tossed onto a wood-burning grill, such as tofu, ice cream, or liquids (along with some sandalwood incense and Chanel N°5, if you're Maggie from *Northern Exposure)*. Butter also has some of the same phenols as smoke; try adding it to butter for table service with bread.

The big long evil list of nasty chemicals and ingredients that one would expect to see on a liquid smoke bottle? "Water, smoke." In and of itself, liquid smoke is not artificial. It does not undergo any chemical modifications or refining steps that alter or change the compounds that would have been present in traditional smoking.

In theory, some of the mutagenic compounds (those that cause cancer) normally present in traditionally smoked foods are present in much smaller quantities in liquid smoke, meaning that liquid smoke might actually be somewhat safer for you than traditionally smoked foods. However, be aware that liquid smoke will have some amount of mutagenic compounds present. As a substitute for smoking foods, it should be as safe as traditional smoking, but you probably shouldn't douse a teaspoon of it on your morning eggs every day until further research is done.

In addition to the following two recipes, consider revisiting the Salmon Gravlax recipe from earlier in this chapter (see 275) and adding liquid smoke to give it a cold-smoked flavor.

S'mores Ice Cream

This recipe uses liquid smoke to impart the toasted flavor of campfire-roasted marshmallows. The concept was inspired by a demo by Kent Kirshenbaum of NYU's Experimental Cuisine Collective.

You'll need a standard ice cream mixer, or you can go all-out geek and either make your own (see 92 in Chapter 3) or use liquid nitrogen or dry ice. For the latter options, see the instructions on 377 in Chapter 7.

To create the base, combine in a mixing bowl:

2 cups (475g) whole milk

1 cup (238g) heavy cream

1/3 cup (75g) sugar

1/4 cup (75g) chocolate syrup

3/4 cup (25g) medium-sized marshmallows

15 drops (0.75g) liquid smoke

Proceed with the directions for your chosen method of making ice cream. Once the ice cream has set, stir in:

1 cup (60g) graham crackers, toasted and chopped into pieces

Serve with hot fudge or chocolate syrup—whipped cream, cherries, and nuts optional.

Oven-Cooked Barbeque Ribs

In a large baking pan (9″ × 13″ / 23 cm × 33 cm), place:

2 pounds (1kg) pork baby back ribs, excess fat trimmed off

In a small bowl, create a dry rub by mixing:

1 tablespoon (15g) salt

1 tablespoon (15g) brown sugar

1 tablespoon (9g) cumin seed

1 tablespoon (9g) mustard seed

20 drops (1g) liquid smoke

Cover ribs with spice mix. Cover baking pan with foil and bake at 300°F / 150°C for two hours.

In a small bowl, create a sauce by mixing:

4 tablespoons (60g) ketchup

1 tablespoon (15g) soy sauce

1 tablespoon (15g) brown sugar

1 teaspoon (5g) Worcestershire sauce

Remove foil from baking pan and coat ribs with sauce. Bake for 45 minutes, or until done.

Note

- *Experiment with other savory spices in the dry rub, such as chilies, garlic, or paprika. Also, try adding items such as onions, garlic, or Tabasco to the sauce.*

Making Liquid Smoke

The smells that we associate with that smoky, barbeque goodness result solely from the chemical reactions that occur during pyrolyzation (burning) of wood. The flavor that you think of as "smoky" does not come from a chemical interaction between the food and the smoke. This lucky quirk means that the chemicals in smoke can be isolated, so the stage of generating smoke flavor can be separated from the step of applying that flavor to food.

You can make your own liquid smoke for about $20 worth of supplies and a few hours of your time. For day-to-day uses, you're way better off buying liquid smoke from the grocery store, but it's rewarding to see how straightforward it is to make, and the process touches on some elementary chemistry techniques as well.

Liquid smoke is made by heating wood chips to a temperature high enough for the lignins in wood to burn, condensing the resulting smoke, and then dissolving it in water. The water-soluble components of smoke remain dissolved in the water, while the non-water-soluble components either precipitate out or form an oil layer that is then discarded. The resulting product is an amber-tinted liquid that you can brush onto meats or mix in with your ingredients.

What actually happens when you burn wood? Wood is primarily made of cellulose, hemicellulose, and lignin, which during burning convert to several hundred different chemical compounds. The aromatic molecules that provide smoke flavoring are generated by the lignin, which breaks down at around 752°F / 400°C. Cellulose and hemicellulose break down at lower temperatures (480–570°F / 250–300°C), but they generate compounds that both detract from the flavor and are mutagenic. This is why, when grilling, you should make sure you have a hot fire, which will guarantee that the lignins, and not just the celluloses, break down.

Making your own liquid smoke can be a little tricky because of the high heat required to properly burn the lignins and the difficulty in correctly capturing the resulting lignin-based compounds, not to mention the need for proper chemistry lab equipment for creating a closed system and heating it safely.

Kent Kirshenbaum demonstrates making liquid smoke during a talk at NYU's Experimental Cuisine Collective (see http://www.experimentalcuisine.org). Here, he burns hickory chips using a propane blowtorch. The smoke is then piped through a water flask (on right), which traps the water-soluble particulate in suspension.

Start by placing wood chips—either hickory or cedar—into a vessel that can be sealed (to create a closed system) and heated with a burner or blow torch. Run a line from the closed system into a container of water, so as to filter the smoke vapor through the water. Heat the vessel, making sure to get it hot enough for the lignins to burn. Because the "tasty molecules" of the smoke are water soluble, the water will end up capturing those flavors, becoming your liquid smoke. Discard any solids that precipitate out or oils that separate and float to the top. Theoretically, something like this could be done with a pipe on a charcoal grill, with the pipe sealed on one end and copper tubing running from the other end into a water container, but it's definitely not up to lab safety protocols.

If you do manage to make your own liquid smoke—it does make for a fun experiment—you'll probably find that it's a lot more work than it's worth. Still, understanding that liquid smoke is nothing more than smoke particles captured in water removes most of the mystery about what's in the bottle at your grocery store.

7

Fun with Hardware

IF YOU'RE ANYTHING LIKE ME, THIS IS THE FIRST CHAPTER YOU'LL FLIP TO WHILE PERUSING THIS BOOK IN THE BOOKSTORE. And, might I add, you have *excellent* taste.

While this chapter is designed such that a foodie-geek can jump right in, really, it does assume that you're up to speed with pairing flavors, that you understand various cooking and baking techniques, and that you're familiar with some of the chemistry concepts covered in earlier chapters. So, don't judge this book solely by this chapter.

Modern commercial kitchens, probably including the high-end ones in your town, use many tools that consumers rarely encounter but that can help create some absolutely stellar meals. We'll cover a few of the commercial and industrial tools used in preparing foods, and throw in a few, uh, "crazy" (and fun!) things that you can do as well.

Time and temperature really are the two key variables in cooking (see Chapter 4). Under normal circumstances, cooking is performed with these variables at moderate values: roasting potatoes for half an hour at around 350°F / 177°C, baking pizza at 450°F / 230°C for 10 minutes, or churning ice cream at –20°F / –29°C for an hour or so. But what happens when you move one of these variables to an extreme?

Cooking at extreme temperatures isn't as uncommon as it might sound at first. Potatoes, for example, wrapped in foil and roasted in the coals of a campfire are in an environment that reaches well above 800°F / 425°C. With this in mind, it shouldn't be too much of a stretch to imagine baking thin-crust pizza in 45 seconds at 900°F / 480°C (the result is amazingly good!). And making ice cream in 30 seconds with liquid nitrogen isn't just fun; this technique actually makes *great* ice cream because the water crystals don't have time to form large aggregates, resulting in a smoother texture.

It's also possible to move cooking times to extreme values. *Sous vide* cooking, the topic of the first half of this chapter, enables this by precisely controlling the temperature of the cooking environment, a water bath, so that it is equal to the target temperature of the cooked food. This allows time to run to extreme values without any risk of overcooking (at least in the conventional sense).

Beyond sous vide, other techniques can be used to produce new culinary creations—or at least, to return an iota of sanity to the life of the commercial chef by making some preparations vastly easier than they are with traditional methods. Filtration makes easy work of creating stocks, clear juices, and consommés. Cream whippers can "whip" air into liquids, allowing for the quick creation of not just whipped creams, but also mousses, foams, and even cakes. And for extreme temperature variations, we'll take a look at blowtorches and high-heat ovens on the hot side and liquid nitrogen and dry ice on the cold side. We'll talk about all of this in the second half of this chapter.

Unfortunately, many of these techniques involve tools that you're unlikely to find at your nearest shopping mall. Expect to do some online sleuthing or to break out the wire cutters and soldering iron, and be willing to try, try, and try again. This chapter is all about experimentation. As with the modern food additives section in Chapter 6, the "recipes" here are really only simple examples to give you a sense of where to start with your experiments. Use your creativity and imagination to create your own dishes!

Brownies in an Orange

Using food as a serving bowl is nothing new: stew in a bread bowl, sliced fruit in half a cantaloupe, and now, brownies in an orange.

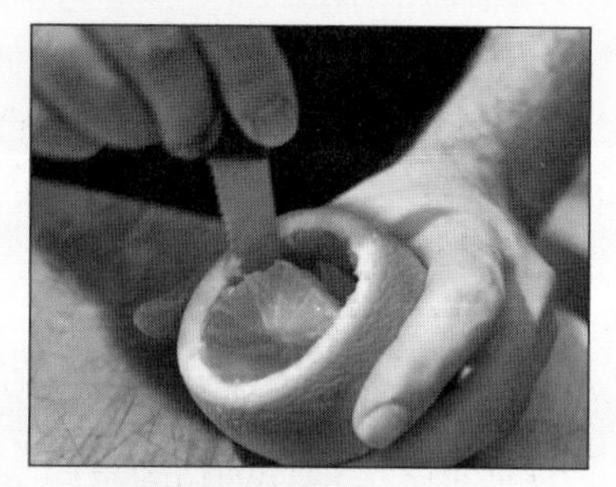

Cut the top off and trim out the center.

Fill with brownie mix (guilty pleasure).

Bake until a toothpick inserted 1" / 2.5 cm deep comes out clean. Dust with powdered sugar.

Sous Vide Cooking

With a name like "sous vide," this cooking technique sounds foreign, and for good reason: the French chef George Pralus introduced it to the culinary world in the 1970s. While foreign in origin, it is certainly not complicated or mysterious. At its simplest, sous vide cooking is about immersing a food item into a precisely temperature-controlled water bath, where the temperature is the same as the target temperature of the food being cooked. Translation? Ultra-low-temperature poaching. And since the temperature of the water bath isn't hotter than the final target temperature, the food can't overcook. Sous vide cooking essentially locks the variable of temperature in the "time * temperature" formula.

The temperature of the water bath is chosen to trigger chemical reactions (e.g., denaturing, hydrolysis) in some compounds in the food while leaving other compounds in their native state. It is one of the biggest culinary revolutions to hit the commercial cooking scene in the past few decades, but has appeared in the U.S. only recently. If I could pick only one new cooking method out of this entire book for you to try, sous vide would be it, hands down. The reason sous vide is so, well, amazing is that foods cooked this way have no gradient of doneness and the associated overcooked outer portion. Instead, the entire piece of food has a uniform temperature and uniform doneness.

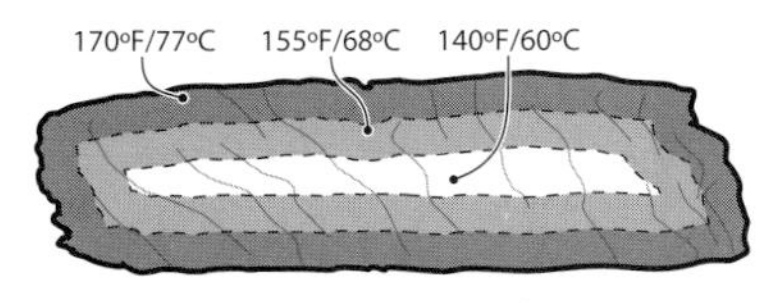

Gradient of doneness for traditional cooking methods

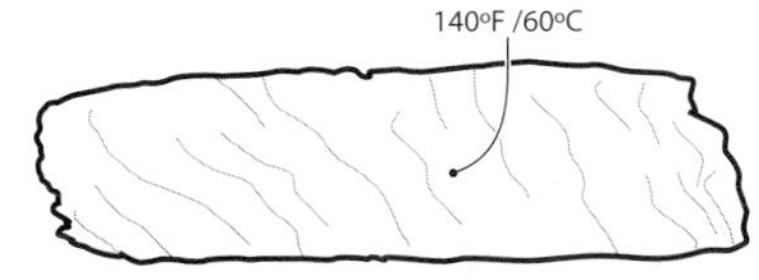

"Gradient" of doneness for sous vide cooking method

Foods cooked sous vide have no temperature gradient, meaning that the entire portion of food is cooked to a consistent level of doneness.

The name sous vide (meaning "under vacuum") refers to the step in the cooking process where foods are placed in a vacuum-pack plastic bag and sealed. Using a vacuum bag—a plastic bag that is sealed after all the air in it has been removed—allows the water in the bath to transfer heat into the food while preventing the water from coming into direct contact with it. This means the water does not chemically interact with the food: the flavors of the food remain stronger, because the water is unable to dissolve and carry away any compounds in the food. (Sous vide is a funny name; I think it should have been called "water bath cooking," because the actual heat source is usually a bath of water. Bain-marie was already taken, I suppose. Still, as with the name "molecular gastronomy," once something gets popularized, it tends to stick.)

The steak tip on the left was cooked sous vide at 140°F / 60°C; the one on the right was pan-seared. Note that the sous vide steak has no "bull's eye" shape—that is, it's consistently medium-rare, center to edge, while the seared steak is well-done on the outside and rare in the middle.

Sous vide cooking doesn't have to be done with a sealed bag in water. A few items don't need to be packed at all. Eggs, for example, are already sealed (ignoring the microscopic pores), and when using this technique for secondary applications like preheating vegetables such as bok choy for steaming, there's no benefit to sealing the food in plastic.

You can also use other fluids instead of water: oil, for example, or even melted butter. And because meats don't absorb fat the same way that they can water, when using one of these as the liquid medium some applications can skip sealing. This can be extremely useful for those foods that might be difficult to seal. Chef Thomas Keller, for example, has a recipe for poaching lobster tails in a bath of butter and water (*beurre monte*, melted butter with water whisked in, which has a higher burning temperature than butter alone).

Temperature-controlled air would technically work as well, but the rate of heat transfer is much, much slower than for water—roughly 23 times slower. Given the low temperatures involved, something like chicken in a 140°F / 60°C "air bath" would take so long to come up to temperature that bacterial growth would be a serious concern. Using a liquid such as water ensures that heat can penetrate the food via conduction—liquid touching plastic touching food—rather quickly. Water is cheap and easy to use, so you'll almost always see it called for, but some chefs do occasionally use other liquids.

The classic example given to explain how sous vide cooking works is cooking an egg. Since different egg proteins denature and coagulate at slightly different temperatures (most are in the range of 144–158°F / 62–70°C), holding an egg at various temperatures within that range will result in varying consistency of egg white and yolk. (Refer back to the discussion of egg proteins setting at different temperatures on page 181 in Chapter 4.)

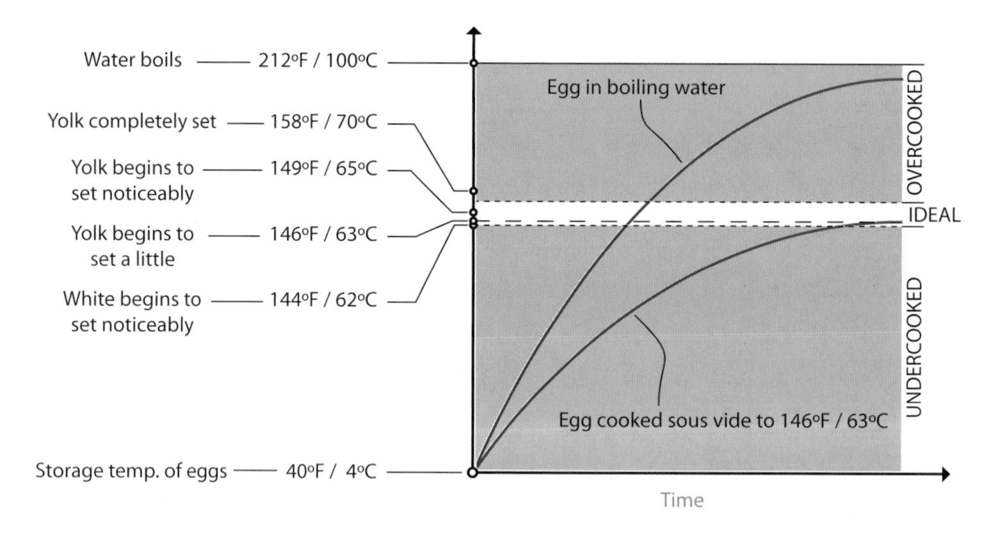

To some, a "perfect" soft-cooked egg should have a slightly runny, custard-like yolk and a mostly set white. Cooking an egg in water brought to a boil can result in an overcooked end result, because the temperature of the egg ramps up to boiling point until it is pulled out. In sous vide, the temperature of the egg reaches only the ideal temperature of the cooked egg, so it cannot overcook.

Cooking eggs in a sous vide bath at 144.5°F / 62.5°C.

By immersing the egg in a water bath held at that temperature you ensure that the egg cannot get any hotter, so in theory, those proteins that set at a higher temperature will remain in their native form. In reality, most chemical reactions in cooking aren't specific to a particular temperature, but are dependent on time-at-temperature. In practice, though, this simple model is accurate enough to explain how sous vide cooking works.

For a "perfect" soft-cooked egg, try cooking it sous vide at 146°F / 63°C for one hour. Because eggs contain many proteins that set at slightly different temperatures, you can experiment by adjusting the temperature up or down a few degrees to suit your personal preferences.

For other foods, consider the compounds they contain, determine the temperatures at which the compounds undergo their different transformations, and pick a temperature high enough to trigger the reactions you do want, yet low enough not to trigger the ones you don't.

Tip: after cooking an egg sous vide, crack open and drop the egg (without shell!) into a pot of just-boiled water. Then fetch the egg out immediately. The hot water will rapid-set the outside of the egg for better appearance and easier handling.

Sous vide cooking isn't a magic bullet, though. For one thing, the textures of some foods break down when held at temperature for any extended period of time. Some types of fish will break down due to enzymatic reactions that normally occur at such a slow rate that they are not noticeable in traditional cooking methods. Sous vide also doesn't reach the temperatures at which Maillard reactions or caramelization occur; meats cooked sous vide are commonly pan seared or even blowtorched briefly after cooking to introduce the flavors brought about by these browning reactions. The largest drawback, however, is the requirement to pay serious attention to food safety issues and pasteurization.

Pasteurization reduces bacterial levels to a point where food can be considered reasonably safe. If it is stored improperly after pasteurization, bacteria can reproduce above safe levels.

Sterilization completely eliminates the target bacteria.

Foodborne Illness and Sous Vide Cooking

Sous vide cooking, when done properly, can safely create amazingly tender chicken, a perfect soft-cooked egg, or a succulent steak. However, it's also possible to set up a perfect breeding ground for bacteria if the food is mishandled. The heat involved in sous vide cooking is very low, so if you start with, say, a very large piece of frozen meat, it will take a long time to come up to temperature and will spend too much time in the breeding ranges of common foodborne pathogens.

With sous vide cooking, it is possible to cook meats to a point where they are texturally done—proteins denatured—but have not had sufficient time at heat for bacteria and parasites to be rendered nonviable. For these reasons, sous vide cooking has run afoul of some restaurant health inspectors: without proper procedures and clear guidelines, pathogens such as listeria and botulism are valid concerns when the food is mishandled. These concerns can be addressed with a clear understanding of where the risks are and what factors mitigate them. With the popularity of sous vide cooking on the rise, health inspectors are creating new guidelines, and depending upon where you live, they might already be comfortable blessing restaurants that have demonstrated proper handling procedures.

With low-temperature cooking, it's possible to violate the "40–140°F / 4–60°C danger zone" rule (see 160 in Chapter 4) and its derivative rule:

> *Thou shalt pasteurize all potentially contaminated foods.*

In the FDA's *Bad Bug Book*, the highest survival temperature listed for a foodborne pathogen at the time of this writing is 131°F / 55°C, for *Bacillus cereus*, which is relatively uncommon (you're 50 times more likely to get ill from salmonella) and, while unpleasant, has caused no known fatalities. The next highest survival temperature listed by the FDA is 122°F / 50°C, which gives you an idea of how much of an outlier *B. cereus* is.

Why is the danger zone a problem for dishes cooked sous vide, even at temperatures high enough to kill bacteria? The issue is that food cooked sous vide takes longer to come up to temperature than food cooked via other methods, and during that time heat-stable toxins can form. A number of foodborne illnesses are brought about by toxins and spores produced by bacteria. Even if the bacteria are killed, there could be sufficient time for them to produce enough toxins to be harmful in overly large cuts of meat. *To be safe, make sure that the core temperature of your food product reaches temperature within two hours.*

The temperatures in the danger zone rule build in a safety cushion, and for a broad, simple rule for all consumers, this is a good thing. If you are going to violate the temperature rules—e.g., cooking fish to only a rare temperature—be aware that you risk contracting a foodborne illness. For sous vide dishes that go from fridge to plate in less than two hours *and* where the danger zone rule is violated, the risks are equivalent to eating the raw item. If you are comfortable eating beef tartar or raw tuna in sushi, foods cooked sous vide are no worse when properly handled. Still, if you are cooking for someone who is in an "at-risk" group, you should avoid serving these foods just as you would avoid serving raw or undercooked items, especially as you can prepare a number of dishes sous vide that pasteurize the food and taste fantastic.

Sous vide cooking methods can be grouped into two general categories: cook-hold and cook-chill. In *cook-hold,* the food is heated up and held at that temperature until it is served. In *cook-chill*, the food is heated up, cooked, then *rapidly* chilled in the fridge or freezer for later use. (Use an ice-water bath to shock the food.) With the cook-chill approach, a greater amount of cumulative time is spent in the danger zone: first while the food is being heated, then while it's being chilled, and then while it's being heated again. Since it's the cumulative time in the danger zone that is of concern, I find it easier to use the cook-hold method, so that I simply don't have to worry about the cumulative time.

For the home chef concerned about food safety (that *is* all of you, right?), there is an easy way to remain safe (well, *safer*; it's all about risk mitigation and relative risks). When cooking sous vide, give preference to the cook-hold method, and be aware of the minimum temperatures required for pasteurization. This is an oversimplification, but it's an easy rule to follow. The better boundary guideline is to make sure to get the food above 136°F / 58°C—the lowest temperature given in the FSIS food guidelines—within a two-hour window and to hold it above that temperature for long enough to pasteurize it.

You can hold food above 140°F / 60°C for as long as you want; it's actually safer than storing food in the fridge.

Pasteurization is *not* an instantaneous process. For food to be pasteurized, it *must* be held at the target temperature for a sufficient length of time for the appropriate reduction of bacteria to occur. Consumer guidelines for cooking meats such as poultry specify temperatures of 165°F / 74°C because at that temperature the bacterial count will be reduced so quickly that there is no need to address the concept of hold time, and slight errors in temperature measurement and thermometer calibration will not be of concern.

With meats such as chicken breast, the required hold time at 140°F / 60°C for enough bacteria to die can be half an hour or more, meaning the food needs to reach 140°F / 60°C and then sit at that temperature for at least the prescribed amount of time. We'll cover the necessary hold times for different kinds of foods later in this chapter, because the hold times vary depending upon the composition of the food.

Cooking in the...Dishwasher?

Invariably, some people raise their eyebrows when I first start to describe sous vide cooking. The idea of cooking in a water bath is just plain foreign at first. But remember: cooking is about the application of heat, regardless of the source of that heat. Sous vide cooking is not the same as boiling food (that'd require the water to be around 212°F / 100°C). It's not even like simmering or poaching, in which the liquid environment is often hotter than the target temperature. Sous vide is the application of a very low, controlled temperature, in some cases as low as 116°F / 47°C.

Consider a piece of salmon cooked to medium doneness, which is an internal temperature of around 126°F / 52°C. On the grill, you'd cook the salmon until the core temperature reached 126°F / 52°C, but by that point, the outer portions of the fish would be hotter. In a water bath at 126°F / 52°C, the entire piece of fish would reach that temperature—but no higher. A ¾" / 20 mm-thick fillet of salmon will cook to medium in about 30 minutes at 126°F / 52°C.

Note that this temperature does *not* pasteurize the salmon. Handle it like raw/undercooked fish.

If you're anything like me, at some point, the following thought will occur: *wait a second, my tap water is about that hot...hmm...* I've tried it, and it *does* work: place your fish (sealed in an airtight bag) in a container in your sink, flip open the hot water tap, and keep a slow, constant trickle running. Check the temperature with a thermometer, and set your timer. It's not exactly energy efficient, even at a slow trickle, but it *does* work, sorta, at least for fish. Foods like chicken and beef require water hotter than what your tap delivers, and even if you did manage to get a stream of 140°F / 60°C water out of your tap, long cook times (e.g., 24-hour brisket) would make it impractical.

So, the next thought a geek might have would be: *wait a sec, did you say 140°F / 60°C? That's about how hot dishwashers get!* Search online for "dishwasher recipes" and yup, it has been done. People have cooked fish and even vegetarian lasagna in their dishwashers. If you try it, just remember to keep the time from fridge to plate at two hours or less, and treat the food as potentially raw or undercooked.

Sous Vide Hardware

Sous vide cooking requires very little in the way of hardware: a heater to keep the water bath at temperature, and a vacuum sealer to package foods so that they can be placed in the water bath without coming into direct contact with the water. While the commercial tools can still set you back many hundreds of dollars, a DIY version can easily be made for less than $100, and resealable plastic bags can be used in lieu of a standard vacuum system.

Water heaters

One difficulty of sous vide cooking is maintaining a water bath at a precise temperature, +/– 2°F / 1°C. The early days of sous vide cooking used laboratory equipment designed to hold water baths at the precise temperatures needed for controlling chemical reactions, but as you'd imagine, the lab gear has the drawback of being expensive. We're right on the cusp of a wave of new products targeted at the home chef who wants to cook sous vide, and while the prices might still be out of your reach, they'll surely come down until at some point the "three-in-one rice cooker" (steamer and slow cooker, too!) will become a four-in-one rice cooker.

Industrial circulators. These are lab-grade units that either are designed to be submerged into a container of water (e.g., hotel pans) or are enclosures with built-in containers.

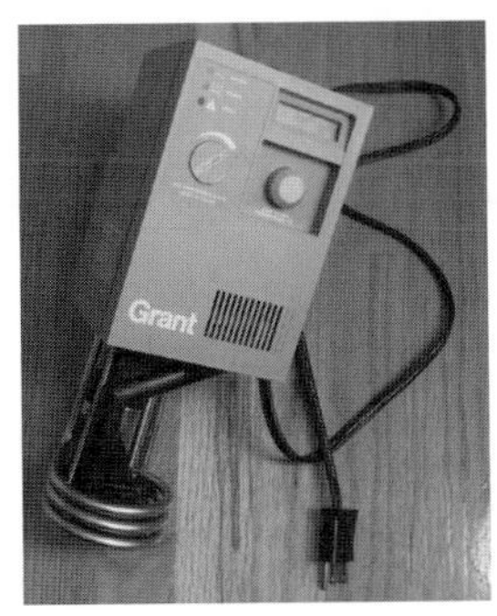

PolyScience is the most common manufacturer (*http://www.cuisinetechnology.com*), with new units costing around $1,000. Grant is also a common maker (*http://www.grantsousvide.com*). With luck, you can pick up a used unit at an online auction site for considerably less, but be aware that you'll have no idea what chemicals or pathogens a used unit might have been exposed to. If you do go this route, a three-step wash seems to be the standard cleaning regimen suggested: run it in a bath of vinegar, then one of bleach, and finally one of rubbing alcohol.

Consumer sous vide products. With the popularity of sous vide rising, a number of consumer products have recently been released or are in development that bring the cost of the hardware down to the $400 range, such as the Sous Vide Supreme. While still on the expensive side, as a piece of consumer kitchen equipment, it's not unreasonable, and prices will inevitably fall. Given the versatility and usefulness of the cooking technique, definitely consider looking at this category of products. See *http://www.cookingforgeeks.com/book/sousvidegear/* for suggestions on current products.

DIY sous vide. Other commercial products supply the "sous vide logic" but are BYOHS (bring your own heat source). Appliances like slow cookers contain the necessary cooking parts already: they hold a reservoir of liquid, have a heating element, and are designed to run

Make Your Own Sous Vide Setup

If you're the type inclined to fiddle with electronics, you can build your own sous vide rig by ordering a few parts online and spending a few hours tinkering.

The actual electronics necessary to maintain a water bath at a set temperature are simple enough: a basic slow cooker, a thermocouple, and a simple thermostat controller to switch the heat source on and off.

First, the slow cooker. The slow cooker will serve as the brawn, holding the water and providing the heat source. Snag a cheap slow cooker—you need one that will turn back on after losing power. Look for one that has a physical knob; the digital ones reset and stay off after power has been cut and then restored.

Next, the thermocouple. If you have a standard kitchen probe thermometer (which you really should), the probe—long braided cable, metal probe—is a thermocouple. For a sous vide rig, you'll need a type J thermocouple, which is made of materials that give it good sensitivity in the temperature ranges of sous vide cooking. This should cost around $15 to $20; search online for "type J probe" or search for part 3AEZ9 on on *http://www.grainger.com*.

Finally, the temperature controller. Just about any thermocouple-based temperature switch will work; look for one that runs off 12 volts DC, such as Love Industries' TCS-4030, which runs about $75. Snag a 12-volt wall wart (AC/DC power adaptor) while you're at it.

Once you have all the parts on hand, it's a relatively straightforward procedure to perform the lobotomy on the slow cooker: hook the thermocouple up to the probe inputs on the switch and connect the 12-volt power supply to the switch, then snip the slow cooker's electrical cord and run one side of it through the switch. Create a small hole in the lid of the slow cooker and poke the thermocouple through. Make sure you use enough water in the slow cooker that the thermocouple makes contact with the water when the lid is on!

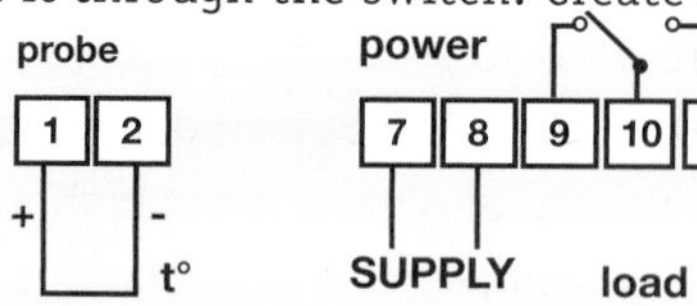

for extended periods of time. You can repurpose them for sous vide cooking by adding an external controller that switches the slow cooker on and off to keep it near a target temperature. See the sidebar on making your own sous vide rig for details.

Vacuum packers

Regardless of what type of vacuum packer you use, make sure that the plastic bag you're using is heat-stable.

Commercial in-chamber vacuum sealers. The industrial vacuum sealers create a (mostly) air-free chamber (a true vacuum). Unfortunately, they cost thousands of dollars. Fortunately, you don't need one. While there are a number of handy applications for them (*mmm*, watermelon steak), sous vide doesn't require this level of vacuum seal.

Consumer vacuum food sealers. These devices suck the excess air out of a plastic bag and then seal the bag by means of melting and fusing the opening of the bag. They don't create a true vacuum (in the sense that the food isn't subjected to a reduction in atmospheric pressure), but they do pull out most of the excess air. This is perfect for sous vide, because the function and purpose of the bag is just to allow rapid heat transfer from the water bath to the food via convection currents. Air in the bag would both slow the rate of heat transfer and also cause the bag to float in the water, preventing the face-up side of the bag from absorbing heat.

Resealable plastic storage bags. Sealable sandwich and storage bags (e.g., Ziploc bags) are not safe for *boiling* food. The concern with boiling applications is the potential for the plastic to leach into the food. The melting point of the type of plastic used in these bags is only a few degrees higher than the boiling point of water. BPA (bisphenol A) contamination would also be of concern, especially if you are adding oils inside the food. Check to make sure the manufacturer of the bags you have does not use BPA.

SC Johnson, the manufacturer of Ziploc-branded bags, does not use BPA.

Sous vide cooking, however, does not boil the water. 170°F / 75°C is about the highest temperature you might use in sous vide applications; 140°F / 60°C is about the warmest that's commonly encountered. Is 170°F / 75°C safe? SC Johnson is on record claiming that, yes, "[Ziploc] bags can be safely heated to 170 degrees Fahrenheit [76°C]."

Remember, the function of the bag is to allow heat to pass quickly from water to food via convection, so if you do use a plastic bag, make sure to remove as much air as possible. You can submerge most of the bag, leaving just the sealing strip at the top above water, and then seal it. Adding a bit of olive oil or marinade helps, because it'll better conform to the shape of the food.

Douglas Baldwin on Sous Vide

Douglas Baldwin is an applied mathematician at the University of Colorado at Boulder who, failing to find a good guide to sous vide, created his own, "A Practical Guide to Sous Vide Cooking," available at http://www.douglasbaldwin.com/sous-vide.htm. *He is also the author of* Sous Vide for the Home Cook *(Paradox Press).*

How did you hear about sous vide, and how did you get involved in it?

I was reading an article in the *New York Times* by Harold McGee, and he mentioned sous vide. While I knew quite a bit about cooking, I had never heard the term before and was intrigued. So I did what any good geek does: I went to Google and did some research. There was some information but not enough to meet my curiosity. So I turned to the academic journals and found a wealth of information.

It took me three or four months to collect and distill the 300 or so journal articles I found and publish the first draft of my guide online. I also did some calculations to figure out how long it takes things to cook and how long it takes to make them safe.

Safety is one of the big topics that comes up with sous vide and I'd love to talk about that in a moment. But first, what turned out to matter more than you expected when cooking sous vide?

People always worry about the vacuuming process, but that's really the least important part, even though the name sous vide means "under vacuum." It's really the precise temperature control that is important.

Long-term precision is important, because you don't want slow drifts when you're cooking for days to cause your meat to be overcooked. But short-term fluctuations in temperature really aren't that important because they will only affect the very outer portion of the meat. As long as the heat is oscillating less than one or two degrees Fahrenheit and the mean temperature is constant, you should be fine.

Wow! Cooking meats for days? What sorts of meats actually need cooking for that length of time?

Well, my favorite is beef chuck roast cooked for 24 hours at 130°F / 54.4°C. It's delicious. It transforms one of the least expensive cuts of beef into something that looks and tastes like prime rib.

It's all about the conversion of collagen into gelatin. This conversion is pretty rapid at higher temperature, taking only 6 to 12 hours at 175°F / 80°C to completely convert everything—well, almost everything. But at lower temperatures like 130–140°F / 54.4–60°C, it can take 24 to 48 hours for the same conversions to occur.

When I look at something like brisket being cooked at 130°F / 54.4°C for 48 hours, alarm bells go off in my head. Isn't there a potential bacterial risk here?

Well, certainly there's no risk at 130°F / 54.4°C. The pathogen that determines the lowest cooking temperature is *Clostridium perfringens*. Its highest temperature reported in literature is 126.1°F / 52.3°C. So as long as you're above that temperature, there won't be any food pathogens growing.

Now, there is the possibility of spoilage or beneficial microorganisms growing at these lower-cooking temperatures. That's one of the reasons that some people will sear ahead of time or drop the package of vacuum-sealed food in a pot of boiling water for a couple of minutes to kill off any thermophilic microorganisms that might be in there, like lactobacilli. But, in terms of safety, there's no concern whatsoever.

How about things like salmon, which are cooked at even lower temperature ranges than 130°F / 54.4°C?

If you would be fine eating the salmon raw, then cooking it for a couple of hours at a very low temperature, say 113°F / 45°C, isn't going to be a problem. If you wouldn't be comfortable eating it raw, then you probably shouldn't be cooking it at anything less than pasteurization temperatures and times.

Most food scientists and food safety experts agree that you should pasteurize fish. Even though it may not taste the same, or possibly quite as good, at least you'll feel a little more safe.

Food safety is about controlling both the actual and the perceived risk. Many people perceive the risk of fish to be much less than the risk of pork, but in many ways it's probably the other way around.

In our modern agro-industrial complex, we don't really know where things come from. With this decrease in knowledge of where our food came from, what field, how it was processed, and how it finally gets to our table, I tend to take the attitude of "pasteurize everything and hope for the best." Though it may not be what everyone wants or likes to hear.

What are the risks and what can somebody in the kitchen do to partially mitigate those risks?

When you're trying to deal with food safety, especially when it comes to pathogens, it is about three things. First, starting with a low initial level of contamination, which would mean buying, for example, very good and very fresh fish for which you know the origin. The second is to prevent the increase in the level of contamination and is frequently accomplished with cold temperatures or acids. The third is reducing the level of contamination, usually by cooking.

The problem is that if you're cooking fish sous vide at only 113°F / 45°C, then you won't reduce the pathogens to a safe level. So either pasteurize your fish by cooking it at 140°F / 60°C for about 40–50 minutes or make sure that very few pathogens grow and that the fish has a very low amount to begin with by buying from a trusted source.

Can one reduce the level of parasites by freezing?

Parasites, certainly. Though freezing fish at home will affect the quality of fish, because consumer freezers just can't freeze the fish fast enough to prevent large ice crystals from forming. Now, it's completely possible that you can buy already frozen, high-quality fish, or simply find out from your fishmonger whether or not it has already been frozen for a sufficient amount of time to kill any parasites.

But freezing won't kill the different bacterial food pathogens that one might be concerned with, and there's always the concern of chemical contamination, especially with shellfish that are harvested from questionable waters.

How do you know if something will work when you go to sous vide it?

I never really know, but I like to really scour the research journals for clues to the underlying processes involved. I first look to see if anyone else has already done it. With the wealth of scientific knowledge now available to us through the Internet, it's very likely that someone has asked and answered a closely related question. Then I just try and adapt it to the home kitchen.

It always surprises me how often I can take things directly from an academic journal and apply them in the kitchen.

Cooking with Sous Vide

While the general principles of sous vide cooking are the same regardless of the food in question, the exact temperatures required to correctly cook and pasteurize it depend upon the specifics of the item at hand. Different meats have different levels of collagen and fats, and denaturation temperatures for proteins such as myosin also differ depending upon the environment that the animal came from. Fish myosin, for example, begins to denature as low as 104°F / 40°C, while mammalian myosin needs to get up to 122°F / 50°C. (Good thing, too, otherwise hot tubs would be torture for us.)

Because meats can be grouped into general categories, we'll cover them in broad categories. We'll look at beef and other red meats together, for example, but keep in mind that variations between different red meats will mean that very slight changes in cooking temperature can yield improvements in quality. Data for the graphs in these sections are from Douglas Baldwin's "A Practical Guide to Sous Vide Cooking"; see the interview with him on the previous page for more information.

Beef and other red meats

There are two types of meats, at least when it comes to cooking: tender cuts and tough cuts. Tender cuts are low in collagen, so they cook quickly to an enjoyable texture; tough cuts require long cooking times for the collagen to dissolve. You can use sous vide for both kinds of meat; just be aware of which type of meat you're working with.

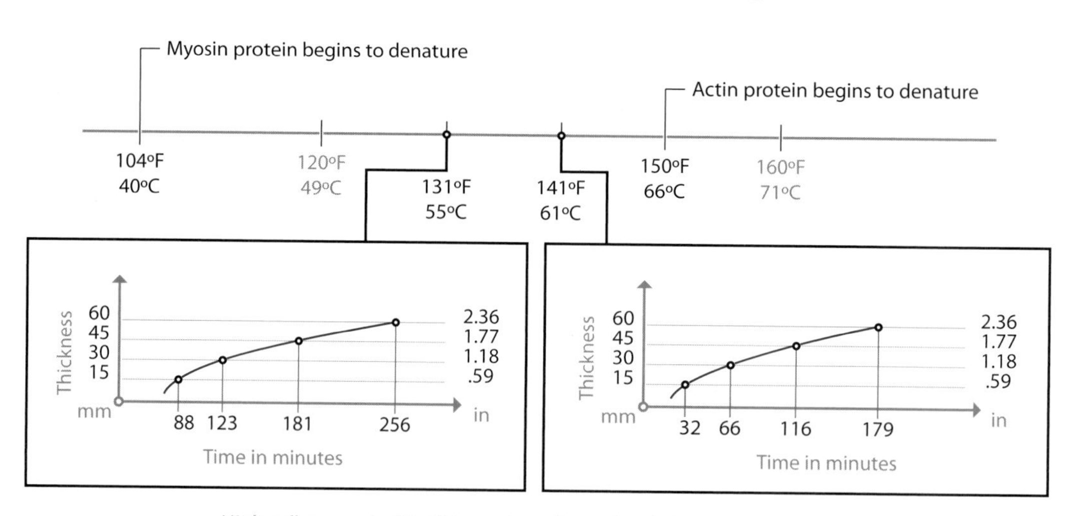

Time at temperature chart for beef and other red meats.

Beef Steak Tips

One of the primary benefits of sous vide is the ability to cook a piece of meat, center-to-edge, to a uniform level of doneness. Beef steak tips are a great way to demonstrate this.

Place in a vacuum bag:

1–2 pounds (~1 kg) steak tips, cut into individual serving sizes (7 oz / 200g)

1–2 tablespoons (15–25g) olive oil

Salt and pepper, to taste

Shake to coat all sides of the meat with the olive oil, salt, and pepper. Seal the bag, leaving space between each piece of meat so that the sous vide water bath will make contact on all sides.

Cook in a water bath set to 145°F / 63°C for 45 minutes. Remove bag from water bath, snip open the top, and transfer the steak tips to a preheated hot pan, ideally cast iron. Sear each side of the meat for 10 to 15 seconds. For a better sear, don't move the meat while cooking each side; instead, drop it on the pan and let it sit while searing.

You can create a quick pan sauce using the liquid generated in the bag during cooking. Transfer the liquid from the bag to a skillet and reduce it. Try adding a dash of red wine or port, a small pat of butter, and a thickening agent such as flour or cornstarch.

Notes

- *In sous vide applications, it is generally easier to portion out the food into individual serving sizes before cooking. This not only helps transfer heat into the core of the food faster (less distance from center of mass to edge), but it also makes serving easier, as some foods—especially fish—become too delicate to work with after cooking. You can still seal all the pieces in the same bag; just spread them out a bit to allow space between the pieces once the bag is sealed.*
- *I find adding a small amount of olive oil or another liquid helps displace any small air bubbles that would otherwise exist in a dry-packed bag. The quantities of oil and spices are not particularly important, but the direct contact between the spices and food does matter. If you add spices or herbs, make sure that they are uniformly distributed throughout the bag; otherwise, they will impart their flavor only to the pieces of meat they are touching.*

Some chemical reactions in cooking are a function of both time and temperature. While myosin and actin proteins denature essentially instantly at sufficient temperatures, other processes, such as collagen denaturation and hydrolysis, take noticeable amounts of time. The rate of reaction increases as temperature goes up, so while collagen begins to break down at around 150°F / 65°C, duck legs and stews are often simmered at or above 170°F / 77°C. Even at this temperature, the collagen still takes a matter of hours to break down.

The drawback to cooking high-collagen meats at this temperature, though, is that actin also denatures. While the fats in high-collagen cuts of meats can mask this, there is still a certain dryness to the finished dish. Since collagen begins to break down at a lower temperature than actin, though, it's possible to avoid this. The catch is that the rate of reaction is so slow that the cooking time stretches into days. With sous vide, though, this isn't a problem, if you don't mind the wait.

48-Hour Brisket

Seal in a vacuum bag:

1–2 pounds (0.5–1 kilo) high-collagen meat, such as brisket, chuck roast, or baby-back pork ribs

2+ tablespoons (30g) sauce, such as barbeque sauce, Worcestershire sauce, or ketchup

½ teaspoon (3g) salt

½ teaspoon (3g) pepper

Cook for 24 to 48 hours at 141°F / 60.5°C. Cut bag open and transfer the meat to a sheet pan or baking dish and broil to develop browning reactions on outside of meat, one to two minutes per side. Transfer liquid from bag to a saucepan and reduce to create a sauce. Try sautéing mushrooms in a pan in a bit of butter until they begin to brown and then adding the sauce to that pan and reducing until the sauce is a thick, almost syrupy liquid.

Notes

- *If your meat has a side with a layer of fat, score the fat to allow the marinades to contact the muscle tissue underneath. To score a piece of meat, drag a knife through the fat layer, creating a set of parallel lines about 1" / 2 cm apart, then a second set at an angle to the first set to create a diamond pattern.*
- *For additional flavors, add espresso, tea leaves, or hot peppers into the bag, along with whatever liquid you use. Liquid smoke can give it a smoky flavor as well.*
- *If your sous vide setup does not have a lid, be careful that water evaporation doesn't cause your unit to burn out or auto-shut off. One technique I've seen is to cover the surface of the water with ping-pong balls (they float); aluminum foil stretched over the top works as well.*

Fish and other seafood

Fish cooked sous vide is amazingly tender, moist, and succulent. Unlike fish that has been sautéed or grilled—cooking methods that can lead to a dry and rough texture—sous vide fish can have an almost buttery, melt-in-your-mouth quality. Other seafoods, such as squid, also respond well to sous vide cooking, although the temperatures do vary.

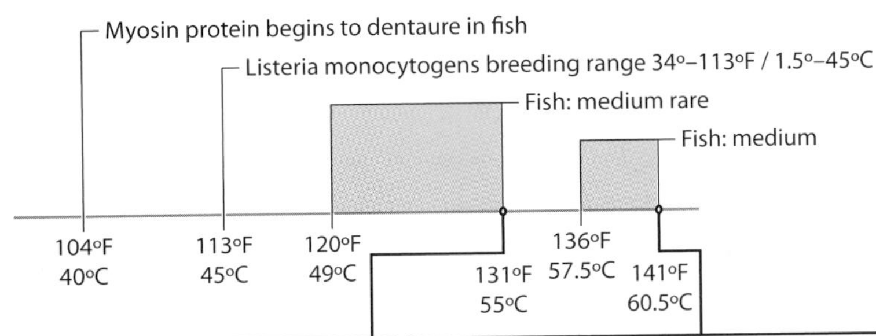

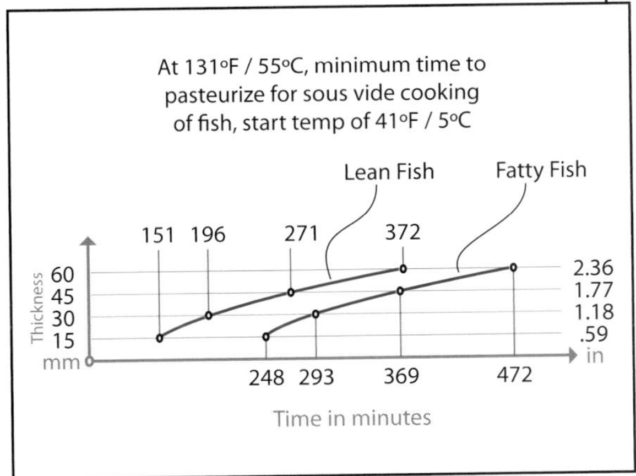

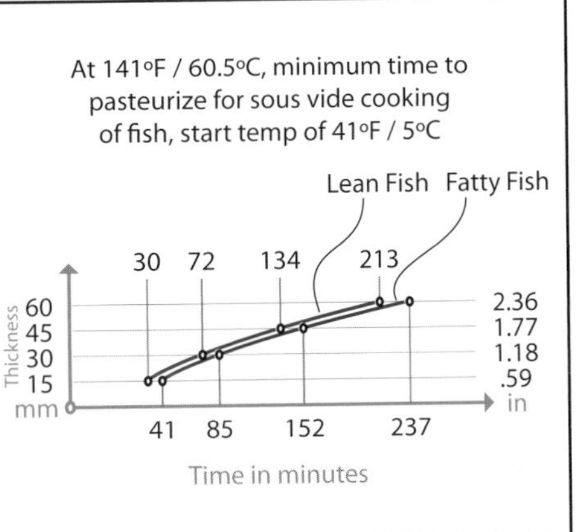

Time at temperature chart for fish and seafood.

If you are going to be using sous vide cooking in any professional setting, I highly recommend consulting Chef Joan Roca's book *Sous Vide Cuisine* (Montagud Editores).

Cooking fish sous vide is so straightforward that you don't need a recipe to understand the concept. The following tips should help in your experimentation with sous vide fish:

- Fish cooked to a doneness level of medium rare (131°F / 55°C) or more undergoes pasteurization by being held at temperature for a sufficient length of time (see the time-for-thickness graphs provided for lean and fatty fish).
- Lean fish, such as sole, halibut, tilapia, striped bass, and most freshwater fish, require less time to cook and pasteurize than fattier fishes, such as arctic char, tuna, and salmon.
- For fish cooked to a doneness level of only rare (i.e., cooked in a water bath set to 117°F / 47°C), pasteurization is not possible. Thus, if you are poaching salmon at 117°F / 47°C, be mindful that it will not actually get hot enough to kill all types of bacteria commonly implicated in foodborne illnesses. (Salmonella, fortunately, is not prevalent in fish.) Cooking fish at 117°F / 47°C for less than two hours presents *no worse* an outcome than eating the fish raw, so the usual recommendations for fish intended to be served raw or undercooked apply: buy sashimi-grade, previously frozen fish to eliminate parasites (see "How to Prevent Foodborne Illness Caused by Parasites" in Chapter 4), and don't serve the fish to at-risk individuals.

The FDA's 2005 Food Code excludes certain species of tuna and "aquacultured" (read: farm-raised) fish from this requirement, depending upon the farming conditions (see FDA Food Code 2005 Section 3-402.11b).

- If your fish comes out with white beads on the surface (coagulated albumin proteins), brine it in a 10% salt solution for 15 minutes before cooking. This will "salt out" the albumin via denaturation.

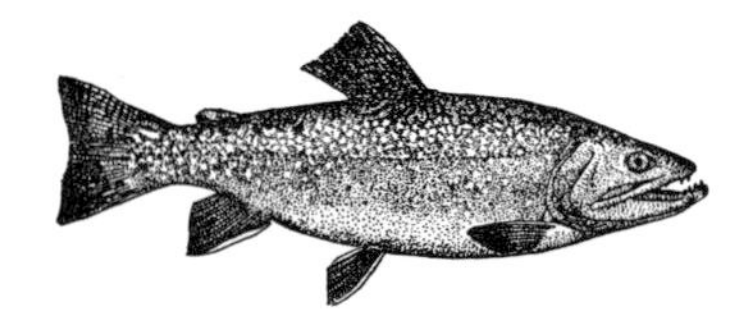

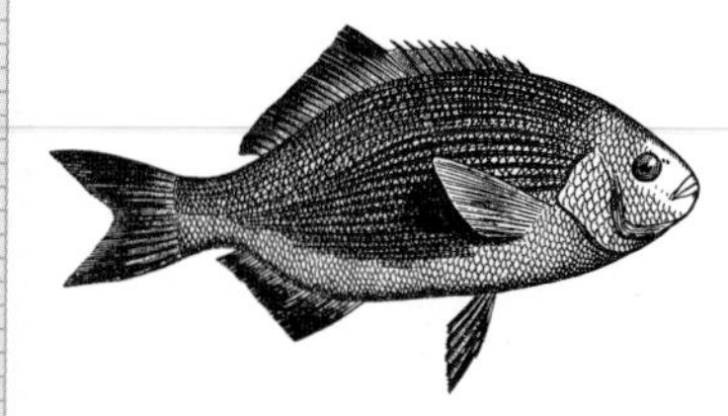

Sous Vide with Prepackaged Frozen Fish

The grocery stores where I live sell frozen fish in vacuum-packed bags. In some cases, the fish, which has been cut into individual portions, is frozen in marinade, making it the perfect sous vide–ready food: it's already vacuum-packed; it has been frozen per FDA standards, thus killing common parasites; and it has been handled minimally, having been frozen and sealed shortly after catch, reducing chances of bacterial cross-contamination. The time is ripe for sous vide to catch on big time: the food industry is already selling food in sous vide–ready packaging!

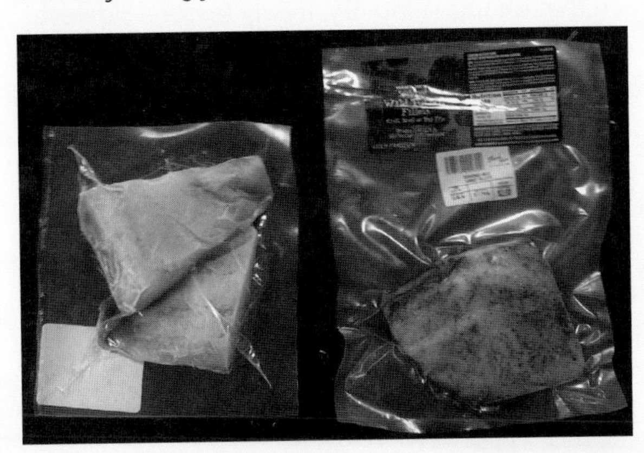

My favorite use of sous vide—well, besides making so many foods just plain delicious and easy to prepare for dinner parties—is using prepackaged frozen fish to make my daily lunch. My routine is fast, easy, cheap, and yummy:

Fill sous vide container (a pasta pot, in my case—I have an industrial circulator) with hot water from the tap. Using hot water means I don't have to wait for the immersion circulator to heat up the water.

Drop frozen vacuum-packed fish in the water, as-is, straight from the freezer. Because it's a single portion, the amount of time it'll take to thaw is relatively short. Just remember: pasteurization times start once the core of the food has reached the target temperature. With a frozen item, you'll have a hard time knowing when this occurs. I cook a single portion of fish for long enough to ensure that both thawing and pasteurizing have occurred. And because sous vide cooking is forgiving of longer cook times, for most types of fish leaving them in the water bath for an extra half hour won't affect the quality.

Go for a run. Go to the gym. Do some errands. Write a section in a book about cooking frozen fish in a bag.

Fish out the bag, cut it open, drop the fish on a plate with some steamed veggies and brown rice, and voilà: lunch.

If you're preparing yourself a meal ahead of time, you can drop the cooked fish into a container with some frozen veggies, which do double duty by acting as ice cubes to rapidly cool the fish down.

The quality of frozen fish can really vary. Frozen salmon from one store can turn out mushy and unappetizing, while the same type of salmon from a different chain can come out moist, succulent, and perfect. This is mostly likely due to differences in freezing techniques: rapid freezing does less damage to the tissues by limiting the amount of time ice crystals have to aggregate and form larger, dagger-like shapes that can pierce cell walls. If you've had bad results with frozen fish, blame the freezing technique, not the fact that it has been frozen.

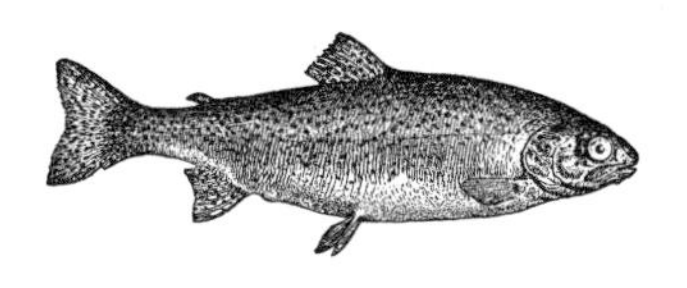

Chicken and other poultry

One of the greatest travesties regularly foisted upon the American dinner plate is overcooked chicken. Properly cooked chicken is succulent, moist, and bursting with flavor—never dry or mealy. True, the potential of contracting salmonellosis from undercooked chicken is real; besides, raw chicken is just gross. But I'm not suggesting undercooking chicken—just cooking it correctly.

The "problem" with cooking chicken "correctly" is that, from a food safety perspective, ensuring pasteurization (sufficient reduction of the bacteria that cause, say, salmonella) requires holding the chicken at a high enough temperature for a sufficiently long period of time. "Instant" pasteurization can be done at 165°F / 74°C, but at this temperature the actin proteins will also denature, giving the chicken that unappealing dry, mealy texture. However, pasteurization can be done at lower temperatures, given longer hold times. Sous vide is, of course, extremely well suited for this: so long as you hold the chicken for the minimum pasteurization time required for the temperature you're cooking it at, you're golden. Even if you hold it too long, as long as it's below the temperature at which actin denatures, the chicken will remain moist. Another win for sous vide!

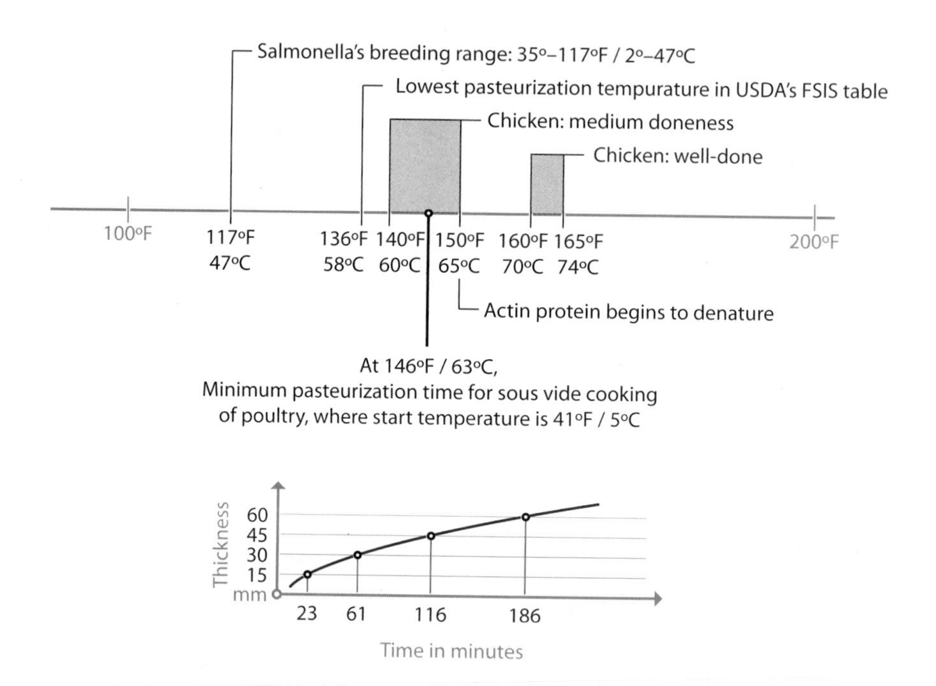

Time at temperature chart for poultry.

Sous vide chicken breast

As with fish, you don't need a recipe in the traditional sense to try out sous vide cooking with chicken. Here are some general tips:

- Chicken has a mild flavor that is well suited to aromatic herbs. Try adding rosemary, fresh sage leaves, lemon juice and black pepper, or other standard flavors in the bag. Avoid garlic, however, because it tends to impart an unpleasant flavor when cooked at low temperatures. When adding spices, remember that the items in the bag are held tightly against the meat, so herbs will impart flavors primarily in the regions that they touch. I find that finely chopping the herbs or puréeing them with a bit of olive oil works well.
- As with other sous vide items, allow space between the individual items in the vacuum bag to ensure more rapid heat transfer, or place individual portions in separate bags.

Slow Cooker Versus Sous Vide

"Wait a sec," you might be thinking, "this 'sous vide' thing...how's it different from a slow cooker?" I thought you'd never ask!

They're not actually that different. Both hold a reservoir of liquid at a high-enough temperature to cook meat but not boil water. Sous vide cooking has two advantages over traditional slow cooking, though: the ability to dial in a particular temperature, and to minimize the amount of variance that occurs around that temperature.

With a slow cooker, your food cooks somewhere in the range of 170–190°F / 77–88°C. The exact temperature of your food and the extent to which that temperature fluctuates aren't so important for most slow-cooked dishes. This is because slow cooking is almost always done with meats that are high in collagen, and as discussed in Chapter 4, these types of meat need longer cook times in order for the collagen to denature and hydrolyze and transform into something palatable.

However, this isn't true for cuts of meat that are low in collagen, such as fish, chicken breast, and lean cuts of meat. For these low-collagen items, cooking needs to denature some proteins (e.g., myosin) while holding other proteins native (e.g., actin). The difference in temperature at which these two reactions occur is only 10°F / 5°C, so precision and accuracy are important. Sous vide wins hands down. It's not even close.

Try cooking ducks legs both ways. Seal up two legs and cook them sous vide at 170°F / 77°C. Meanwhile, prepare a second set of legs in a slow cooker. Cook for at least six hours and then examine the difference.

Sous vide duck legs.

Slow-cooked duck legs.

Vegetables

The geeky way to think about cooking is to consider the addition of heat to a system. Adding heat isn't a spontaneous thing: there will always be a heat gradient, and the difference between the starting and target temperatures of the food will greatly affect both the cooking time and the steepness of the gradient.

This is one reason to let a steak rest at room temperature for 30 minutes before grilling: 30 minutes is short enough that bacterial concerns are not much of an issue, but long enough to lower the temperature difference between raw and cooked steak by a third. You can use a water bath to the same effect for vegetables: reduce the heat delta by holding them in a moderate-heat water bath (say, 140°F / 60°C) for 15 to 20 minutes, and then steam or sauté them.

Yes, you can cook vegetables sous vide, too, but because vegetables don't begin to cook until relatively high temperatures—typically above 185°F / 85°C—and even then take a while, it's easier to cook them with traditional techniques.

I often cook steak tips at the same time that I preheat bok choy, Swiss chard, or other hearty greens, using the same water bath for both the steak and the veggies. This works because the veggies don't actually cook at the temperature that the meat is cooking at.

This technique works great for small dinner parties. I bag and seal the steak tips just before my guests show up, and once they arrive, I drop the bag of steak tips and some bok choy into a water bath set to 140°F / 60°C. Thirty minutes or so later—after catching up with my guests, sharing a beer or glass of wine, and noshing on cheese and bread—I pull the steak tips out and let them rest for a few minutes, during which time I quarter the bok choy and steam it in a hot frying pan.

Because the bok choy is already warm, it reaches a pleasant cooked texture in two to three minutes, at which point I transfer it to the dinner plates. Reusing the same frying pan, I quickly sear the outside of the steak tips, which I then cut and transfer to the plates. Total time spent while guests wait? Five minutes, tops. Number of dirty dishes? One, plus plates. And it's delicious!

Enhancing texture

Ever wonder why some vegetables in canned soups are mushy, textureless blobs, but others aren't? Some vegetables—carrots, beets, but not potatoes—exhibit a rather counterintuitive behavior when precooked at 122°F / 50°C: they become "heat resistant," so they don't break down as much when subsequently cooked at higher temperatures. Holding a carrot in a water bath at around 120°F / 50°C for 30 minutes causes *enhanced cell-cell adhesion*, science lingo for "the cells stick better to each other," which means that they're less likely to collapse and get mushy when cooked at higher temperatures.

During the precooking stage, calcium ions help form additional "crosslinks" between the walls of adjoining cells, literally adding more structure to the vegetable tissue. Since "mushy" textures occur because of ruptured cells, this additional structure keeps the vegetable tissue firmer by reducing the chance of cellular separation.

The normal solution to mushy vegetables is to refrain from adding them until close to the end of the cooking process. This is why some beef stew recipes call for adding vegetables such as carrots only in the final half-hour of cooking.

For industrial applications (read: canned soups), this isn't always an option. In home cooking, you're unlikely to need this trick, but it's a fun experiment to do. Try holding carrots at 140°F / 60°C for half an hour and then simmering them in a sauce mixed in with a batch of sliced carrots that hasn't been heat-treated. (You can cut the heat-treated carrots into slightly different shapes—say, slice the carrot in half and then half-rounds, versus full-round slices—if you don't mind your experiment being obvious.)

Chocolate

Tempering chocolate—the process of selectively melting and solidifying the various forms of fat crystals in cocoa butter—can be an intimidating and finicky process. The chocolate must first be melted to above 110°F / 43°C, then cooled to around 82°F / 28°C, and then heated back up and held between 89°F / 31.5°C and 91°F / 32.5°C. Once tempered, you must play a thermal balancing act: too warm, you lose the temper, and too cold, it sets.

It's not exactly correct to describe chocolate as something that "melts," because chocolate is a solid sol, a colloid of two different solids: cocoa powder and cocoa fats. The cocoa powder itself can't melt, but the cocoa fats that surround it can. Cocoa butter contains six different forms of fats, and each form melts at a slightly different temperature.

The six forms of cocoa fat are actually six different crystalline structures of the same type of fat. Once melted, the fat can recrystallize into any of the six forms. It's for this reason that tempering works at all—essentially, tempering is all about coercing the fats to solidify into the desired forms.

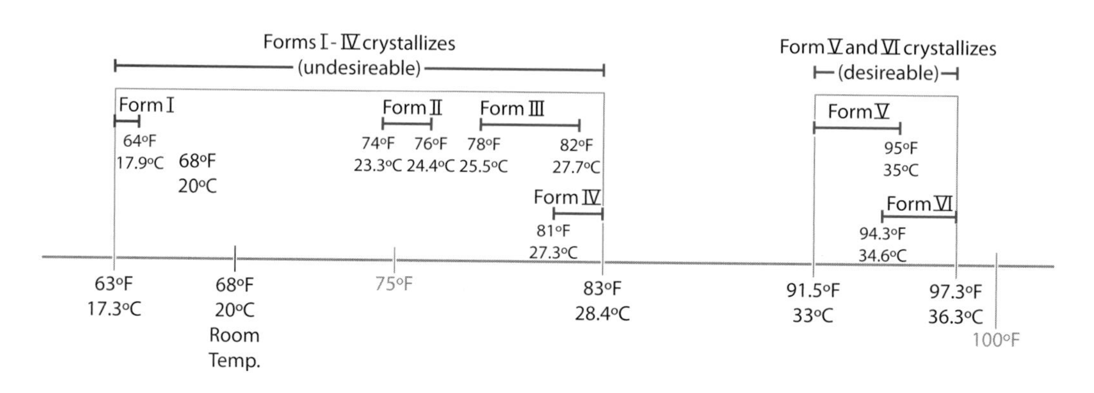

Melting points of the six polymorphs of cocoa fat.

How do scientists tell when something is melting? Two common techniques are used: differential scanning calorimetry (DSC) and x-ray diffraction. In DSC, energy is added to a closed system at a controlled rate, and the temperature of the system is monitored. DSC picks up phase changes (e.g., solid to liquid) because phase changes require energy without a temperature change. X-ray diffraction looks at how x-rays scatter when passed through a sample: with each phase change, the x-ray pattern changes.

It's not a matter of different types of fats; it's the structure that the fat takes upon solidifying that determines its form. Two of these forms (Forms V and VI) link together to create a metastructure that gives chocolate a pleasing smoothness and firm snap when broken. Chocolate with a high number of Form V structures is said to be *tempered.* The other primary forms (I–IV) lead to a chalky, powdery texture. Form VI occurs in only small quantities, due to the temperature range at which it crystalizes. Chocolate that has been exposed to extreme temperature swings will slowly convert to Forms I–IV. Such chocolate is described as having *bloomed*—the cocoa particles and cocoa fats separate, giving the chocolate both a splotchy appearance and a gritty texture.

To further complicate things, the fats in cocoa butters don't actually melt at an exact temperature, and the composition of the fats varies between batches. The ratio of the different fats determines their exact melting point, and the ratio varies depending upon the growing conditions of the cocoa plant. The fat in chocolate from beans grown at lower elevations, for example, has a slightly higher melting point than chocolate from beans grown at higher, cooler elevations.

Still, the temperature variances are relatively narrow, so the ranges used here generally work for dark chocolates. Milk chocolates require slightly cooler temperatures, because the additional ingredients affect the melting points of the different crystalline forms. When looking at chocolate for tempering, make sure it does not have other fats or lecithin added, because these ingredients affect the melting point.

Luckily for chocolate lovers worldwide, chocolate has two quirks that make it so enjoyable. For one, the undesirable forms of fat all melt below 90°F / 32°C, while the desirable forms noticeably melt around 94°F / 34.4°C. If you heat the chocolate to a temperature between these two points, the undesirable forms melt and then solidify into the desirable form.

The second happy quirk is a matter of simple biology: the temperature of the inside of your mouth is in the range of 95–98.6°F / 35–37°C, just above the melting point of tempered chocolate, while the surface temperature of your hand is below this point. Sure, a certain sugar-coated candy is known to be made to "melt in your mouth, not in your hands," but with properly tempered dark chocolate, the sugar coating isn't necessary (it is necessary for milk chocolate, though, which melts at a temperature lower than that of your hand).

M&Ms were developed in 1940 by Frank C. Mars and his son, Forrest Mars, Sr. During the Spanish Civil War (1936–1939), Forrest saw Spanish soldiers eating chocolate that had been covered in sugar as a way of "packaging" the chocolate to prevent it from making a mess.

How does all of this relate to sous vide cooking? Traditional tempering works by melting all forms of fat in the chocolate, cooling it to a low enough temperature to trigger nucleation formation (i.e., causing some of the fat to crystallize into seed crystals, including some of the undesirable forms), and then raising it to a temperature around 90°F / 32.2°C, where the fats crystallize to make Form V crystals.

This three-temperature process requires a watchful eye and, during the second step, constant stirring to encourage the crystals to form while keeping them small. Water baths allow for a shortcut in working with chocolate: already tempered chocolate doesn't need to be tempered *if* you don't get it any hotter than around 91°F / 32.8°C. The desirable forms of fat won't melt, so you're good to go. To melt already tempered chocolate, seal it in a vacuum bag and submerge it in a water bath set to 91°F / 32.8°C. (You can go a degree or so warmer; experiment!) Once it's melted—which might take an hour or so—remove the bag from the water, dry the outside, and snip off one corner: instant piping bag.

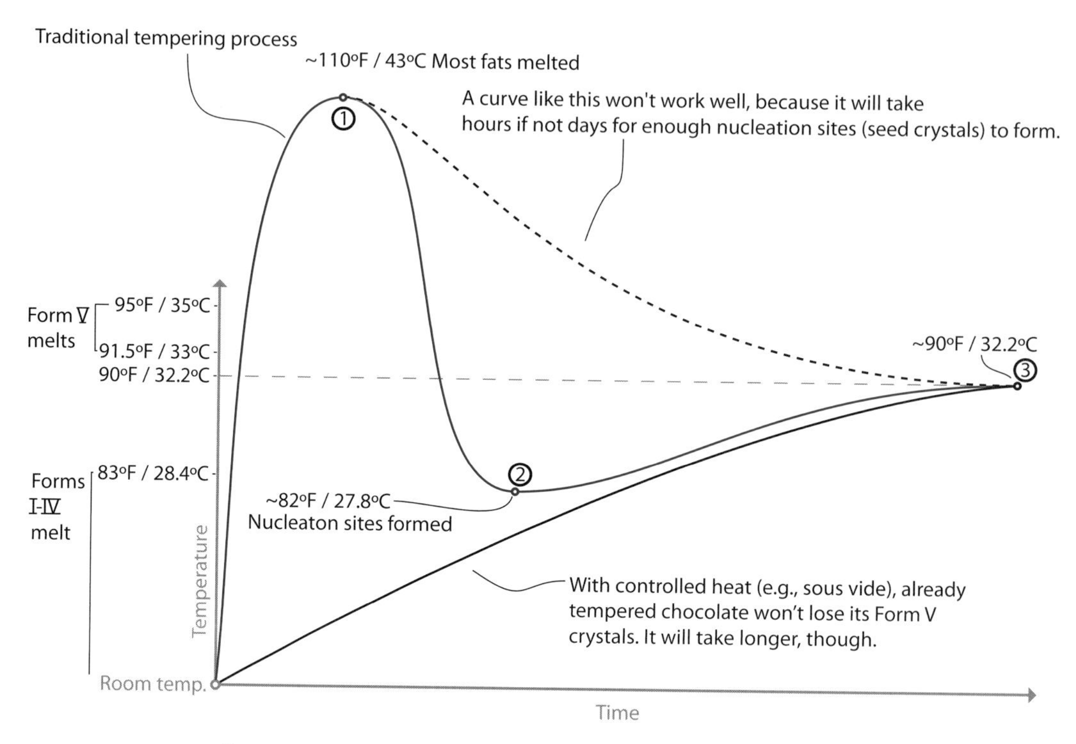

Temperature versus time chart for melting and tempering chocolate.

If you're going to be working with chocolate on a regular basis, the sous vide hack will probably get tiring. It works, but if you have the dough to spend, search online for chocolate tempering machines. One vendor, ChocoVision, sells units that combine a heat source, a motorized stirrer, and a simple logic circuit that tempers and holds melted chocolate suitable for everything from dipping fruit to coating pastries to filling chocolate molds. Of course, if you have a slow cooker, thermocouple, and temperature controller…

Chocolate Almond Bars

My local grocery store recently started carrying specialty bars of chocolate infused with unusual ingredients: curry powder and coconut; plums, walnuts, and cardamom; even bacon bits. These exotic chocolate bars also carried exotic price tags, so I thought: how hard can it be to make these? With sous vide, it turns out it's downright simple.

Place tempered chocolate in a vacuum bag. Use chocolate in bar form; chocolate chips might not work if they aren't as well tempered.

Add your flavorings. Try almonds or hazelnuts (at about a 1:2 ratio—one part nuts to two parts chocolate by weight). Your ingredients should be dry. Any water in them will cause the chocolate to seize up.

Seal, drop in a water bath set to 92°F / 33.5°C, and wait for chocolate to melt, which may take an hour or two.

After the chocolate is thoroughly melted, work the bag to distribute the chocolate and flavorings. You can use a rolling pin to work the fillings around if using something like nuts.

Let bag rest on counter to cool.

Once cooled, snip the bag open and peel it off the chocolate. You can break the bar up into pieces.

Try using coffee beans (yum), candied grapefruit rind, dried fruits such as cranberries, or a mix of toasted nuts (almonds, pistachio, and pecans, and maybe a pinch of cayenne pepper).

Flash Pickling with a Vacuum Sealer

...or How to Void Your Warranty in Three Easy Steps

Once you have a sous vide setup, you also have most of the tools needed to do flash pickling. In the culinary world, *flash pickling* refers to submerging a food item in a liquid-filled container, evacuating the container, and then repressurizing the container. Unlike traditional pickling, which requires time (or heat) to coerce the pickling liquid into the food, flash pickling is instant, hence its name.

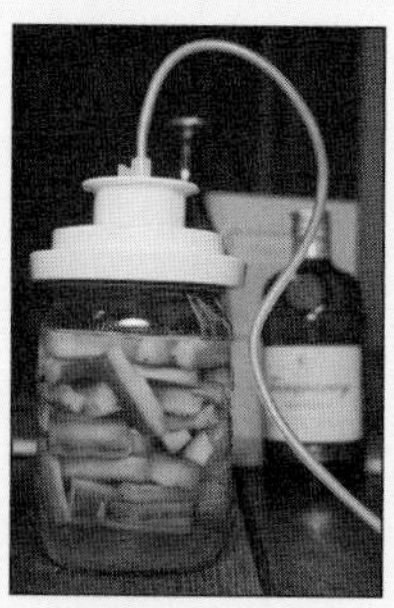

Cucumbers being flash-pickled in gin using a consumer jar sealer attachment.

Under vacuum, microscopic air pockets in foods like sliced apples and cucumber wedges lose their air. Upon returning to atmospheric pressure, the food expands back out to its original shape, a bit like a sponge. But because the food is submerged, liquid is pulled back in, instead of air. Why bother? Because "Manhattan apples" (use whiskey) or "martini pickles" (use gin) are just plain *awesome*. See *http://video.nytimes.com/video/2007/12/04/magazine/1194817116911/the-edible-martini.html* for a video of Dave Arnold talking about the process using a commercial vacuum sealer.

The pros, who use commercial vacuum chambers, can just drop the food into the liquid bath and clamp down on the lid. For the rest of us, though, generating a sufficiently strong vacuum isn't so easy. But if you have a vacuum food sealer and don't mind voiding your warranty, there is a way.

I should have just written an entire chapter called "Voiding Your Warranty."

Consumer vacuum sealers have a pressure switch that triggers them to stop pumping and start sealing, meaning that they stop short of creating a strong enough vacuum to create a good pickle. But if you disable the pressure switch, the unit should continue to pump indefinitely, or until the motor burns out.

To make a DIY flash-pickling system, start with a consumer vacuum sealer. You'll need a toggle switch and an extra piece of wire, along with a screwdriver and wire cutters.

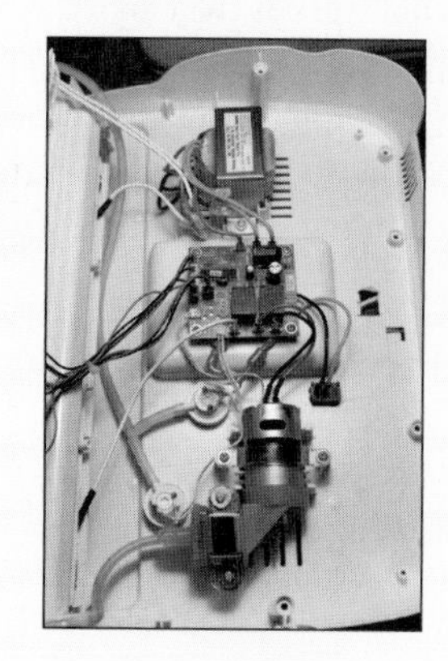

Start by popping open the vacuum sealer. It should look something like this.

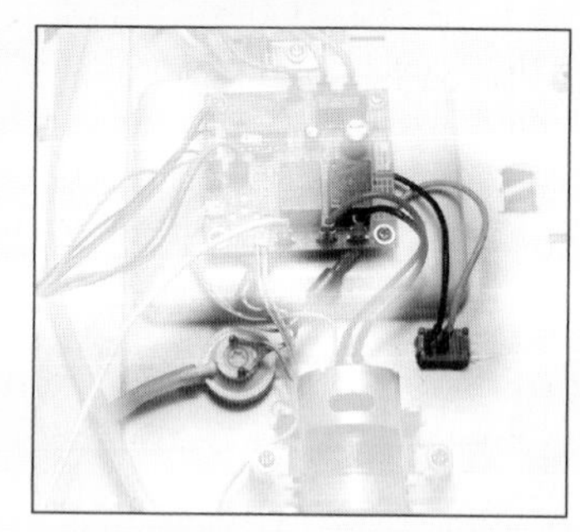

Locate the pressure switch (highlighted on left). Cut one of the wires that runs back to the circuit board and interpose a toggle switch (highlighted on right). Cut a small hole in the plastic and mount the toggle switch so that you can flip it from the outside.

Commercial Hardware and Techniques

What goes on behind those two-way swinging doors leading to the commercial kitchen? More and more restaurants are sharing with the public what they're doing, even going so far as to blog their thoughts and recipes for all the world to see. Why? Well, for one, it serves as great publicity for the restaurants. And secondly, so much of what's done in the high-end modernist restaurants requires so much work that it's probably cheaper for a home chef to go and eat at the restaurant than it would be to try undertaking one of their recipes anytime soon.

Even if you're not going to attempt a full 26-course dinner, you can learn a lot by seeing how the pros approach food and the lengths to which they go to in their quest for a truly fantastic and delightful meal.

Since the techniques in this section are not, in and of themselves, going to put dinner on the table, you might wonder how to work them into your cooking. Think of this section like knife skills for modernist cuisine: a few pointers for what's happening behind those swinging doors. For inspiration and ideas of what to do with these skills, try turning to the Internet. Here are a few blogs worth checking out (most of these are associated with interviewees in this book as well):

Cooking Issues *(http://cookingissues.com)*
: Nils Norén, Dave Arnold, and other members of the French Culinary Institute blog about their investigations into cooking phenomena, giving good explanations of how to use new technologies.

eGullet.org *(http://forums.egullet.org)*
: The mother of all forums related to food, eGullet is home to many threads covering almost any topic you can imagine related to the creation of food, including the infamous sous vide thread.

Ideas in Food *(http://blog.ideasinfood.com)*
: Alexander Talbot and Aki Kamozawa blog about their work with food, sometimes including insightful recipes and tips.

Playing with Fire and Water *(http://www.playingwithfireandwater.com)*
: Linda Anctil's blog posts give an evocative and creative approach to food.

In this section, we'll take a look at a few techniques that are common in commercial restaurants and examine ways that they can be useful to the home chef. This isn't by any means a complete list. Rather, this should be enough to get you started thinking outside the box (or, harking back to the functional fixedness concept discussed in the opening chapter, getting to see the box in a different way).

3D Printing and Mold Making

Many aspects of "playing with your food" are beyond the reach of most commercial restaurants, either because they're not worth the time or require a geek to do it.

For a few high-end restaurants, spending the time involved in making custom molds allows them to create innovative and unusual experiences. Working with fabricators, they'll create custom silicone molds ranging in shapes of everything from vegetables to eggs, using them to mold asparagus puree set with gelling agents or for signature desserts.

Then there's the geek side of things. If you happen to have access to a CNC (computer numeric control) printer, such as MakerBot's Cupcake, try printing your own molds and cookie cutters. Here's an example, using none other than that famous penguin, Tux. (Tux is the Linux kernel's official mascot.) You'll need a cookie cutter, sugar cookie dough, and frosting.

Create the cookie cutter. This is the hardest part (second hardest, if you're the type to eat all the cookie dough before getting to the end). Assemble a MakerBot CNC printer and print a Tux cookie cutter, following the STL and G code files at *http://www.cookingforgeeks.com/book/cookie-cutter/*.

Bake the cookies. Using the cookie cutter (shown on the left in the photo below), create your Tux cookies and bake. Allow the cookies to cool before frosting.

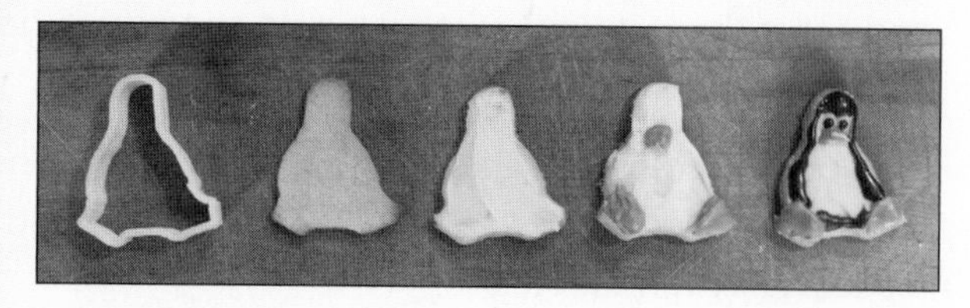

Frost. Until MakerBot comes out with a Frostruder that supports multiple colors, you'll have to do this by hand. Prepare a batch of frosting (see page 245 in Chapter 5 for a simple frosting recipe) and divide it into three bowls, putting most of the frosting in the first bowl. Add yellow food coloring to the second bowl; you'll use this for Tux's yellow feet and beak. Add red and blue food coloring to the final bowl; when mixed together, this will make an almost-black frosting.

To frost, take a first pass using the white frosting, covering the entire cookie in a single full layer of white frosting. Using a dinner knife, take a second pass, lightly smearing the yellow frosting for his beak and feet. For the third pass, transfer the black frosting to a plastic sandwich bag, snipping off the corner to make a piping bag, and carefully dot the two eyes and black edge.

Filtration

Filtering is a common technique for separating solids from liquids in a slurry. Filtering is usually done to remove the solids—for example, to create a clear broth free of particulate matter or a juice free of pulp. Other times, the solid matter, such as browned butter solids, is the desired item.

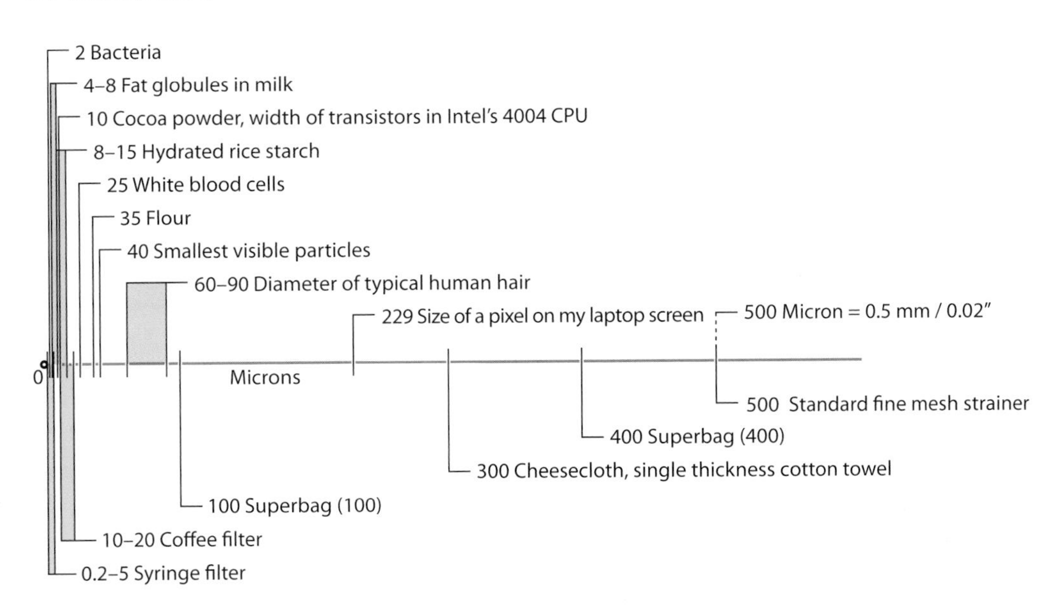

Sizes of common items (top portion) and common filters (bottom portion).

Besides filtration, which we'll talk about here, additives can be used to separate out some types of solids. Some manufacturers use isinglass, a collagen derived from fish bladders, in beer and wine making. The isinglass binds with yeast and causes it to precipitate out. (Sorry, vegetarian beer lovers.) And consommé is traditionally clarified using egg whites, which, like isinglass, bind to small particulates and then coagulate into a large mass that's easily removed. Mechanical filtration, in contrast, has the advantage of being fast and easy.

Reasons for filtering in the kitchen can range from aesthetic (including traditional broths like consommé) to practical (needing particulate-free liquid to work with in cream whippers, as described later in this chapter—the particulate would potentially clog the system).

Which type of filter to use depends on the size of the solids. A *chinois*—a conical strainer—is fine for straining out spices and solids from a broth and is the standard go-to item for filtration. To mechanically mash foods and give them a finer texture, you can push them through a perforated sheet of steel. Traditional European soups, such as vichyssoise (potato and leek), pass through these to ensure a smoother mouth-feel.

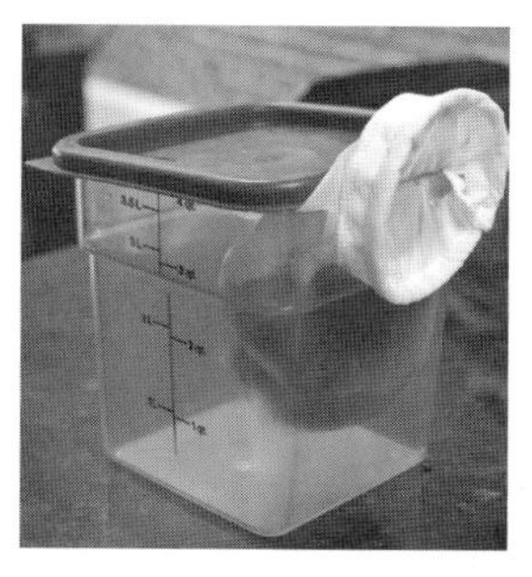

A standard modern technique for making clarified liquids such as consommés is to freeze the liquid and drip-thaw it through a filter, such as a Superbag.

High-end chefs use finer filtration to achieve other effects. Straining out the solids in tomato juice to get a clear, transparent tomato water requires a much finer filter. You can also use hydrocolloids: create a gel with gelatin (e.g., stocks) or agar (e.g., Dave Arnold's lime juice on page 368), and pass the gel through a filter. The gel will hold on to most of the solids, while the filter will hold on to the gel.

International Cooking Concepts sells a filter bag called a "Superbag" that's dishwasher safe, reusable, and highly durable. For a tenth of the price, McMaster-Carr sells mesh filter bags that are FDA compliant and rated to 220°F / 104.4°C. Search for part 6805K31on *http://www.mcmaster.com*.

If I were to write a "cooking geek purity quiz," one question would definitely be "How many things have you ordered from McMaster-Carr?"

The McMaster-Carr product uses a stiffer material and doesn't drain as quickly as the Superbag, however. With this size of filtration, you can quickly create flavored liquids such as nut milks (purée presoaked almonds, drop in filtration bag, squeeze liquid out) or fruit juices (purée cantaloupe, drop in filtration bag, squeeze liquid out). Try other things, such as asparagus and olives.

These finer filters can also be used for drip filtration, where the solids are rested in the filter bag and the liquids are given time to percolate out slowly. Purée tomatoes, drop them in a fine (~100 micron) filter sleeve, clamp in a storage container, and let drip overnight in the fridge to create semiclear tomato water.

Stock, broth, and consommé

Stock, broth, consommé—what's the difference? *Stock* and *broth* are both liquids made by simmering vegetable and/or animal matter. Traditionally, stocks are made with bones, which have collagen. Most of this collagen breaks down and converts to gelatin, which gives the stock a lubricious mouth-feel and, at sufficient concentrations, causes the stock to turn into a gel when cooled. The cans of "stock" that you find in the grocery store are really broth—they don't have the same level of gelatin that a proper stock should have.

If the canned "stock" carried in grocery stores had gelatin, it would be gelled like Jell-O.

Stocks are generally more of an ingredient—not highly seasoned, usually added to a soup or dish. Broth is a finished product, and strictly speaking broths should be made without bones; they contain no gelatin and so are comparatively much thinner than stocks. From a practical perspective, in home cooking you can treat them as the same thing in most cases. Just don't try to make a dish such as aspic that relies on gelatin using broth.

Both stock and broth contain fats and solid particulate matter from the vegetable and animal products they're made with, giving them a cloudy appearance. A *consommé* is a clarified version of either stock or broth, from which the particulates and some of the fats have been filtered out. The traditional method for clarification involves creating an egg-white "raft" that is gently stirred while the broth is simmered. It's time-consuming, and while you should try it sometime, it's not likely to be an everyday cooking technique. An easy modern method involves using the gelatin present in a true stock to trap the particulate matter. Freeze the stock, and as it thaws, the gelatin will hold on to the particulate matter; thaw it in a filter that's fine enough to hold onto the gelatin, and the resulting liquid that passes through the filter will be consommé.

"Filtering" by Evaporation

Okay, this isn't really filtering in the true sense of the word, but you *can* separate a liquid from any compounds dissolved in it by boiling off the liquid. Think sea salt: saltwater is allowed to evaporate, leaving behind just the salt.

A rotary evaporator (*rotovap*) is nothing more than a fancy (and unfortunately very expensive) tool for replicating what happens in a salt flat, but it's designed to enable better control and to allow capturing of both parts (i.e., the salt *and* the water after separating). It separates a solvent from a liquid or solid by gently boiling it away under a precise vacuum and temperature and then condensing the vapor in a flask, a process known as *distilling*. It's like boiling water on the stove and collecting the steam that condenses on the lid, but with far more precision.

Distilling under a vacuum lowers the boiling point of the solvent (usually water or ethanol), meaning that any compounds that are heat-unstable remain undisturbed. With a rotovap, alcohol or water can be boiled off without the changes in flavor that normally come about from cooking. Chefs have made flavorings using everything from common vanilla to offbeat items such as "sea" (using sand) and "the woods" (damp dirt from the forest). Rotovaps can also be used to remove solvents from a food—removing alcohol to make whiskey essence, water to increase the concentration of fresh-squeezed juices, or both alcohol and water to make sauces such as port syrup.

Unfortunately, commercial rotovaps are *expensive,* and the process of distilling foods is heavily regulated. For all the details, see the Cooking Issues blog writeup at *http://www.cookingissues.com/primers/rotovap/*.

Basic White Stock

In a large stockpot (6 qt / 6 liter), add the following and sweat the vegetables until they begin to soften, about 5 to 10 minutes:

2 tablespoons (25g) olive oil

1 (100g) carrot, diced

2 (100g) celery ribs, diced

1 medium (100g) onion, diced

Add:

4 pounds (2kg) bones, such as chicken, veal, or beef

For bones, look for "chicken backs" in your grocery store.

Cover with water and bring to a slow boil. Add aromatic herbs and spices, such as a few bay leaves, a bunch of thyme, or whatever suits your taste. Try star anise, ginger root, and cinnamon sticks for something closer to the stock used in Vietnamese *Phở*.

Simmer for several hours (two to three for chicken bones; six to eight for thicker and heavier bones). Strain and cool; transfer to fridge.

If you're worried about leaving the stove on, use a slow cooker.

For a bit of overkill, here's what straining a batch of white stock in various ways yields, starting with the coarsest straining and going progressively finer. (I removed the bones and vegetable matter with a ~5,000-micron spider strainer before running the stock through the 500-micron filter.)

500 micron: stuff caught by a chinois or fine strainer

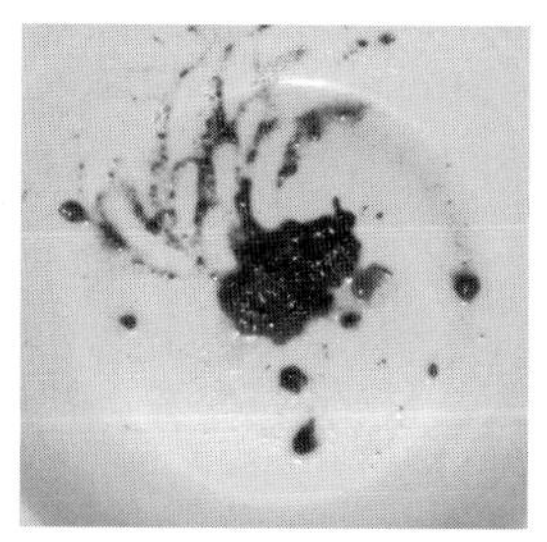

...then 300 micron: stuff caught by a cotton towel

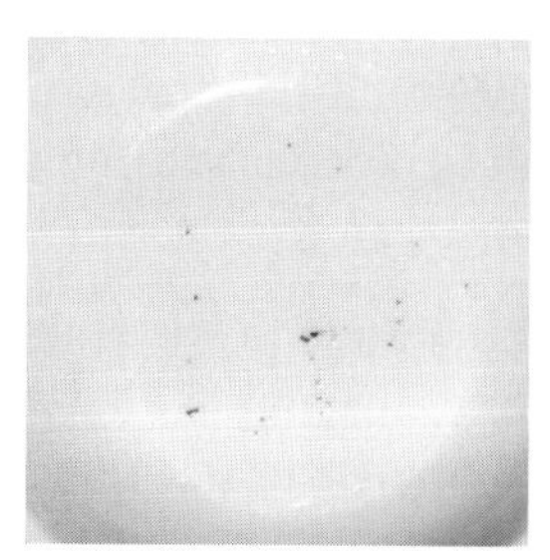

...then 100 micron: stuff caught by a Superbag mesh filter.

Drip-Filtered Consommé

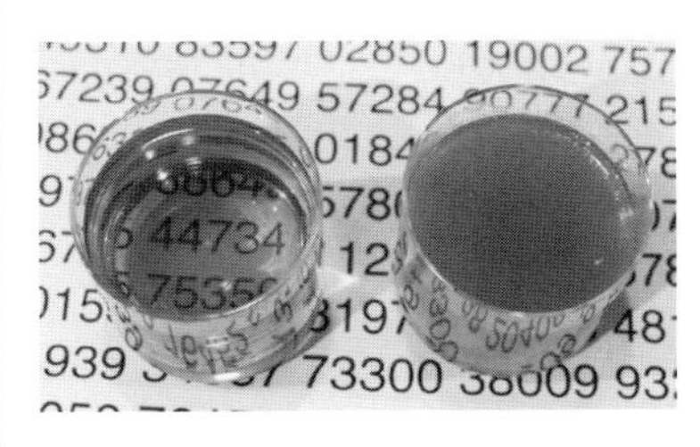

Consommé made via drip-thawing stock (left), compared to the original stock (right) filtered at 100 microns. Note the transparency of the consommé—it looks like filtered apple juice.

To make a drip-filtered consommé, start with a proper stock. The gelatin is a necessary component, because it serves the same function as the egg-white raft in the traditional method.

Once the stock has cooled and gelled (leave it overnight), transfer the gelatin to the freezer and let it freeze solid. As the water in the stock freezes, it will push the impurities into the gelatin. After it's frozen, put the stock into a filter bag or strainer lined with a cotton towel and let it drip-thaw on the counter for an hour, or in the fridge overnight. The filter or towel will hold on to the gelatin, and the gelatin will hold on to the smaller particles.

Make sure the container you freeze the stock in is smaller than the filter bag you use; otherwise, you won't be able to fit the frozen block into the filter.

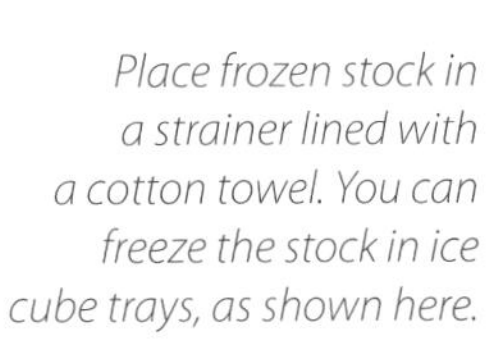

Place frozen stock in a strainer lined with a cotton towel. You can freeze the stock in ice cube trays, as shown here.

After an hour or two, the stock will have thawed, with the consommé in the pan and the cotton holding on to these weird blob shapes of gelatin.

Dave Arnold on Industrial Hardware

PHOTO OF DAVE ARNOLD USED BY PERMISSION OF JEFF ELKINS, *HTTP://JEFFELKINSPHOTO.COM*

Dave Arnold teaches at the French Culinary Institute in New York City, where he instructs students about modern techniques and equipment. He also contributes to the excellent Cooking Issues blog at http://www.cookingissues.com.

How do you get someone to make the mental leap, to think analytically, and to think outside the box, while in the kitchen?

For people who don't naturally think this way, you can't expect them to start organically. You just want to give them another set of tools to work with in the kitchen. So we take something that they take for granted, like cooking eggs, and then break it into a zillion little components. We set up grids where we manipulate single variables. This means that we look at two variables at once in a grid format—for example, time against temperature—and manipulate one variable to see how it affects the other.

One of the classic examples is coffee. The variables are knowable, but why is so much coffee, specifically espresso, terrible? There are plenty of people that have machines that are good enough. It's good to think analytically. If you're messing around with coffee and you're changing x, y, and z, it's the equivalent of standing in front of a big control board with a bunch of dials and then just spinning the dials. To teach someone to make good coffee, you have to teach them how to lock down all their variables and then alter them one at a time. When you're making espresso, most people choose to alter their grinds as their variable. They find that it's easier to lock in the temperature, the dosing, the pressure, and then manipulate grinds. It teaches them how to manipulate variables and think analytically about something.

If we're trying to figure out the variable of temperature with eggs, we'll just do it. I'll use a circulator to cook 10 eggs at very precise temperatures. We'll do it multiple times and we'll crack them and see what the behavior is. Or we'll teach people how to make grids to test two different variables in order to figure out something like the effect of heat on searing meat. We'll set up a tasting grid and they can taste it. I think this helps people to pick up that skill. It's all about control and the ability to observe.

What sort of hardware have you repurposed for the kitchen?

Basically, a chef is going to want to steal anything that can help them heat differently or homogenize or blend differently. Most of what we use that has been repurposed aren't necessarily our own ideas. You can crib things off of other people. Everyone is using liquid nitrogen now, which is fantastic stuff.

Even stuff normally found in the kitchen we just use in a different way. A lot of people are doing interesting work with pressure cooking nowadays. We use ultrasonic cleaners and rotovaps a lot. We've been running some experiments on torches recently. Why do things that are hit with torches taste like torch? I'm beginning to think that it's the component added to gases to make them smell so that you can tell when you have a leak. I think the torch flavor is due to not fully combusting all of the stinky stuff. I wanted to crisp something big, and so I fired up the roofing torch with propane, and it didn't taste bad. I tried to shoot a regular torch through a screen to see whether we could combust any of the torch smell by capturing it on the screen and blowing it through. That also works.

How do you balance experimenting with safety?

Teach yourself as much as you can about the risks involved with any potential new endeavor. The Internet is also good for that, because there are plenty of people who've already hurt themselves. Do a lot of research; read a lot of things. There're a lot of opinions out there, and what one person says may not necessarily be true. It doesn't take too much Googling around to find out that someone has already tried to carbonate something by sticking dry ice in a soda bottle and gotten a bunch of shattered plastic in his face as a result.

You don't want to stifle anyone's creativity or their desire to hack around and do things, because that's the fun of it. But it has to be tempered with a certain amount of base knowledge. Things are dangerous under three circumstances: one, if you don't know the procedure at all. That's what happened to the soda bottle guy. He didn't know the procedures. Two, you're completely frightened of something, a piece of equipment or a knife. If you decide to use it anyway, you're more likely to get hurt. Three, when you become complacent. If you're an inherently cautious person and you don't become complacent, that's the safest way to do these kinds of experiments.

What about the safety of used equipment, such as lab gear?

When I got my centrifuge, we bleached and pressure-cooked any parts that would touch any food. When I got my rotovap, I soaked that sucker in a bleach solution and then in boiling water, and then boiling water and bleach. You have biological contaminants and you have poisonous contaminants—all sorts of contaminants. I feel pretty okay that with stainless and glass I can get rid of most bad inorganic stuff, but you just have to pray that you wash enough to get rid of all the organic stuff. From a biological hazards standpoint, you're worried about prions, you're worried that someone has been blending up cow brains doing Creutzfeldt–Jakob research or something like that. You can't cook it away, they're heat-stable. Then you're counting on mechanical washing.

I'm curious, what do you do with a centrifuge?

A lot of people buy centrifuges because they think they're going to get awesome results with a centrifuge. What you really need to do is borrow someone else's first. All a centrifuge does is separate things based on density.

If you're cooking, you want a lot of product, because you want to serve a lot of people. It's not often feasible. Unilever donated a centrifuge to us, and I had more time just to play around. Now we're doing a lot of things like making our own nut oils, or clarifying things like apple juice, where we're spinning it down to increase our yield. Also, you can blend olives, cured ones like kalamata, and then you spin them. It breaks into three layers. You have the best olive brine ever for a dirty martini, hands down. You have a completely flavorless middle layer you throw away. Then you have a really interesting layer of olive oil from cured olives. That's kind of fun. Expensive, though.

We're taking things into the kitchen that aren't from the kitchen, not just laboratory equipment. There's a whole group of people that make their own chocolates. They use a stone grinder from India that's used to grind dahl. We've taken that, and we're making things that have the textural properties of chocolate, which aren't related to chocolate at all, like ketchup and mustard. Most stuff in the kitchen is going to be equipment-based, but it's not necessarily new technology or lab technology. Sometimes it's just learning new techniques. It's more of an attitude.

I'll give you another example: how are you supposed to cook mushrooms? You're not supposed to soak mushrooms. They always tell you to wipe off your mushrooms.

I usually just do a quick wash. My take has always been that it doesn't actually absorb that much water.

It actually does. Mushrooms are little sponges, but here is the thing: our contention has always been that it's just going to take longer to cook. Which is true. We did a test where not only did we soak the mushrooms in slice form but then crowded the pan—all the things that you're not suppose to do with a mushroom.

The amazing thing was not that it didn't make a difference in cooking them, but that the ones that we had soaked and crowded were *better*. The reason is because while the soaked mushrooms are sitting there giving off their water and stewing in their own juices, they're collapsing. It's no longer a sponge to soak up oil, so by the time all the water had boiled off and they started sautéing they had already collapsed, and they weren't absorbing the oil. The non-soaked mushrooms, at the end of sautéing, had soaked up all of the oil and in fact wanted more oil. The ones that had been soaked hadn't even absorbed all of the oil. Some of the oil was still left in the pan.

So just by normal observation, because we had measured things and were trying to figure out what was going on, we realized that everything that they teach you about mushrooms is wrong. You're not going to measure every time, but you would never pick up on stuff like that unless you were really thinking analytically about what's going on.

I think it's actually the key to a lot of this. I think there is a certain something that drives some people to go to lengths, when other people just kind of shrug their shoulders and end up not being as curious.

Right, and that's why Harold McGee's website is called "The Curious Cook." A lot of it is about curiosity and then after curiosity—and here's where the real geek thing comes in—is the ability and willingness to actually do something about the curiosity. Go the stupid extra length. Just see whether you can do it.

The Easier, Cheaper Version of "The $10,000 Gin and Tonic"

The UK magazine Intelligent Life *did a piece on Dave Arnold, including the "stupid extra length" he went to to make a perfect gin and tonic. Dave explains:*

It was called the $10,000 gin and tonic because there was all this equipment and time, and rotary evaporation, and the PSI measured carefully, and clarifying juices, etc. I was redistilling lime essence to create a clear lime juice so that I could add that to my quinine simple syrup and gin and get the water level exactly where I wanted it and carbonate it. The reason you want it clear is because gin and tonic should be clear, and should have enough bubbles, and the right alcohol content. So I was able to break out every single variable and recombine them exactly the way I wanted.

The idea of the original recipe, *Bottle Strength G 'n T*, was to produce a gin and tonic shot at bottle strength (80 proof). To do it, we distilled lime juice and gin to capture the fresh volatiles from the juice and increase the proof of the gin. We then added acids back to the distillate to recreate the flavor of the lime juice, along with sugar and quinine, the bitter part of tonic water. Why all this? Adding sugar, acid, etc. lowered the proof of the gin. If we wanted to serve bottle-strength gin and tonic shots, we had to raise the initial proof. Plus, distilling the lime volatiles gave us a perfectly clear drink that carbonates well (pulp is a carbonation killer). We no longer serve this version, because it only tastes good at around 0°F / –18° C. Served any warmer than 5°F / –15° C, and it tastes unbalanced; any colder than –9°F / –23° C, and it is painful going down. It was hard to get people to drink quickly enough, when the shots were at the right temperature.

This same technique, when watered down to 15–20% alcohol by volume, produces our perfect G 'n T, and it's much easier to do. I use what we call *simple agar clarification* on lime juice. I can do it in 20 minutes on a camp stove and I don't need the high-end equipment to make it. It's back to a normal cost in terms of equipment, except for a carbonation rig. The good news is that it's very inexpensive to get a real carbonation rig at home. The whole carbonation rig costs well under $200. A single 20-pound tank of CO_2 costs about $20 to refill, and it makes 200 to 400 gallons of seltzer or liquor. Everyone should have one in their house. Everyone.

Clarified lime juice. Squeeze the juice from 10 limes into a container, running through a sieve to remove pulp. Weigh the juice; it should be around 500g. Set aside.

In a pan, create an agar gel using water and agar. Measure out a quarter of the amount of lime juice in water, roughly 125g of water, and create a 10% agar gel, around 12g (this will result in a 2.0% concentration once mixed with the lime juice). Once the agar has melted, remove from heat and pour the water-agar mixture into the container with the lime juice and let it rest for half hour or so, until set.

Once the lime gel has set, use a whisk to break the gel into pieces. Take the whisk and make zigzag slashing cuts; don't actually whisk the gel.

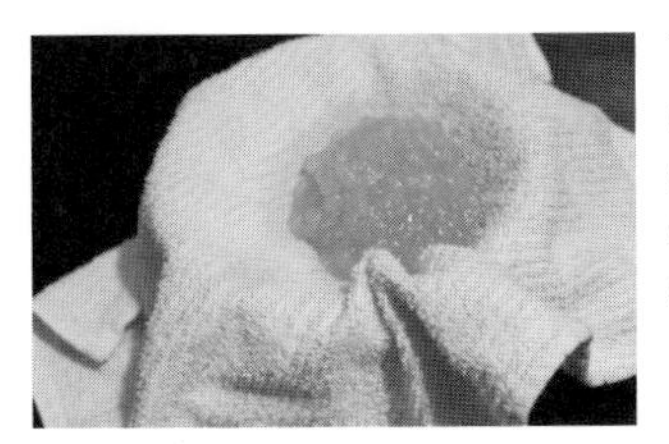

Transfer the broken gel to a cheesecloth (real cheesecloth, not the loose mesh stuff) or towel. Fold the cloth up into a ball.

Hold the balled cloth above a coffee filter and squeeze it with your other hand, massaging it to force out as much liquid as possible. (The coffee filter will catch any small chunks of agar that happen to leak through).

Simple syrup with quinine. Create a simple syrup (2 parts sugar, 1 part water), then add diluted quinine sulfate. Be careful! Quinine in anything other than minute quantities is poisonous! The legal limit is 83 parts per million of quinine, which is 0.083 grams of quinine sulfate per liter of liquid. You will need considerably less than this. Quinine goes from pleasantly bitter to extraordinarily bitter rather quickly. Make a solution of 1g quinine sulfate in 500 ml of water (or gin) and use no more than 40 ml of this solution per liter of finished product and you will be okay. You will probably like less than half that amount.

To assemble:

4 oz gin
2 oz clarified lime juice
Simple syrup with quinine to taste
Salt to taste

Chill in freezer. Carbonate to 40 PSI.

Cream Whippers (a.k.a. "iSi Whippers")

We're all familiar with whipped cream in a can. A *cream whipper* is a reusable version of the can, without the cream, that you fill with cream or whatever else you like. They're a simple yet clever design: pour your contents into the container, screw on the lid, and pressurize using a small, bullet-like cartridge that provides either nitrous oxide or carbon dioxide to the can through a one-way valve.

Cream whippers take their name from their primary purpose: making whipped cream. With a whipper, you can control the quality of the ingredients and the amount of sugar used. Fill it up, store in the fridge, and there's no functional difference between a whipper and the more familiar whipped cream in a can.

The obvious extension is to create flavored whipped cream. Toss some orange zest and maybe a bit of vanilla sugar into a pint of organic cream, and spray away. Try tea-infused cream: steep some Earl Gray in cream and transfer it to the whipper, or go smoky and use Lapsang Souchong. Just remember to strain the tea leaves out before filling the canister of the whipper! You can also spike the cream—make amaretto cream to go on your coffee with 4 parts heavy cream, 2 parts amaretto liqueur, and 1 part powdered sugar.

But the real fun with cream whippers (besides whipped cream fights) is passing other liquids through them. You can whip any liquid or mixture that has the ability to hold air—that is, anything that can be turned into a foam (sometimes called an *espuma* in menu speak), including foamed "waters" flavored like carrots or desserts like chocolate mousse. You can even put pancake batter in a cream whipper (hence the whole "pancakes in a can" thing). Because the contents are ejected under pressure, small, pressurized bubbles come along for the ride and expand, leading to mechanical injection of air into the liquid. This is why cream turns into whipped cream, although the foam that's generated isn't as stable as manually whisked whipped cream.

The most common brand of cream whipper used in the food industry is made by iSi (it's not uncommon to hear a cream whipper simply called an "iSi"). Regardless of manufacturer, basic models run $40 to $60 dollars; cartridges are about $0.50 each in bulk.

Don't use chargers made for BB guns. They aren't food grade.

This might be more than you want to spend upfront for just whipped cream, but if you're a regular user of the canned stuff, the long-term savings alone will make it worthwhile. If you want to play around with textures and flavors in the kitchen, it's downright cheap.

Cream whippers also come in a "thermal" style (i.e., built like a Thermos) that's useful for keeping contents cold if you're working onsite. The thermal versions can't be used in water baths, though, making it harder to do hot foams or to partially poach the contents à la sous vide for egg-based custards.

A few things to keep in mind when working with a whipper:

- Make sure the gasket is properly seated and the threads on the lid are clean when screwing on the lid, unless you want chocolate cake batter, cream, or pancake mix sprayed 10 feet in a random direction.
- Always run your liquid through a strainer (~500 micron is fine) to remove any particulates that might clog the nozzle. You can skip straining things like plain cream, of course.
- When working with heavier batters, you can double-pressurize the canister. After pressurizing with one cartridge, remove it and pressurize with a second one. You'll find that the pressure decreases as you run through the contents, because the airspace in the whipper increases as the contents are ejected.
- If your liquids fail to foam correctly, try adding some gelatin, which provides structure. If you don't mind taking a shortcut, try using flavored Jell-O.

You can also use a whipper as a source of pressure. One technique uses an adaptor from McMaster-Carr to connect the spray nozzle of the whipper to a length of plastic tubing. Fill the tubing with a hot liquid with agar or other gelling agent, let it set, and use the whipper to force-eject the "noodle."

Another thing to try is using a CO_2 cartridge to create "whipper fizzy fruit"—fruit that has been carbonated, giving it a fizzy texture. Try popping grapes, strawberries, or sliced fruit such as apples and pears into the canister and pressurizing it. Let rest for an hour, depressurize, and remove fruit. Not exactly haute cuisine, but fun to do as a party trick. Fizzy raspberries make a great basis for a mixed drink.

Chocolate Mousse

This creates a very light chocolate mousse, almost the complete opposite of the dense chocolate mousse based on agar from Chapter 6 (see page 311).

Heat to a temperature hot enough to melt chocolate (130°F / 55°C):

1 cup (250g) heavy cream

Remove from heat and whisk in to melt:

6 tablespoons (60g) bittersweet chocolate
¼ teaspoon (0.6g) cinnamon

Transfer to whipper canister and chill. Make sure the liquid is completely cold—fridge temp—before using. Otherwise, the cream won't whip.

Pressurize and dispense into serving glasses or on a plate, as desired.

Notes

- *You can dump the canister in a plastic container filled with half ice, half water to chill it quickly.*
- *The cream really does need to be completely chilled. If it's not, instead of getting a light, airy chocolate mousse foam, you'll get a jet of chocolate-flavored heavy cream.*

Foamed Scrambled Eggs

This egg foam is something like a whipped mayonnaise, but incredibly light. Try it with steak and fries. This recipe is based on a recipe by Alex and Aki of http://www.ideasonfood.com *fame.*

Measure out into a bowl:

4 large (240g) eggs
5 tablespoons (75g) heavy cream
½ teaspoon (2g) salt
½ teaspoon (2g) sriracha sauce

Using an immersion blender, thoroughly purée the ingredients. Strain into a nonthermal whipper and screw lid on, but do not pressurize. Place whipper in a water bath at 158°F / 70°C and cook until the mixture is partially curdled, around 60 to 90 minutes. Remove from bath, check that eggs are just partially set, and pressurize. Dispense into small bowls and garnish, or use as a component in a dish.

Notes

Try using the small strainer from a loose-leaf teapot when filtering liquids—it's easier to hold above the container while pouring in the mixture.

30-Second Chocolate Cake

In a microwave-safe bowl, melt:

4 oz (113g) chocolate (bittersweet preferably)

Add and thoroughly whisk together:

4 large (240g) eggs

6 tablespoons (80g) sugar

3 tablespoons (25g) flour

Pass the mixture through a strainer to remove any lumps and to filter out the chalazaes (the little white strands that attach the yolk to the egg white). Transfer to whipper and pressurize.

Spray mixture into a greased glass, ramekin, or whatever microwave-safe container you will cook it in, leaving at least the top third of the container empty. The first time you do this, I recommend using a clear glass so that you can see the cake rise and fall as it cooks.

Microwave for 30 seconds or until the foam has set. Flip onto a plate and dust with powdered sugar.

Powdered sugar is the bacon of the pastry world. It goes well with almost everything and is great for covering up things like tears or holes—in this case, covering up the Nutella filling.

For better-tasting results, try adding Nutella or Fluff: spray a thin layer of cake batter, drop a spoonful of filling into the center, and then spray more cake batter on top of and around the filling.

After cooking, cover in chocolate and do a small loopy white icing thing on the top, and you've got something close to commercial cream-filled cupcakes.

Notes

- *Try spraying a thin layer of the batter onto a plate and cooking that. Peel it off the plate, coat the top with a layer of jam or whipped cream, and roll it up to create a log-shaped chocolate treat.*
- *To be fair, you can do a close approximation without using a whipper. Search online for "microwave chocolate cake." I find that the iSi Whipper version produces a more uniform, spongier cake, though.*

Unbaked: nonwhipped (left) and whipped (right).

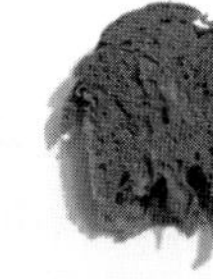

Baked: nonwhipped (left) and whipped (right).

"Cooking" with Cold: Liquid Nitrogen and Dry Ice

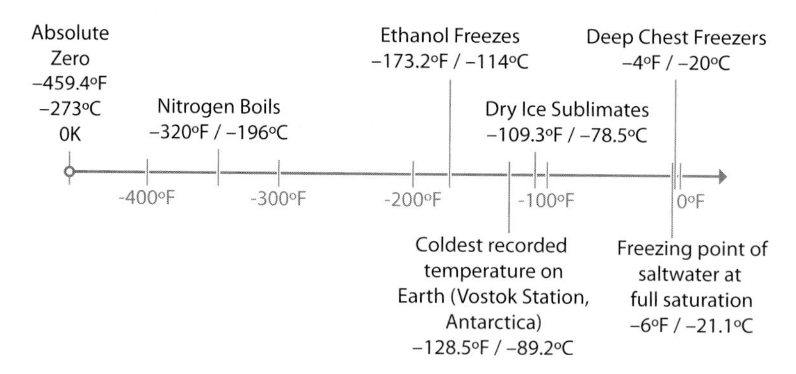

Common and uncommon cold temperatures.

Okay, strictly speaking, cooking involves the application of heat, but "cooking" with cold can allow for some novel dishes to be made. And liquid nitrogen and dry ice can be a lot of fun, too!

If there's one food-related science demo to rule them all, ice cream made with liquid nitrogen has got to be the hands-down winner. Large billowy clouds, the titillating excitement of danger, evil mad scientist cackles, and it all ends with sugar-infused dairy fat for everyone? Sign me up.

While the gimmick of liquid nitrogen ice cream never seems to grow old (heck, they were making it over a hundred years ago at the Royal Institution in London), a number of more recent culinary applications are moving liquid nitrogen (LN_2, for those in the know) from the "gimmick" category into the "occasionally useful" column.

Dangers of liquid nitrogen

But first, a big, long rant about the dangers of liquid nitrogen. Nitrogen is mostly inert and in and of itself harmless, making up 78% of the air we breathe. The major risks are burning yourself (frostbite burn—it's cold!), suffocating yourself (it's not oxygen), or blowing yourself up (it's boiling, which can result in pressure buildup). Let's take each of those in turn:

- *It's cold.* Liquid nitrogen boils at −320°F / −196°C. To put that in perspective, it's further away from room temperature than oil in a deep-fat fryer: seriously cold. Thermal shock and breaking things are very real concerns with liquid nitrogen. Think about what can happen when you're working with hot oil, and show more respect when working with liquid nitrogen. Pouring 400°F / 200°C oil into a room-temperature glass pan is *not* a good idea (thermal shock), so avoid pouring liquid nitrogen into a glass pan. Splashes are also a potential problem. A drop of hot oil hitting your eye would definitely not be fun, and the same is true with a drop of liquid nitrogen. Wear closed-toed shoes and eye protection. Gloves, too. While the probability of a splash is low, the error condition isn't pleasant.

- *It's not oxygen.* This means that you can asphyxiate as a result of the oxygen being displaced in a small room. When using it, make sure you're in a relatively well-ventilated space. Dorm rooms with the door closed = bad; big kitchen space with open windows and good air circulation = okay.
- *It's boiling.* When things boil, they like to expand, and when they can't, the pressure goes up. When the pressure gets high enough, the container fails and turns into a bomb. Don't *ever* store liquid nitrogen in a completely sealed container. The container *will* rupture at some point. Ice plugs can form in narrow-mouthed openings, too, so avoid stuffing things like cotton into the opening.

"Yeah, yeah," you might be thinking, "thanks, but I'll be fine."

Probably. But that's what most people think until they're posthumously (post-humorously?) given a Darwin Award. What could possibly go wrong once you get it home? One German chef blew both hands off while attempting to recreate some of Chef Heston Blumenthal's recipes. And then there's what happened when someone at Texas A&M removed the pressure-release valve on a large dewar and welded the opening shut. From the accident report:

> *The cylinder had been standing at one end of a ~20′ × 40′ laboratory on the second floor of the chemistry building. It was on a tile-covered, 4–6″ thick concrete floor, directly over a reinforced concrete beam. The explosion blew all of the tile off of the floor for a 5′ radius around the tank, turning the tile into quarter-sized pieces of shrapnel that embedded themselves in the walls and doors of the lab... The cylinder came to rest on the third floor leaving a neat 20″ diameter hole in its wake. The entrance door and wall of the lab were blown out into the hallway. All of the remaining walls of the lab were blown 4 to 8″ off of their foundations. All of the windows, save one that was open, were blown out into the courtyard.*

Do I have your attention? Good. End rant.

Okay, I promise to be safe. Where do I get some?

Look for a scientific gas distributor in your area. Some welding supply stores also carry liquid nitrogen.

You'll need a *dewar*—an insulated container designed to handle the extremely cold temperatures. Depending upon the supplier, you may be able to rent one. Dewars come in two types: nonpressurized and pressurized. Nonpressurized dewars are essentially large Thermoses. The pressurized variety has a pressure-release valve, allowing the liquid nitrogen to remain liquid at higher temperatures, increasing the hold time.

Unless you're renting dewars and having them delivered to your location, stick with a nonpressurized one. Small quantities of liquid nitrogen in nonpressurized dewars don't require hazmat licenses or vehicle placarding when properly secured and transported in a private car. It's still considered hazardous material though, because handled improperly, it can cause death. Transportation falls under "material of trades" and it is your responsibility to understand the regulations. For example, New York State defines anything under 30 liters / 8 gallons as a small quantity. (For details, see *https://www.nysdot.gov/divisions/operating/osss/truck/carrier/materials-of-trade.*)

> Standard lab safety protocols for driving small quantities of liquid nitrogen around usually state that two people should be in the car and that you should drive with the windows down or at least cracked.

When it comes to working with liquid nitrogen, I find it easiest to work with a small quantity in a metal bowl placed on top of wooden cutting board. Keep your eyes on the container, and avoid placing yourself in a situation where, if the container were to fail, you would find yourself getting splashed.

Don't sit at a table while working with it. Standing is probably a good general rule to reduce chances of injury. And remember: it's cold! Placing a noninsulated container such as a metal bowl directly on top of countertops, especially glass ones, is not a good idea.

> I once cracked a very nice countertop with an empty but still cold bowl during a demo at a large software company whose name begins with the letter M. I'm *still* sheepishly apologizing for it.

One final tip: when serving guests something straightaway after contact with liquid nitrogen, check the temperature (using an IR thermometer) to make sure the food is warm enough. (As a guideline, standard consumer freezers run around –10°F / –23°C.)

Making dusts

One of the classic "silly things you can do with liquid nitrogen" tricks is to freeze a leaf or a rose and then whack it against something to shatter it. Unlike traditional methods of freezing, liquid nitrogen freezes the water in the plant so quickly that the ice crystals do not have time to aggregate into crystals large enough to pierce the cell walls and destroy the tissue, meaning the leaf or flower won't wilt when thawed.

In culinary applications, you can use this same property to create "dust" from plant material. Lavender flowers, for example, can be rapidly frozen, crushed with a mortar and pestle (which needs to be chilled in a freezer to keep the frozen plant material from thawing), and then allowed to thaw back out. Some chefs have frozen larger items—beets, for example—causing them to shatter in an organic pattern that couldn't be obtained with a knife.

Making ice cream

The standard formula for LN_2 ice cream goes something like this: cream + flavoring + liquid nitrogen + whisking / mixing = 30-second ice cream.

Be sure to take the necessary safety precautions!

While you can make ice cream with a small quantity of alcohol using traditional methods, those versions have only a mild flavor brought by the alcohols, which are used more as extracts or flavorings than as actual components of the body. With liquid nitrogen, however, you can make a scoop of ice cream with an entire shot of alcohol. Calories are no longer the biggest problem with this type of ice cream; hangovers are.

As with anything you make with liquid nitrogen that's served cold, check to make sure that it's not *too* cold before serving it. An IR thermometer is a handy tool for this. Spot-check your ice cream, and if it's too cold, let it warm up to normal freezer temperature.

Cocoa-Goldschläger Ice Cream

In the metal bowl of a stand mixer, mix:

1 cup (256g) milk
1 cup (240g) heavy cream
¾ cup (180g) Goldschläger (cinnamon liqueur)
¼ cup (80g) chocolate syrup
½ cup (80g) bittersweet chocolate, melted
2 tablespoons (25g) sugar
½ teaspoon (1g) salt
½ teaspoon (1g) cinnamon

Taste the mixture to check the balance (try not to drink it all at this point), and adjust accordingly. Once frozen, the mixture will not taste as strong, so an overly strong mixture is desirable.

Turn your stand mixer on and (carefully! with goggles and gloves!) slowly pour in liquid nitrogen. I find it takes about a 1:1 ratio of mixture to liquid nitrogen to set the ice cream. If you don't have a stand mixer, you can also do this in a metal bowl and stir with a whisk or wooden spoon.

Note

- *To melt the chocolate, microwave the milk and then add the chocolate to the hot milk. Let it rest for a minute, so the chocolate warms up, and then mix to combine. You can nuke the chocolate directly as well, but I find it easier and less likely to burn doing it this way.*

Playing with Dry Ice

Dry ice—solid carbon dioxide—is easier to work with than liquid nitrogen. For one thing, it's solid, so you don't need specialized equipment to handle it. A Styrofoam cooler or even a cardboard box is sufficient. And secondly, it's much more readily available. Just make sure to ask for *food-grade* dry ice!

A few words of warning: like liquid nitrogen, dry ice expands into a much larger volume as it sublimates. Do not store dry ice in a sealed container. Also, dry ice and ethanol form a wet slurry that is *very* dangerous. It's not cold enough to generate the Leidenfrost effect, the phenomenon where a liquid generates a vapor barrier around a much-hotter item. Dry ice and ethanol can wick through clothing and stick to skin.

Besides sticking a chunk of dry ice in a cup of coffee and pretending not to notice while drinking from it (the chunk will sink to the bottom), what else can you do?

Quick-freeze berries. Industry lingo for this is *IQF* (individually quick frozen), in which large blast freezers rapidly freeze individual peas, raspberries, and chicken breasts. You can toss some dry ice in a Styrofoam cooler and mix in a roughly equal amount of berries or veggies, wait until the dry ice has sublimated away, and then bag 'em and stick them in the freezer.

Make ice cream. Works just like liquid nitrogen, only alcoholic flavors probably won't set quite as nicely. Take your food-grade dry ice, place it between two towels, and give it a few whacks with something like a rubber mallet or the back of a frying pan to create a powder. Whisk the powder into the ice cream base until set.

As with LN_2 ice cream, it'll take a little less dry ice if you start with a base already at freezing temperatures.

Create "fizzy fruit." Drop some grapes, bananas, strawberries—really, any moist fruit—into a pressure cooker, toss in some dry ice, and slap on the lid. As the dry ice sublimates, the chamber of the pressure cooker will hold the carbon dioxide (and bleed off any over-pressure amount), and the fruit will absorb some CO_2. Wait 20 to 30 minutes, release the pressure, pop off the lid, and munch away.

DIY "Anti-Griddle" Using Dry Ice

If a griddle cooks foods by adding heat, it should follow that an anti-griddle "cooks" foods by removing heat. PolyScience, known by many chefs for its sous vide recirculating units, makes a product that does exactly that: its anti-griddle cools down whatever you put on the griddle surface.

You can make a do-it-yourself version by using dry ice, ethanol, and a sheet of stainless steel. You'll need a solid chunk of stainless steel (plan on ordering a piece from a distributor such as McMaster-Carr). I have a 6″ × 6″ / 15 cm × 15 cm slab that normally lives in my freezer; it's handy for those times when you want to cool down a small item quickly. Here's how it works:

1. Rig up a bed of crushed dry ice. Try using a cookie sheet placed on top of a wooden cutting board. The cookie sheet will hold the dry ice/ethanol slurry, and the cutting board will provide insulation between the extremely cold cookie sheet and your countertop. Alternatively, if you have the lid to a Styrofoam container, using the inside, indented part can serve both purposes.

2. Pour a small amount of ethanol onto the bed of crushed dry ice—enough to create a level top. (You can use rubbing alcohol or cheap vodka.) The ethanol will remove any air gap between the pieces of dry ice and the stainless steel griddle, and it won't cause the dry ice to froth in billowy clouds like water would.
3. Plop the square of stainless steel on top of the ethanol-topped dry ice. It should be a complete contact fit, just like a heat sink on top of a CPU.
4. Spray or coat the top surface of the stainless steel with a nonstick cooking spray, butter, or oil.
5. Drop your food to "cook" on the surface, smoothing it out into a pancake shape if desired. After 10 seconds or so, use a spatula to flip it and set the other side. As a starter, try whipping some cream up in a bowl with a bit of sugar and chocolate syrup. Try using a cream whipper with flavored foams or the chocolate mousse recipe from earlier in this chapter.

Fun with Hardware

Windell Oskay and Lenore Edman: Electrocuted Hot Dogs and Apple Pie

*Windell Oskay and Lenore Edman blog about DIY and open source hardware projects on their website (*http://www.evilmadscientist.com*) and occasionally dive into the food arena with their "Play with your food" posts.*

You are unique in the sense that you do some really extreme stuff both in hardware and in food. How much of what you do is happenstance versus planned?

Lenore: Most of the time we are brainstorming. Any topic is fair game for a project. We may have talked about doing an "Apple" apple pie long, long ago, and then we're at the cooking store and looking at all of the different round-cornered square pans thinking, "This might work," or "Oh, wow! This is perfect!"

Windell: That also brings up one of the most important methods we have for solving problems, which is to put it on a list and wait for a really long time until we think of a solution.

How do you know if a project is going to work?

Lenore: You try it. The apple pie went through several iterations before we figured out a way to make it aesthetically pleasing.

Windell: There were a couple of different pie shells. We're not running a commercial kitchen. It actually takes us time to make a pie shell and let it chill. We don't have an extra one just sitting ready.

Why do you do these projects?

Lenore: That's a good question. Why do you breathe?

Windell: What else are we supposed to be doing? We like to do cool stuff. We have a chance to, so why not?

Lenore: It's rewarding to see other people enjoying our projects, so publishing them is rewarding. And you've got to eat, right? So you might as well eat something interesting. You have to wear clothes.

Windell: So wear something interesting.

If you knew the world were going to end tomorrow, what would you want your last meal to be?

Lenore: I don't know… We eat a lot of good food, so it's not like there's one thing in particular that I would regret not having eaten.

Windell: If the world were going to end tomorrow, the last thing I think I would do is sit down and have a nice big comfy meal. It doesn't really seem like that's going to happen.

Lenore: I guess if you knew you couldn't do anything about it, that might be a fine way to end. Sit down and have a big comfy meal. It seems unlikely, but…

Windell: Water and hard tack in the bunker.

Electrocuted Hot Dog

Since heat is a form of energy (heat = kinetic energy of molecules in a system), adding energy to a system can cause it to heat up, which is why a hot dog gets hot when electricity runs through it. (Hot dogs happen to be made of materials—proteins, fats, a little bit of salt—that are conductive enough for this to work.)

But the potential for killing yourself on a live wire is high enough that it's not even funny to joke about doing it. If you really want to electrocute your dogs, search *http://eBay.com* for "Presto Hotdogger."

Visit *http://www.evilmadscientist.com/article.php?story=hotdogs* for more information.

P.S. LEDs light up when "plugged in" to a hot dog!

PHOTOS OF ELECTROCUTED HOT DOG USED BY PERMISSION OF WINDELL OSKAY

"Apple" Apple Pie

You too can make an *Apple* apple pie. Lenore and Windell used their laser cutter and a square springform pan, but with care, you can use a knife to cut the logo and a square glass pan to bake the pie. (If you're not a purist shooting for an edible replica of a Mac Mini or Apple TV, a standard round pie will taste just as good.) For details, see *http://www.evilmadscientist.com/article.php?story=ApplePie*.

PHOTOS OF APPLE PIE USED BY PERMISSION OF LENORE EDMAN

Fun with Hardware

Cooking with (a Lot of) Heat

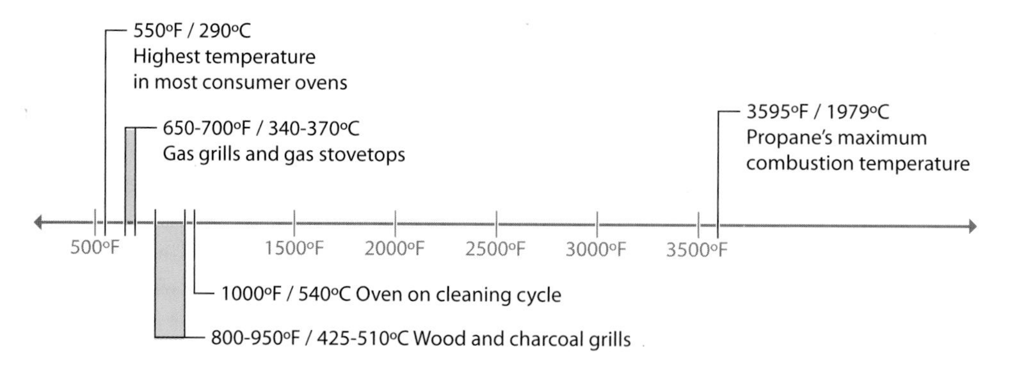

Common and uncommon hot temperatures.

If cooking at 400°F / 200°C produces something yummy, surely cooking at 800°F / 425°C must produce something twice as yummy.

Well, okay, not quite—and by now, hopefully your mental model of how heat is transferred to food and the importance of time and temperature for gradients of doneness should have you slamming this book shut while muttering something about software engineers not understanding hardware. (Guilty as charged.)

But there are some edge cases—just as with "cooking" with cold—where extremely high heat can be used to achieve certain effects that are otherwise difficult. Let's take a look at a few dishes that can be made by transferring *lots* of heat using blowtorches and high-temperature ovens.

Blowtorches for crème brûlée

Blowtorches can be used to provide very localized heat, enabling you to scorch and burn just those parts of the food at which you aim the flame. Torching tuna sushi, roasting peppers, and browning sous vide–cooked meats are all common uses, but creating the sugary crust on crème brûlée is the canonical excuse for a blowtorch in the kitchen. You can also use a blowtorch to prerender the fatty side of meats—try scoring and then torching the fatty side until it begins to brown before roasting.

When it comes to buying a torch, skip the "gourmet" torches and head to a hardware store to pick up a propane blowtorch—not a MAPP gas one, though. The smaller torches sold by kitchen specialty shops burn butane and work okay, but they don't pack the same thermal punch as the hardware-store variety, which have larger nozzles and thus larger flames.

Quinn's Crème Brûlée

Prepare six ramekins for baking by placing them in a large glass baking dish; set aside. Preheat oven to 325°F / 160°C.

In a bowl, separate out five large egg yolks, saving the egg whites for some other dish (see the section on egg whites on page 252 in Chapter 5 for suggestions). Whisk the egg yolks until light and frothy; set bowl aside.

In a saucepan, measure out:

2 cups (475g) heavy cream

½ cup (100g) sugar

Heavy cream and whipping cream are essentially the same thing in the United States. Heavy cream usually has a slightly higher percentage of fat while whipping cream typically has a stabilizer such as carrageenan added, but you can usually use either one regardless of what is called for.

Cut a vanilla bean lengthwise and use the edge of a spoon to scrape out the seeds. Add both seeds and bean to saucepan. Set the burner to medium heat and cook the cream, sugar, and vanilla for 10 minutes, stirring continuously. Meanwhile, in a separate pot, bring to a boil enough water to partially fill the glass baking dish holding the ramekins.

After the cream mixture has been cooked for 10 minutes, fetch out the vanilla bean and discard it. Strain the mixture through a ~400 micron filter (cheesecloth works fine) into a measuring cup or other container that's easy to pour from.

Set the bowl with the egg yolks on the counter, where you can whisk the yolks with one hand and hold the saucepan with the other. Slowly drizzle the hot cream mixture into the egg yolks, whisking the entire time to prevent the hot cream from cooking the egg yolks. Too slow is okay; too fast, and you'll end up with scrambled eggs. (Sweet, tasty scrambled eggs, to be sure.)

Ladle the mixture into the six ramekins, taking care to not transfer any foam that you may have whisked up. (The foam will float and set on top of the brûlée.) Add the boiling water into the baking dish—enough to reach halfway up the sides of the ramekins—and transfer to oven.

Bake until the centers of the custards jiggle just a little when shaken, about 30 to 35 minutes. They should reach an internal temperature of 180°F / 82°C. Remove ramekins from baking dish and chill in fridge until cold, about three hours. (You can store them longer, of course.)

You can create a quick work surface for blowtorching by flipping a cookie sheet upside down and setting the ramekins on top.

Once cold, sprinkle a thin coating of sugar over the top of the custard. Using a blowtorch, melt and caramelize the sugar, sweeping the flame slowly across the surface until you're happy with the color and appearance. Keep in mind that darker sugar will be more bitter; also make sure to at least melt all of the sugar, as otherwise the granulated, unmelted sugar will give an odd mouth-feel.

Transfer ramekins to fridge and store for 10 minutes to allow the sugar to cool; then serve. You can hold the torched brûlée for up to an hour before the sugary crust begins to get soggy.

Note

- *Try infusing other flavors into the cream as you cook it, such as orange, coffee, cocoa powder, or tea.*

You can "upgrade" Bananas Foster—a simple and tasty dessert where the bananas are cooked in butter and sugar, spiked with rum, and then served over vanilla ice cream—by sprinkling sugar on the cooked bananas and then using a blowtorch to caramelize the sugar. To create a work surface, flip a cast iron pan upside-down, line it with foil, and set the bananas on that.

Practice using a blowtorch by melting sugar sprinkled on a sheet of aluminum foil on top of a metal cookie sheet or cast iron pan. Don't get the flame too close; this is the most common mistake when cooking with a blowtorch. The blue part of the flame is hottest, but the surrounding air beyond the tip will still be plenty hot. You'll know you're definitely too close when the aluminum foil begins to melt—around 1220°F / 660°C.

High-heat ovens and pizza

A serious—some might even say OCD—discussion of pizza is clearly a must-have for a cookbook for geeks. I've tried to restrain myself from dwelling too much on pizza, having already given it plenty of airtime in Chapter 5, but it covers so many variables in cooking: flavor combinations, Maillard reactions, gluten, fermentation, temperature. We've covered the first four elsewhere in the book, but we haven't yet talked about temperature and pizza.

If you want to make a crispy thin-crust pizza, a high-heat oven is critical. It takes a sufficiently hot environment to set the outer portions of the pizza dough quickly enough to create the characteristic crispiness and flavors. How hot is hot? The coldest oven I've found acceptable for flat-crust pizza was a gas-powered brick oven at 550°F / 290°C, where the pizza was dropped onto the brick floor of the oven.

The better flat-crust pizza I've had is cooked either in wood-fired brick ovens or on a grill over wood, at 750°F / 400°C, with parts of the oven pushing 900°F / 480°C. For comparison, my local normal "thick-crust" pizza place runs its oven at 450°F / 230°C in the winter, 350°F / 175°C in the summer. (The oven can't be run any hotter in summer without the kitchen becoming unbearable.)

By trying various temperatures, I've found 600°F / 315°C to be the lower limit for getting a crispy, flavorful crust. At 700°F / 370°C, the crust becomes noticeably better. And at 950°F / 510°C? It takes 45 seconds to cook a pizza. But how can you get these temperatures? Most of us don't have ovens that normally reach 950°F / 510°C, let alone 700°F / 370°C, and few of us have brick ovens, either. What's a thin-crust-pizza-loving geek to do? If only there were a flow chart for this...

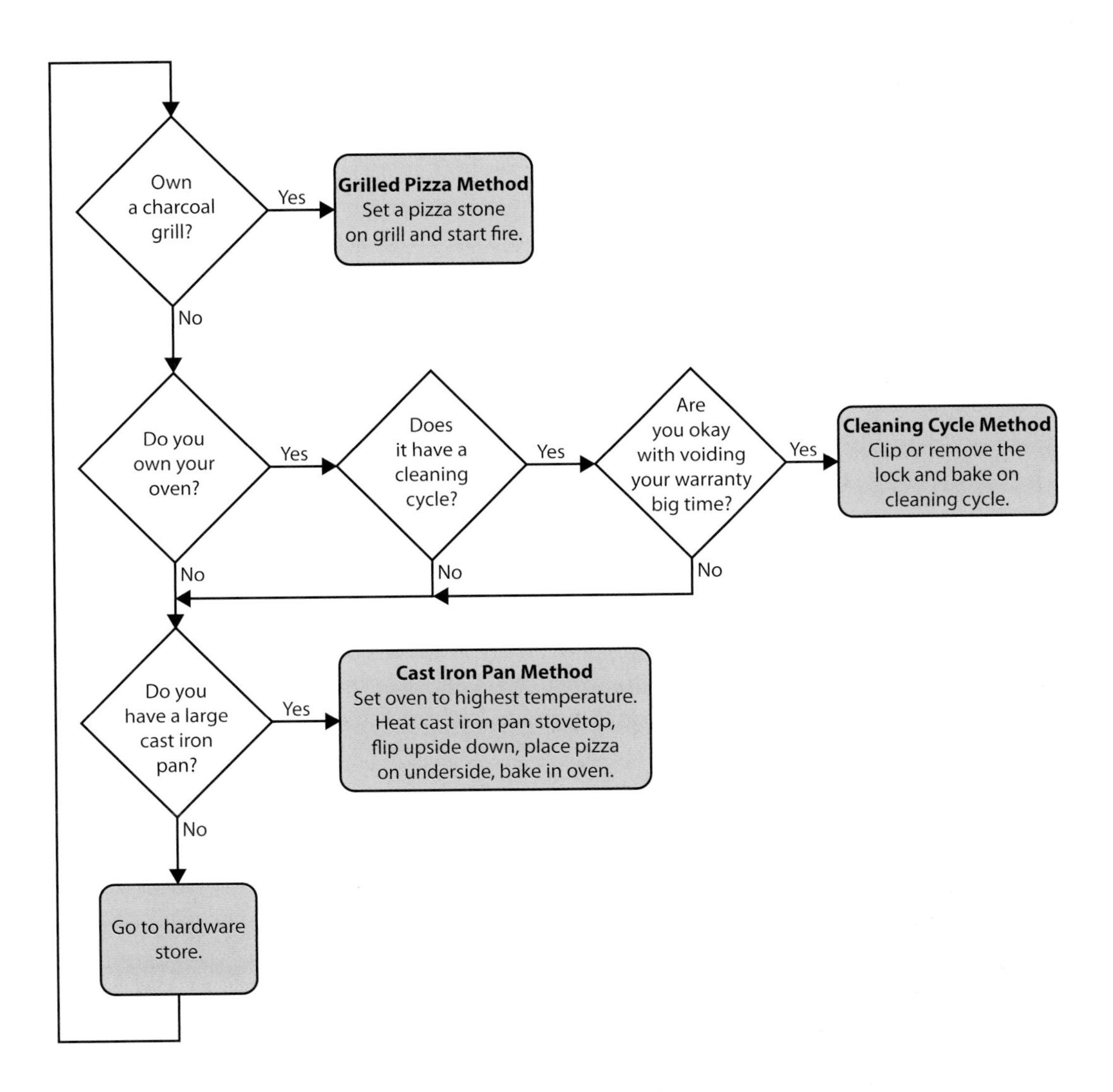

Decision tree for how to cook a pizza.

Fun with Hardware

High-Heat Methods for Pizza

Charcoal or wood grill method. This is by far the easiest method. Grills fueled by charcoal or wood get hot, easily up into the 800°F / 425°C temperature range. (Propane grills tend to run cooler, even though propane itself technically burns hotter.)

See the No-Knead Pizza Dough recipe and accompanying pizza-making instructions on page 238 in Chapter 5.

Place a pizza stone on top of the grill and light the fire. Once the grill is good and hot, use a pizza peel (a piece of cardboard works just as well) to transfer the pizza with toppings onto the grill. Depending upon the size of your grill and the size of your pizza, you might be able to cook the pizza directly on top of the grill, sans stone—give both a try!

Superhot cast iron pan method. What if getting a grill isn't an option for you, as is the case for many apartment dwellers? There are still a few ways left to get up to sufficiently hot temperatures. While most consumer ovens reach only 550°F / 290°C, both the oven's broiler and the stovetop can reach higher temperatures. Leave an empty cast iron pan on a burner at full throttle and it'll reach 650°F / 340°C in 5 or 10 minutes. And the infrared radiation from a broiler is even hotter.

Preheat oven to 550°F / 290°C, or as hot as it goes.

Superheated cast iron under broiler.

Heat up cast iron pan on stovetop at maximum heat for at least five minutes.

Place cast iron pan upside down in the oven under a broiler set to high and par-bake the pizza dough until it just begins to brown, about one to two minutes.

Transfer dough to cutting board and add sauce and toppings. Transfer back to cast iron pan and bake until toppings are melted and browned as desired.

If you don't have a broiler, you can try a doubled-up cast iron pan approach:

Doubled-up cast iron.

Heat up two cast iron pans on maximum heat.

Par-bake the dough, flip it onto a cutting board, add toppings, and return it to the hot cast iron pan.

Cover the first cast iron pan with the second one, preferably using a larger pan so that it doesn't touch the pizza toppings.

Cleaning cycle method (a.k.a. "oven overclocking"). As we've discussed, one of the key variables for good thin-crust pizza is an extremely hot oven. Consumer ovens just don't get hot enough; 550°F / 290°C is still a good 150–200°F / 80–110°C away from where the "real" thin-crust pizzas are cooked. If only there were a way to hack an oven to get it that hot! It turns out that there is, but it's dangerous, voids your warranty, and, given that the alternative ways of getting this kind of heat are far, far easier, is really not worth doing. Still, for the sake of my readers, I tried this method, conceived by Jeff Varasano. (See the interview with him on 236 for details.)

Ovens get a lot hotter—a lot, lot hotter—when they run in the cleaning cycle. The problem is that consumer ovens mechanically lock the door, preventing you from slipping a pizza in and out at those temperatures, and leaving a pizza in for the entire cleaning cycle will result in a most unpleasant burnt taste, to say the least.

Cut or remove the lock, however, and ta-da! You've got a superheated oven. After a bit more fiddling and testing, I had an oven that I measured at over 1,000°F / 540°C. The first pizza we tried took a blistering *45 seconds* to cook, with the bottom of the crust perfectly crisped and the toppings bubbling and melted.

However, the middle of the pizza—the top portion of the dough and the bottom portions of the sauce—never had a chance to cook, so the 1,000°F / 540°C pizza wasn't quite right (too hot). Another attempt at around 600°F / 315°C resulted in the opposite outcome: the pizza was good, but it didn't capture the magic of the crispy thin crust and toasty-brown toppings (too cold). Around 750–800°F / 400–425°C, however, we started getting pizzas that were darn good (just right).

Ovens aren't designed to have their doors opened when running in the cleaning cycle. Honestly, I don't recommend this approach. I broke the glass in my oven door and had to "upgrade" it, although it is cool to have bragging rights to an oven sporting a piece of PyroCeram, the same stuff the military used for missile nose cones in the 1950s.

There's also the issue of how hot the surrounding countertop and cabinetry can get. Commercial stoves are designed for these sorts of temperatures and as a result require a large air gap between the appliance and any combustible materials. Given that an upside-down cast iron pan under a broiler or a wood-fired grill turn out delicious flat-crust pizzas, I'm afraid I have to recommend that you skip the oven overclocking, even if it is fun.

Nathan Myhrvold on Modernist Cuisine

PHOTO USED BY PERMISSION OF NATHAN MYHRVOLD

Nathan Myhrvold, formerly CTO of Microsoft, is among many things an avid cook. He has been working on a book covering the techniques of modernist cuisine.

Tell me about your background with food and how you came to be so interested in it.

I've been interested in food as long as I've been alive. When I was nine years old, I announced to my mother I was going to cook Thanksgiving dinner. I went to the library, got a bunch of cookbooks, and I did. Amazingly, she let me do it, and even more amazing, it worked out!

In 1995, while I was working at Microsoft as a Senior Vice President, I decided that I wanted to go to cooking school. I took a leave of absence and went to a school in France, *L'école de la Varenne*. I went through an intensive professional program. After retiring from Microsoft, I started my own little company, but I'd been interested in food and so decided to write a book. There were lots of big, thick books on cooking, teaching you how to do classical cooking, but there was no modern technique within those books; they were all about the techniques of the past. I got the notion that there really was an opportunity to write a book about modernist cuisine—something that would be encyclopedic for the techniques of modern cuisine.

If I didn't do it, it's not clear that anyone else would, at least not for a very long time. I decided that this was my way to make a contribution to the food world. I could create a book many years sooner than anybody else because of the time, energy, and money involved. It could do something unique in terms of bridging the gap between the understanding of science and the practice of cooking in an accessible way.

What's your definition of modern cooking? The term that would come to many people would be molecular gastronomy.

I deliberately don't use that name. The term that I'm using is *modernist cooking*. I call it modernist because it's analogous to what modernist architecture and modern art did in that it is a somewhat self-conscious attempt to break with the past. It has all of the intellectual hallmarks of modernism.

That happened 100 or 50 years ago in art and architecture but not in cooking. There are chefs who take offense to it if you call it *molecular gastronomy*. It's not a terrible name per se, but it means so many different things to different people. Modernist is a more inclusive term.

Can you give me an example of something that's surprised you in studying these techniques?

There's a cooking technique called *confit* that means "preserved" in French. You cook the meat in oil or fat at a relatively low temperature for a long period of time, like 8 or 12 hours. Any chef would tell you that confit is a cooking technique that involves cooking in fat, which has a characteristic effect on the meat.

One day we were discussing this, and I said, "How can this possibly work? How can cooking meat in oil actually change the meat? That makes no sense to me at all. The molecules are actually too big to penetrate into the meat. It's got to be on the outside and so on and so forth."

So we did a bunch of experiments, and it doesn't really have the effect that you would think. If you steam meat without any oil and you put oil in at the end, you can't actually tell the difference.

Presumably you can't do it just in a water bath with no fat.

We did that, too. You can't tell the difference! You can tell the difference if you cook it at a different temperature or for a different period of time. But if you're cooking at the same temperature and time, whether it's sous vide or steamed or cooked confit, you really can't tell the difference afterward. That was a big shock to us.

There's a bunch of other things that have been quite surprising in determining how techniques work. People will frequently drop meat into ice water to stop the cooking. It's called *shocking*.

Suppose you're cooking a big roast or something that's got some thickness to it. A lot of books will say take it out and then plunge it in ice water to really stop the cooking. It doesn't work at all! The temperature at the core of the meat will not be affected by you dumping it in ice. You will cool the whole thing by dumping it in ice water, but it's not actually going to affect the maximum temperature the core reaches.

Heat and cold "travel" at the same speed. It's not exactly correct, but if you think about a wave of heat going from the outside in, shocking it is going to put a wave of cold, a "negative" wave of heat. But it doesn't go faster, and the hot wave that started before will hit the center before the cold wave does.

Wow, that makes a lot of sense. Are there other examples of processes that you've discovered that apply to the way that most people cook on a day-to-day basis?

One of the things that we've spent a bunch of time on in the book is explaining the role of humidity in cooking. Most food is wet. When you heat wet things, they give off water and that takes a tremendous amount of energy to do. The rate at which the water evaporates depends on what the humidity is.

If you cook something in Aspen in the winter when the humidity outside is really low, and you cook that same thing in Miami in the summer when the humidity is very high, you actually get radically different results. It can make a 10 degree difference in the temperature that the food is experiencing, particularly at the onset.

We went through a whole bunch of examples like this. It turns out that humidity is a huge factor in how cooking actually happens. A convection steam oven controls the humidity, and that's its huge advantage. One of the advantages of sous vide is you seal the food up in a plastic bag where there is no variation in humidity. But if you're cooking out in the open air, humidity actually makes a big difference. That's one of the reasons that people don't have their recipes turn out quite like they thought.

Is that something that's important to absolutely every cook in America? I can't tell you that it is. I think it's kind of cool; it certainly will matter to professional chefs. Every chef has had the situation where they try the recipe in the book and it doesn't work, or the chef travels and the food doesn't quite turn out right. This is one of the reasons. If you're not controlling humidity, it's a free variable, and it will make a big difference.

People don't generally understand how much energy it takes to boil water. This dramatically affects cooking. If you just look at the latent heat of vaporization of water, it takes four joules of energy to move a gram of water one degree Celsius, 400 joules to take it from just above freezing up to the edge of boiling, and 2,257 joules to boil it. That's why steam engines work. All kinds of things are driven off this one fact.

How do you think what you've learned will change the approaches of chefs and amateur cooking enthusiasts?

What we're hoping to do is enable chefs to use a broad range of techniques to make the kinds of food they want to make. Right now there is a set of chefs who are using these very modern techniques. There are a lot of others who don't.

It's very hard to learn all of this stuff. We're hoping that we can give chefs and amateurs an accessible way to understand how it works. If we can do that, I think that we can really make a difference in how folks cook. That's not world peace; it's not solving global warming or something like that, but it is something that, within the cooking world, I think people are going to find tremendously exciting and empowering.

Any parting words of wisdom that you would give somebody learning to cook?

Learning to cook is a wonderful thing to do and I highly recommend it to folks. The message in a lot of recipes is, "Don't worry about how it works, just do this, this, and this, and the right thing will happen."

When it works, that's okay. When it doesn't work, you don't really know why. I always feel cheated when that's the case. I want to find out why. I'm still learning how to cook. I think even the best chefs in the world are still learning how to cook, and it's that learning and that exploration that makes it interesting.

APPENDIX

Cooking Around Allergies

I LOVE THE CHALLENGE OF COOKING WITH CONSTRAINTS. With allergies, the challenge is to prepare a meal with a certain set of ingredients considered off-limits.

Food allergies are caused by an immune system response to certain types of proteins. In some individuals, the immune system misidentifies certain proteins as harmful and generates a histamine reaction in response to them. Immune reactions can occur within a few minutes to several hours of ingesting the offending food item. Minor reactions include a tingling sensation on the tongue or lips, itchy eyes, runny nose, or skin rashes lasting from a few hours to a day. More extreme reactions include throat constriction, nausea, vomiting, diarrhea, or coughing. Oh, and death.

If you ever encounter a reaction that involves tongue swelling, throat constriction, or restricted breathing—hallmarks of an anaphylactic reaction—call 911 and get to a hospital *immediately*, because the swelling can increase to the point where it cuts off the airway. Those who know that they have particularly strong allergies will often carry an Epipen, a small pen-sized medical device that auto-injects epinephrine to control the allergic reaction. (The injection buys 15 to 20 minutes of time to get to a hospital for further care.)

Since an allergy is a response to a particular protein in food, not the food itself, and because some types of proteins denature below the temperature at which the foods containing them are cooked, certain allergies apply only to uncooked foods. Your guests will be able to tell you their particular constraints.

When shopping for a meal to cook for someone with an allergy, be sure to read the labels on any packaged goods you consider. Also, be careful if you are reusing components or sauces from previous meals, because things like soy and nuts can show up in unexpected places. When in doubt, pick recipes with fewer ingredients to avoid unexpected surprises.

Chef Card

If you have serious food allergies, consider creating a *chef card* that you can hand to a waiter when dining out. A chef card is a small, index-sized card that communicates your allergies explicitly, quickly, and clearly. One chef I know commented that "they're very helpful. It's nice when a customer with allergies gives a server one, so they can bring it in the kitchen and I can read it out to the whole staff."

WARNING! I am severely allergic to ____________________

In order for me to avoid a **life-threatening reaction**, I **must avoid** all foods that contain these ingredients:

Please ensure that my food does not contain any of these ingredients, and that any utensils and equipment used to prepare my meals, as well as prep surfaces, are thoroughly cleaned prior to use. **THANK YOU for your cooperation.**

© 2006, The Food Allergy & Anaphylaxis Network, www.foodallergy.org.

Chef Card, courtesy of the Food Allergy & Anaphylaxis Network; for a customizable and printable version, see *http://www.foodallergy.org/page/chef-card1*.

Also, check how sensitive your guests are to their allergies. If they are especially sensitive, you will need to be particularly diligent to avoid cross-contamination while working in the kitchen. It's probably best to avoid using any allergen-containing item in the entire meal, but if a guest has an allergy broad enough that you elect to cook that person a special side dish, you should treat the allergens as you would raw meats: separate them out from the safe foods, and wash *all* items that will come in contact with that side dish (preferably in a dishwasher, as sponges can harbor enough traces to cause cross-contamination).

A friend of mine has learned that cross-contamination of gluten can occur even if she butters her bread and then uses the same knife to slice a pat of butter to drop into a pan; the few micrograms of bread carried back onto the butter knife are enough to trigger an allergic reaction in her child. This is a really extreme case, but do check with your guests about how sensitive their allergies are.

Avoiding cross-contamination can be difficult, because it might occur in many places you'd never think it would. For example, if you're cooking both rice noodles and regular pasta, the residual gluten left on your strainer after running the regular pasta through it might be enough to contaminate the batch of rice pasta. Selecting your recipes with care can help you avoid some of these problems. Again, everyone reacts differently, so your level of vigilance should be adjusted as necessary to match your guests' needs.

Substitutions for Common Allergies

So, you've just found out that someone you're cooking for is allergic to an ingredient in your favorite family dish. What to do?

This section includes a number of suggestions for ingredient substitutions for the eight most common allergies, based on information from Kristi Winkels's website, Eating with Food Allergies (*http://www.eatingwithfoodallergies.com*). Visit her website for additional suggestions and recipes tailored to those with allergies.

This list contains many of the common ingredients and foods to avoid, but you should still check any questionable ingredients with your guests.

Dairy Allergies

Ingredients to avoid

Casein, whey, whey solids, buttermilk solids, curds, milk solids, lactalbumin, caseinate, sodium caseinate.

Foods commonly containing dairy

Milk, buttermilk, chocolate (milk and dark), hot chocolate, "nondairy" creamers, baked goods, spreads including butter and many margarines (even some that say "nondairy" on the label), cheeses, yogurts, frozen yogurts, frozen desserts such as ice cream, sherbets, some sorbets, whipped toppings.

Substitutions

For milk

Soy, rice, potato, almond, oat, hemp, and coconut milk are all possible substitutes for cow's milk. If you aren't dealing with a soy allergy as well, soy milk is a good option; it tastes pretty good and, when fortified, contains roughly the same amount of calcium and vitamin D (two important nutrients, especially for children). Rice milk is also often fortified and, like soy milk, can usually be found at the regular grocery store. Potato milk is available in specialty food stores in powder form.

For margarine

When searching for a dairy-free margarine, be sure to examine the product labels carefully and make sure the ingredient list does not contain "milk derivatives." Also bear in mind that most "light" margarines are not suitable for baking. Look for Earth Balance Light and Fleischmann's Unsalted Margarine brands.

For yogurt

If you're a yogurt fan, check out soy yogurt or coconut milk yogurt. Try using it as a dip for fruit, or buy plain and use it to make a creamy salad dressing.

Egg Allergies

Ingredients to avoid

Albumin, globulin, lysozyme, livetin, silici albuminate, Simplesse, vitellin, meringue, ingredients containing the word "egg" such as egg white, ingredients that begin with "ovo" (Latin for "egg").

Foods commonly containing egg

Baked goods (cookies, cakes, muffins, breads, crackers), desserts (custards, puddings, ice creams), battered foods (fish and chicken nuggets), meatballs, meatloaf, pastas, sauces, dressings, soups.

Substitutions

While dishes like omelets and egg salads are out, you can still achieve reasonable results in baked goods. Eggs provide air and leavening in cakes, add structure to breads and cakes, and supply liquid in cookie doughs, cakes, and muffin batters. Determine which functions the egg provides in the baked item and experiment with using one of the following alternatives.

To replace one egg in baking:

Baking powder, water, and oil

Whisk together until foamy: 1½ tablespoons (20g) oil, 1½ tablespoons (22g) warm water, and 1 teaspoon baking powder.

EnerG Foods Egg Replacer

Whisk with water until fluffy; then add to your mixture. This is a great all-purpose egg substitute.

Unflavored gelatin

Mix 1 teaspoon (4g) unflavored gelatin with 1 tablespoon (15g) warm water. You should be able to find unflavored gelatins in your grocery store near the flavored gelatin (like Jell-O).

Flaxseed meal

Mix 1 tablespoon flaxseed meal with 3 tablespoons warm water; let sit for 10 minutes. It does have a strong flavor, so does not work as an all-purpose egg replacement, but can be useful in cakes, pumpkin bars, oatmeal applesauce cookies, and muffins.

Fruit puree

In some cases, you can use a quarter cup of puréed banana or apple. Experiment!

Fish/Shellfish Allergies

An allergy to fish does not necessarily mean an allergy to shellfish, and vice versa. However, if you are cooking for someone who has an allergy in either category, the safest approach is to entirely avoid fish and seafood, unless your guest has specifically advised you of allowable food items.

Foods commonly containing fish or shellfish

Anything with fish or seafood, including imitation crab meat, Caesar salad, Caesar dressing, Worcestershire sauce, some pizzas, gelatin (sometimes derived from fish or shellfish bones), some marshmallows, some sauces, antipasto dishes.

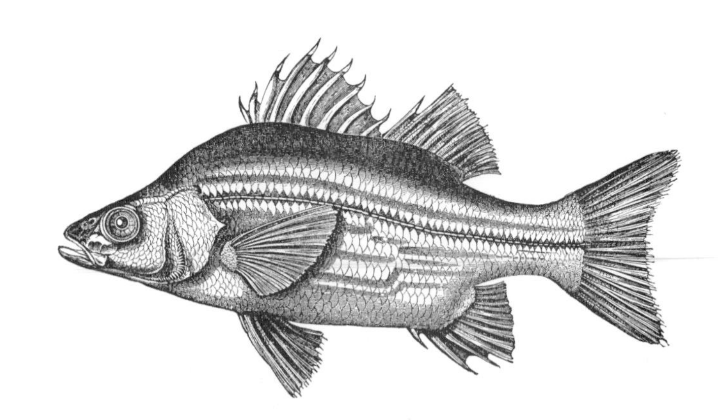

Peanut Allergies

Ingredients to avoid

Peanuts, peanut butter, peanut starch, peanut flour, peanut oil, mixed nuts, crushed nuts, hydrolyzed plant protein, hydrolyzed vegetable protein, vegetable oil (if the source isn't specified), and depending upon the severity of the allergy, anything that states "may contain trace amounts of peanuts."

Foods commonly containing peanuts

Baked goods, baking mixes, chocolate and chocolate chips (many contain trace amounts of peanuts), candy, snacks, nut butters, cereals, sauces (peanuts are sometimes used as a thickener), Asian food (stir fry, sauces, egg rolls), veggie burgers, marzipan (almond paste).

Substitutions

If you have a dish that calls for peanuts directly, you might be able to substitute something else, such as cashews or sunflower seeds. For peanut butter, you can use soy nut butter, almond butter, cashew butter, or sunflower butter, if your guest is not allergic to them (true seeds and soy differ from peanuts).

Tree Nut Allergies

Ingredients to avoid

Almond (butter, pastes such as marzipan, flavoring, extract), brazil nut, cashews (butter, flavoring, extract), chestnuts (water chestnuts are okay as they're not actually nuts), hazelnuts (filberts), hickory nuts, macadamia nuts (Queensland nut, bush nut, maroochi nut, queen of nuts, bauple nut), pecans, pine nuts, pinon (pignoli), pistachios, walnuts, nut meal, nougat, nut paste, Nutella.

Foods commonly containing nuts

Baked goods, snack foods, Asian foods, pesto, salads, candy. Cross-contamination is a major concern, so inspect packages for statements such as "may contain trace amounts of..."

Substitutions

Working around nut allergies can be tricky. As with peanut allergies, your best bet is to select recipes that don't rely on nuts. In salads and snacks, you can use seeds, such as sunflower, pumpkin, or sesame seeds. Sunflower butter can replace nut butters.

Sesame seed allergies are not uncommon, so check with your guest on this substitution.

Soy Allergies

Ingredients to avoid

Hydrolyzed soy protein, miso, shoyu sauce, soy-anything, soy protein concentrate, soy protein isolate, soy sauce, soybean, soybean granules, soybean curd, tempeh, textured vegetable protein ("TVP"), tofu.

Foods commonly containing soy

Baby foods, baked goods (cakes, cookies, muffins, breads), baking mixes, breakfast cereals, packaged dinners like spaghetti or macaroni and cheese, canned tuna packed in oil, margarine, shortening, vegetable oil and anything with vegetable oil in it, snack foods (including crackers, chips, pretzels), nondairy creamers, vitamin supplements.

Substitutions

There are no good substitutes for items like tofu and soy sauce, so choose recipes that don't directly rely on soy-based products. Note that soy is used in an amazing number of commercial products—often in places that you wouldn't suspect, such as pasta sauce—so read labels carefully!

Wheat Allergies

Note that a wheat allergy is *not* the same as a gluten intolerance. Wheat allergy is often confused with celiac disease (gluten intolerance), which is an autoimmune disorder in which the small intestine reacts to the ingestion of gluten. Still, celiac disease is often easier to explain as a severe allergy so that people unfamiliar with the details of it understand the importance of handling food for those with it.

Wheat allergies are triggered by proteins present in wheat specifically, not the gluten. Unlike those who have wheat allergies, individuals with celiac disease must avoid *all* gluten, regardless of source. Be careful to avoid cross-contamination: even a knife used to butter toast might contain sufficient trace amounts of gluten to cause problems, so make sure to carefully wash and rinse utensils, dishes, and hands when cooking for someone with gluten intolerance. For more information on celiac disease, visit *http://www.celiac.org*.

Ingredients to avoid

Wheat (bran, germ, starch), bulgur, flour (graham, durham, enriched), gluten, modified food starch, malt, spelt, vegetable gums, semolina, hydrolyzed vegetable protein, starch, natural flavoring.

Foods commonly containing wheat

Breads (bagels, muffins, rolls, donuts, pancakes), desserts (cakes, cookies, baking mixes, pies), snacks (crackers, chips, cereals), most commercial soups including broths, pastas (noodles, packaged dinners containing pasta), condiments (soy sauce, Worcestershire sauce, salad dressings, barbeque sauces, marinades, glazes, some vinegars), beverages (beer, nonalcoholic beer, ale, root beer, instant chocolate drink mixes), meats (frozen meats that are packaged with broth, lunch meats, hot dogs), gravies and sauces (most likely thickened with wheat flour), flour tortillas, tabbouleh (salad dish), pilafs.

Substitutions

Flour

Replacing wheat flour is tricky, because it contains gluten, which creates bread's characteristic elastic structure and texture. It is difficult to duplicate wheat baked goods (especially bread) without wheat flour. Some nonwheat flours, such as barley and rye flour, do contain the proteins necessary to form gluten.

People with a wheat allergy can usually tolerate those flours while people with celiac disease cannot.

Rice flour and rye flour are easy to find. Check your regular grocery store. You can use either in place of wheat flour in some recipes (substituting at a 1:1 ratio). Tapioca starch, potato starch (use ⅝ cups per 1 cup of wheat flour, a 0.625:1 ratio), potato flour, and sorghum flour can also be used.

You can achieve better results by blending several flours together. For an all-purpose flour mix, combine ¾ cups (120g) white rice flour, ¼ cup (30g) potato starch (not potato flour!), 2 tablespoons (15g) tapioca starch (also called tapioca flour), and, optionally, ¼ teaspoon (1g) xanthan gum.

Pasta

Luckily, there are great alternatives to wheat pasta! Pasta also comes in rice, corn, and quinoa varieties. Take care to not overcook these types of pasta, because they can get mushy and fall apart easily, and remember to make sure the colander is really clean if you've previously used it for wheat pasta.

Snacks

If your guest is more sensitive or has celiac disease, be sure to double-check with the manufacturer about shared manufacturing lines and cross-contamination. Rice cakes, rice crackers, popcorn, and corn and potato chips make for excellent wheat-free snacks (but are not necessarily gluten-free).

Afterword

CURIOSITY AND THE JOY OF DISCOVERING HOW SOMETHING WORKS ARE TWO OF A GEEK'S DEFINING CHARACTERISTICS. I can think of very few other things that have brought me as much joy as learning to cook and providing for others. It scratches the same neurons that solving a puzzle or producing a brilliant piece of code does, but tastes better and often takes less time—not to mention that you can do it for other people and make them happy, too!

Speaking of puzzles, here's how to solve the 12-coin problem I gave in the first chapter. Start with coins 1–4 on the left side and 5–8 on the right side. If the scale registers them as equal, place 9 and 10 on the left and 11 and 1 on the right side. If equal, 12 is the bad one. If not equal, remove 11 and 1 and move 10 to where 11 was. If the scale remains in the same unbalanced position, 9 is the bad one. If the scale is balanced, 11 is the bad one. And if the scale flips to the other side, 10 is the bad one. The trick is to realize that a balance scale can give you not two but three bits of information: <, =, and >, as opposed to = and !=. I'll leave solving the problem of the starting positions of 1–4 and 5–8 being unequal for you.

I hope that by now the puzzle that is learning to cook has been replaced with the joy of understanding the basic mechanics of the system. True, there are still many more puzzles left to understand, but the core principles of cooking can actually be summed up in a single page (see next page).

Whatever your reasons for learning to cook—health, financial, social, giving, romantic—and whatever your style, cooking should be fun. I hope you've found this book useful in showing you ways to bring a certain playfulness to food, both inside and outside the kitchen.

Happy cooking!

Potter's Kitchen Tips

Manage expectations and perceptions. When cooking for someone, expectations and perceptions are just as important as the objective quality of the dish. Only you, as the cook, will know what it was supposed to be. If the chocolate soufflé falls, call it a fallen chocolate cake, toss some berries on top, and ship it.

Use quality ingredients. The number one predictor of a great tasting meal is great-tasting produce and ingredients. Tomatoes should taste like tomatoes, avocados should be soft and creamy, and apples should have their distinctive crisp.

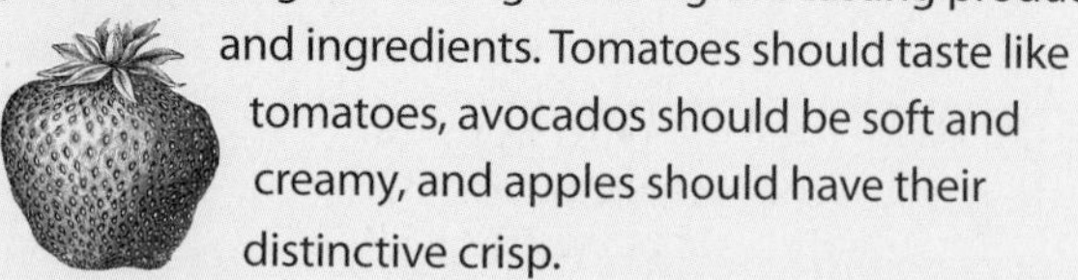

Create harmony and balance. Harmony is found in combining compatible ingredients. Balance is found in adjusting sweetness and sourness (acids) and seasoning correctly with salt. Start with good produce, taste it, and adjust with an acid (vinegar, lemon juice) and salt.

Practice food safety. When working in the kitchen, be mindful of the growing conditions for pathogens. Avoid cross-contamination by washing your hands. Frequently. Foodborne illness isn't fun, but it's usually the secondary complications for at-risk populations that are life-threatening.

Eat whole foods. There's nothing inherently wrong with processed foods, but they tend to be higher in salt, sugar, and fat. Food additives aren't in and of themselves evil, but like anything, too much—or not enough—can be problematic because of what our body does or doesn't do in response.

Measure temperatures, not time. Proteins in meats and starches in grains undergo physical reactions at certain temperatures, regardless of whether they're boiled, grilled, or sautéed. A 4 lb chicken will cook faster than a 6 lb chicken, but both will be done at the same temperature. Timers are useful, but internal temperature tells you a lot more.

Add flavor and aroma with browning reactions. When sugars caramelize (for sucrose, starting at around 340°F / 171°C) and proteins undergo Maillard reactions (starting at around 310°F / 155°C), they break down and form hundreds of new compounds. For some reason, we like the way those compounds smell.

Pay attention to the details when baking. Use weight instead of volume measurements, and pay attention to the various variables in play; gluten levels, moisture content, and pH levels especially. Baking is a great place for A/B experimentation: the ingredients are cheap, relatively consistent, and easy to foist off onto coworkers trying to lose weight (muhahaha).

Experiment! If you're not sure how to do something, take a guess. If you aren't sure which way to do something, try both. One way will probably work better, and you'll learn something in the process. Worst case, you can always order pizza. Have fun, be curious, but use your common sense and be safe.

Index

A

NOTE: **Bolded page numbers** indicate major discussions for that topic. ***Bolded and italicized*** entries indicate recipes.

D

E

F

G

H

I

J

Q

R

S

U

V

W

X

Y

Z

About the Author

Jeff Potter has done the cubicle thing, the startup thing, and the entrepreneur thing, and through it all maintained his sanity by cooking for friends. He studied computer science and visual art at Brown University.

Colophon

The cover, heading, and recipe font is ITC Officina; the body font is Myriad Pro; and the margin note font is Marydale.